Environmental Modelling:
An Uncertain Future?

Uncertainty in the predictions of science when applied to the environment is an issue of great current relevance in relation to the impacts of climate change, protecting against natural and man-made disasters, pollutant transport and sustainable resource management. However, it is often ignored by both scientists and decision makers, or interpreted as a conflict or disagreement between scientists. This is not necessarily the case, the scientists might well agree, but their predictions should still be uncertain and that advice might be important in decision making.

Environmental Modelling: An Uncertain Future? introduces students, scientists and decision makers to:

- the different concepts and techniques of uncertainty estimation in environmental prediction;
- the philosophical background to different concepts of uncertainty;
- the constraint of uncertainties by the collection of observations and data assimilation in real-time forecasting; and
- techniques for decision making under uncertainty.

This book will be relevant to environmental modellers, practitioners and decision makers in hydrology, hydraulics, ecology, meteorology and oceanography, geomorphology, geochemistry, soil science, pollutant transport and climate change.

Keith Beven is Professor of Hydrology and Fluid Dynamics at Lancaster University. While finishing this book he was at Uppsala University in Sweden as Konung Carl XVI Gustafs Gästprofessor i Miljövetenskap 2006/07.

Environmental Modelling: An Uncertain Future?

An introduction to techniques for uncertainty estimation in environmental prediction

Keith Beven

CRC Press
Taylor & Francis Group
Boca Raton London New York

CRC Press is an imprint of the
Taylor & Francis Group, an **informa** business

CRC Press
Taylor & Francis Group
6000 Broken Sound Parkway NW, Suite 300
Boca Raton, FL 33487-2742

Visit the Taylor & Francis Web site at
http://www.taylorandfrancis.com

and the CRC Press Web site at
http://www.crcpress.com

Contents

Figures

Boxes

Preface

This book has had a long gestation. It has its origins in a piece of hydrological modelling work carried out for my PhD about 35 years ago. The results were published much, much later (Beven, 2001) because they were so bad. The modelling was done scientifically and objectively, working to the best standards of the day but the model gave very poor predictions of both the outputs of the small catchment area under study and the internal responses of the hillslopes. At the time, computer limitations were an issue: these were still the days when computer programs existed as boxes of punched cards so that debugging program code and run-time errors was extremely time-consuming. Computer memory and speed were constraints on the element size used in the model and on the periods of simulation that could be run. I remember that we had a small celebration when my simulations finally used less computer time on the University "mainframe" computer than the 12 hours of real time that they were purporting to simulate!

Computer limitations were not, however, sufficient to explain why the results were so bad. My interpretation of the results was that the model did not take adequate account of the nature of the flow processes and heterogeneity of soil characteristics observed in the field, despite the fact that it was based on the best theory available at the time and that the model parameters had been measured in the field (albeit on small samples). Thus, my research career can be summarised as an attempt to cope with this failure and to find ways of simulating environmental systems, in some sense, "properly".

Defining "properly" is, itself, a difficult issue that overlaps into areas of the philosophy of science and this book includes a discussion of some of the philosophical issues involved. It is, of course, quite possible to develop and use environmental models without any explicit underlying philosophy. They are simply useful tools. However, at the risk of making a gross generalisation, I would suggest that most environmental modellers have, at least implicitly, what might be called a pragmatic realist philosophy. I am a hydrologist. I know very well that my computer model is only a set of logical constructs implemented on some complex electronic hardware but I still think of the variables in the model as representing real water. If I make predictions of pollutant transport I think of the concentrations in the model as representing real contaminant. My aim is to improve the models over time, learning from each application to come "closer" to the real quantities of interest.

In the same way, an ecological modeller will hope to represent real populations and communities; the atmospheric modeller real energy and momentum fluxes in the atmosphere; and the volcanologist the properties of real lava flows with their differing mineralogies and gas contents. Their aim will also be to improve their models over time, so that the models become more "realistic", but most environmental modellers will realise that there are limitations on how far we can take this process. The modeller has to be a pragmatist, even if not wishing to be "only" an instrumentalist in retaining any model thought to be "useful" in some sense in prediction.

This book is essentially an exploration of those limitations and their implications for modelling practice and the use of model predictions in decision making. Those limitations mean that there will be inherent uncertainties in the predictive capabilities of environmental models and therefore a risk of being wrong in making a prediction. Thus, it is the thesis of this book that such uncertainties should, wherever possible, be explicitly evaluated and any decision that is based on such predictions should take account of the risk of being wrong.

However, uncertainty estimation has certainly not yet become routine practice, although the research literature is now increasing rapidly. Probability concepts and stochastic models have been used widely but, as will be seen, often in quite a deterministic way. There are good reasons for this, but this does not justify the total neglect of uncertainty that is often evident in environmental predictions. The aim here is to outline the methods that are available and show how they can be used in practice in a way that can be readily understood, especially by those who may need to take uncertainty into account in making decisions. I hope that this might encourage wider understanding in future (and even that in some areas of predicting future changes that the use of uncertainty estimation in environmental modelling should become mandatory!). In a short book covering a wide range of techniques, it is only possible to provide an introduction and guide, but references to more detailed reading are given throughout. A variety of case studies are used to illustrate different methodologies. One of the reasons for the wide range of available methods is that uncertainty estimation for environmental models is still developing and sometimes involves contentious issues. There have been some strong disagreements about appropriate methodologies and the book also tries to reflect some of the current debates.

Environmental models are (or should be) run for a purpose. That purpose might only be as a framework for trying to understand how an environmental system is working, but often they are run to help in some form of decision making. Many people have suggested that decision makers are "not ready" to deal with uncertain predictions from the scientist, or that if the estimate of uncertainty is too large then this may not be helpful in decision making. Both suggestions are, in my view, wrong. Indeed, in my view, it is wrong for a scientist not to associate any prediction that might be used in decision making with a realistic evaluation of uncertainty since this might actually change the decision that is made. Where uncertainties are indeed large, then this might result in a different approach to decision making that is robust to uncertainty and adaptive to learning more about the system. Some of the methods for how to do so are also discussed in this book.

My own research work has been primarily in the areas of hydrological and hydraulic modelling and I have limited experience of other areas of environmental modelling. The examples used in illustrating the techniques presented in the book are therefore biased towards the areas I know best, but I have tried to bring in examples from other disciplines where appropriate. Environmental science is, after all, an interdisciplinary activity within which physical, chemical and biological processes are linked and coupled. Any demarcation of a particular spatial or process domain therefore necessarily involves simplification and, as a result, uncertainty. I hope, therefore, that the techniques presented in this book will be applicable across a wide range of environmental problems and that there is a sufficient range of methods presented that readers might find one or two things that are new to them and that might be useful, particularly in environmental modelling and decision-making applications when it is difficult to make strong assumptions about the various sources of uncertainty. However, the literature on uncertainty estimation is growing rapidly and it is very possible that I have missed some benchmark references from other fields (there is little on uncertainty in geophysical applications and regionalisation techniques, for example). I also realise that not every methodology has been included (the book is already bigger than I originally intended as an introductory text). I can but offer my apologies in advance to those readers who feel that more prominence should have been given to their work.

Over the last 30 years, the ideas that are presented here have been influenced by a wide variety of people, some of whom I have never met but whose writings have caused me to think deeply about how best to do modelling "properly". Citing only the examples of good practice, I should mention first Peter Young at Lancaster University whose attempts to show that science can indeed formally progress by induction have been a consistent inspiration. I also benefited greatly from discussions about modelling with George Hornberger when I worked at the University of Virginia: indeed, our later development of the GLUE methodology for uncertainty estimation is a direct extension of the Monte Carlo sensitivity analysis work that he did with Bob Spear and Peter Young.

Statisticians have tended to claim the realm of uncertainty estimation as their own, but there are many aspects of uncertainty in environmental modelling that have nothing to do with statistics. There are many things about the GLUE methodology that many statisticians do not like, especially subjective likelihood measures, but I have greatly appreciated the open-mindedness of Jonathan Tawn of the Department of Statistics at Lancaster University and Jonathan Rougier, now at Bristol, in discussing some of the issues involved. There have now been applications of GLUE to a wide variety of problems, helped by a large number of graduate students, post-docs, colleagues and visiting scientists at Lancaster. The contributions of Bruno Ambroise, Giuseppe Aronica, Kathy Bashford, Andrew Binley, Sarka Blazkova, Rich Brazier, Kev Buckley, Wouter Buytaert, David Cameron, Hyung Tae Choi, Sarah Dean, Luc Feyen, James Fisher, Stewart Franks, Hannah Green, Barry Hankin, Ion Iorgulescu, Helen Kettle, Rob Lamb, Trevor Page, Florian Pappenberger, Pep Piñol, Renata Romanowicz, Karsten Schulz, Paul Smith, Phil Younger and Susan Zak have all been important. Special thanks are due to Jim Freer for long service in trying out lots of different hydrological modelling and GLUE ideas, enthusiastic promotion of an international community of hydrologists interested in uncertainty in measurements and model predictions, and (not least) keeping a succession of parallel computing machines working at Lancaster.

Tack så mycket också to the group of PhD students from Sweden and Denmark who participated in the course at Uppsala based on a draft of this book and provided some useful feedback, and to Uppsala Universitet and Sveriges Lantbruksuniversitet for the support and warm welcome during my year in Uppsala. Finally, thanks to Monique Romanens, Susanne Abbuehl, Angela Hewitt, Thomas Tallis, Svenska Polskor, and the Arditti, Alban Berg, Borodin, Brodsky, Emerson, Fitzwilliam, Keller, Kronos, Lindsay, Maggini, Uppsala Kammarsolister and Zehetmaier quartets for keeping me saner than I would otherwise have been during the last year of writing.

In the preface of my rainfall-runoff modelling book, I noted that I had written it with Anna in mind and that, should she ever have to read it, I hoped that it would be useful. Totally unexpectedly, she did have to use it as part of an MSc course (with the assessment that "it was not too bad, really"!). So perhaps this time round, I can formally dedicate this book to Anna and the next generation of environmental modellers, in the hope that they might start to implement some of the methods presented here in real applications. After all, some clients are, at last, starting to ask for uncertainty estimates in predictions – so this time I hope it might just possibly be not too bad a guide to the methods available.

Keith Beven
Lancaster, Outhgill and Uppsala

Chapter 1

How to make predictions

As we know, there are known knowns, there are things that we know we know. We also know there are known unknowns, that is to say, we know there are some things we do not know. But there are also unknown unknowns, the ones we don't know we don't know.

> Donald Rumsfeld, former US Secretary of Defense, February 12th 2002

What men really want is not knowledge, but certainty.

> Bertrand Russell, 1964

1.1 The purpose of this book

This book is primarily intended as a discursive examination of the process of uncertainty estimation in environmental modelling as it is done now and how it might be done more "properly", and perhaps more "realistically", in the future. It is intended to be a book that might be useful to students, graduate students, and practitioners interested in increasing their understanding of different methods of uncertainty estimation and how that understanding might be used in decision-making. It is written with both users of environmental models and decision makers in mind; particularly those who have not much previous exposure to uncertainty concepts. It can be read without reference to all the detailed technical material in the Boxes that follow the different chapters. It cannot, of course, be a comprehensive account of uncertainty estimation in all the different disciplines that comprise environmental science. It is intended to be much more a first guide on how to think about the modelling process and choose uncertainty estimation and decision making techniques appropriate to a particular application. It cannot go into all the details and software available for each technique but references are provided to allow the reader to explore further.

It is based on a long experience of trying to cope with predictive uncertainties in one particular branch of environmental science, that of hydrology. As such, it is coloured by the particular problems of hydrology, although these are not so different from many other areas of environmental modelling. Modelling and prediction in hydrology are important in providing information for the practical management of natural resources and natural hazards. Hydrology, however, also has severe limitations as a science resulting primarily from limitations in measurement techniques at the scales at which we want to make predictions. This is particularly the case for the examination of flow processes underground which is where many of the interesting and active hydrological processes take place.

Similar limitations can be found in most areas of environmental science, whether physical, chemical or biological process in the earth, atmosphere or oceans. In all cases it will be difficult to make measurements at the scales at which we wish to make predictions. It will be difficult to define the boundary conditions for a system of interest and, for time-dependent processes, the initial conditions everywhere within the domain of that system. It will also be difficult to specify the physical, chemical and biological characteristics of the domain. Thus, even if we have some understanding of how the system is working, these factors will make prediction difficult. How, then, to make predictions in the face of such difficulties?

One, necessary, answer to this question is "approximately". The very act of prediction involves simplification of the complexity that is the real domain of interest. The discussions that follow are primarily concerned with how to achieve that simplification in a scientifically rigorous way taking proper account of the uncertainties in the modelling process. In the remainder of this chapter we will consider how to approach the environmental modelling process and some of the difficulties involved in trying to model environmental systems "properly".

The reader may also find that there will be some terms in this book that might be new. For clarity, a Glossary of Terms is provided, including a discussion of different types of usage for some of the entries. Words in the text in bold will be found explained in the Glossary. A brief revision of matrix algebra and a guide to sources of software for uncertainty estimation are also provided as Appendices at the end of the book.

1.2 The aims of environmental modelling

The way in which an environmental scientist might choose to make predictions will depend, in part, on the aims underlying the effort. One major aim for any scientist is to show that some of the understanding that has been gained about the controlling processes can be formalised into a system of mathematical relationships that result in verifiable predictions about that system. This is prediction as science to show "that we do, after all, understand our science and its complex interrelated phenomena" (W. M. Kohler, Head of Hydrology at the World Meteorological Organisation, 1969). For many scientists this is a sufficient aim in itself since, if the predictions do not prove to be correct, then it should force a revision of the science that underlies the formal statement of the model used. That this does not always happen will be discussed further later. For now, it is sufficient to note that a secondary aim of developing a scientific model that produces verifiable predictions is to use that model operationally for predictions that will be useful for management and decision-making purposes.

With the growth of computer power and computer modelling capabilities, the aim of producing operational predictions that will be useful for management has been increasingly driven by demand. Now that the results of complex computer simulations of weather systems are routinely shown on television it is perceived that computer predictions should now be possible in many other areas of environmental science, from the transport of toxic immiscible pollutants in groundwater to the impact of climate change on vegetation patterns and floods. This is despite the fact that our knowledge of the properties of specific groundwater aquifers is poor; despite the fact that our ideas about future climates rest on the results of global circulation models that are not yet very secure; and despite the fact that we often complain about television weather forecasts being wrong. In many areas of environmental science the demand for predictions has outstripped the scientific understanding on which predictions must be based. There are certainly some areas in which the answer to the question of how to make predictions should, as yet, be *don't* (or at least don't put too much trust in the model predictions when making decisions).

It is indicative that not many environmental modelling studies show true tests of predictions of the models in the form of post-prediction auditing. Many will show simulations that are compared with past data after some history matching or model calibration has taken place. Some will show similar predictions of periods not used in model calibration as a test of the capabilities of a model. Very few studies have made predictions that have then been verified (or not) by data collected later (something that we have all been taught should be part of the "scientific method").

In fact, experience in this type of post-prediction audit has not been good, at least in the field of groundwater modelling (Konikow and Bredehoeft, 1992). Post-prediction audits made for a variety of different modelling studies showed that, in general, the results were generally poor (see also Anderson and Woessner, 1992). This was often for very understandable reasons, such as wrong assumptions about future boundary conditions, but this does not change the conclusion that the results were poor. What, then, should we conclude about the predictions of the much more complex coupled ocean–atmosphere global circulation models that are being used to predict the expected changes in climate as a result of changing concentrations of greenhouse gases into the future? That their predictions are wrong? Quite possibly, but not necessarily. The more common conclusion is that they are necessarily approximate at present but will be improved as computer power increases and as any mismatches between observed and predicted variables are evaluated and understood. The same would now be true in the case of the groundwater models. In most cases a post-prediction audit would lead to model improvements that would allow better predictions to be made with the benefit of hindsight about, for example, which boundary conditions actually occurred over the predicted period. This means that modellers are rarely forced to admit to false predictions since they can always revise their predictions with hindsight or with a new generation of models and auxiliary conditions. It is worth noting that, viewed in this way, model applications become part of a learning processes, not only about the models but also about the places they are applied to. The idea of modelling as a learning process will be a continuing theme in this book as it is essentially about reducing different forms of uncertainty in making predictions.

Modelling for understanding, modelling for prediction for practical applications and modelling as career are all part of the current practice of environmental modelling. Scientists and practitioners who model and make predictions tend, for the most part, to be pragmatic realists at heart. Their goal is to bring models based on the most comprehensive understanding to bear on prediction problems of operational or practical interest. This would combine the aims of prediction as science, of prediction as practical tool (and of prediction as career). Gradually, as the science progresses, the models used in prediction are expected to evolve to become a more and more realistic description of the real system. This pragmatic realism is one commonly held philosophy of environmental modelling. This is not, however, the only possible philosophical position to take and we will return to discuss this further in Chapter 2 after considering the nature of the modelling process and the different sources of uncertainty that arise in modelling environmental systems.

1.3 Seven reasons *not* to use uncertainty analysis

The issues that are raised by the uncertainty inherent in the application of environmental models have been discussed for two decades and more (e.g. notably Beck, 1987, in the field of water quality modelling). Pappenberger and Beven (2006) have considered why uncertainty estimation is still not yet standard practice in environmental modelling. It remains common to show results without uncertainty bounds to decision makers, at scientific conferences, in refereed publications or in consultancy reports. It seems that there is still significant resistance to the routine use of uncertainty analysis methods by environmental modellers, whether for reasons of expense, understanding of methods, or training in the requisite skills. Yet, the use of uncertainty estimation should be *routine* in environmental science. As yet, despite all of the research on methods of uncertainty estimation that is now available, it is not. Seven of the reasons why not are as follows:

1   Uncertainty analysis is not necessary given physically realistic models.
2   Uncertainty analysis is not useful in adding to process understanding.
3   Uncertainty (probability) distributions cannot be understood by policy makers and the public.
4   Uncertainty analysis cannot be incorporated into the decision-making process.
5   Uncertainty analysis is too subjective.
6   Uncertainty analysis is too difficult to perform.
7   Uncertainty does not really matter in making the final decision.

The reader may well be able to add some other reasons to this list (for example that the whole idea of trying to assess the uncertainties makes his/her head hurt!). Pappenberger and Beven consider each of those seven reasons in turn and suggest that none of them is tenable in many applications, at least where uncertainty estimation is not limited by computational constraints. In particular, they discuss the interaction between scientists and policy and decision makers. The concepts of "uncertainty" and "risk" are perceived and understood in a variety of different ways by different communities and different people. However, it can be shown that when both scientists and public work together this gap may be bridged. For example, several studies have shown that probabilistic weather forecasts can be understood by non-scientist users (e.g. Luseno et al., 2003). Moreover, policy makers derive decisions on a regular basis under severe uncertainties, because the scientific basis is not sufficient at the time. Studies suggest that decision makers actually want to get a feeling for the range of uncertainty and the risk of possible outcomes when it can be provided (e.g. McCarthy et al., 2007). This point is illustrated by the political demand of "handling uncertainty in scientific advice" to the UK Parliament (Ely, 2004). The response may, however, be subject to the type of decision under consideration. Tyszka and Zaleskiewicz (2006) report that people were much less interested in probabilistic information about scenarios when the decision to be made had an ethical dimension.

This is a very important point for the modeller since a misunderstanding of the certainty of modelling results can lead to a loss of credibility and trust in the model and the modelling process (Demeritt, 2001; Lemos et al., 2002). The communication of uncertainty to decision makers, the public and other stakeholders is all important in this process (e.g. Brashers, 2001; Fox and Irwin, 1998; Patt and Dessai, 2005; Faulkner et al., 2007; Stainforth et al., 2007b). Effectively, uncertainty estimation is embedded in the wider decision-making process (e.g. Refsgaard et al., 2005, 2006). The suggestion that scientific uncertainty cannot be understood by stakeholders and decision makers persists (on both sides). There would seem to be little reason why this argument should continue to be made in the future in terms of understanding. This book is, hopefully, a contribution towards easing the communication process and working towards the routine application of uncertainty estimation in environmental modelling. Other initiatives are also helping, such as the more widespread availability of software for uncertainty estimation (see Software Appendix at the end of this book) and the decision tree for uncertainty estimation methods described in Section 1.9 below.

An open scientific discourse on uncertainty would have important implications for the environmental decision process. Uncertainty clearly does matter in the current debate over the significance of future predictions of climate change and its implications for future global policies (and consequent impacts on future water resources management and capital investment). This is an area where the science has not yet matured to the point where an open discourse is possible and expressions of uncertainty are interpreted as simple disagreements amongst scientists. Some disagreements exist, of course, but neither side in the climate change debate has been open in the communication of the uncertainties involved, leading to disputed results rather than risk evaluations. In this book I will try to show that uncertainty estimation need not make the head hurt, and that it can be valuable in policy making and management of environmental systems.

1.4 The nature of the modelling process

1.4.1 From perceptual to procedural models

In the study of any environmental system from the viewpoint of scientific understanding it is possible to perceive much more complexity than it is possible to represent in mathematical form to make quantitative predictions. We know that nature is complex at many levels: we can perceive complexity of processes and the variation of controlling processes over time; we can perceive complexity in defining the boundaries of a system of interest to separate it from its "environment"; we can perceive complexity of external forcing for the system we are interested in; we can perceive complexity of the local characteristics within the system with variations in space and in time. We can describe many of these complexities in qualitative terms, but not necessarily in quantitative terms. But a quantitative, mathematical, description is usually held to be necessary for making management decisions based on firm scientific principles built upon the aim of prediction as science.

It is an extremely important point to note that moving from a model based on our perceptions of the full complexity of a system (the **perceptual model**[1]) to a mathematical description (let us call it a **formal model**) requires the introduction of simplifying assumptions. This is self-evident but, in practice, the complexities that are left out in defining the formal model are often quietly forgotten in model applications. This is fine if the neglected complexities have a negligible effect on the system but this is not always the case; they are certainly important enough to have been *perceived* as potentially important. In many environmental systems the assumptions of the formal model will often involve gross simplifications of the perceptual model (while the perceptual model itself may yet be incomplete because of lack of measurement techniques or other means of identifying significant processes or characteristics of the system). These simplifications are made for good reason. Previous research may not have resulted in an adequate description of some of the perceived complexity at the scale at which predictions are required, while even if a description is available it may be very difficult to estimate the parameters that will allow that description to be used to describe the processes in a particular location or time period.

Such models are examples of what Adam Morton (1993) calls *mediating models*. They mediate between an underlying theory, which is often developed largely in rough qualitative terms (the perceptual model), and the quantitative prediction of system responses. They have the general characteristics revealed by Morton's analysis: they have assumptions that are false *and are known to be false*; they are not, however, arbitrary but reflect physical intuition; they tend to be purpose-specific with different (and possibly incompatible) sets of assumptions and auxiliary hypotheses for different purposes; they have real explanatory power but may never (nor are they necessarily expected to) develop into full theoretical structures. They also have a history, in that successful modelling techniques tend to be refined and inherited by later models.

A further level of simplification may be necessary in solving the equations of the formal model for a particular application. In many environmental modelling problems these equations are nonlinear partial differential equations that do not have general analytical solutions. It is therefore necessary to resort to approximate numerical solutions (finite difference methods, finite element methods, finite volume methods, boundary element methods, . . .) in implementing the formal model as an algorithm or **procedural model** that will run on a computer. Different implementations will, of course, give different predictions depending on the coding and degree of approximation. The procedural model exists as a computer code that provides the quantitative predictions required. It represents a further level of approximation to the processes of the real system. There is a vast literature in both applied mathematics and different application subject areas about how to solve the equations of a formal model accurately and we will not consider this further here. Note only that for many nonlinear problems it is quite possible to implement solutions that give inaccurate solutions to the original equations and any naïve user of a modelling package should be aware that this may be an important source of error if poor algorithms (or poor choices of time steps, spatial discretisations or treatments of boundary conditions) are used.

The differentiation of perceptual, formal and procedural stages in the modelling

---

1 Definitions of all the highlighted terms will be found in the Glossary at the end of the book.

process is a useful one, emphasising the successive level of approximation in moving from qualitative understanding to quantitative prediction. Different terms are used in different fields of modelling to represent these stages.

1.4.2 Parameters, variables and boundary conditions

Every formal model contains both **parameters** and **variables**. These are also words that are used in different ways in different subject areas. For the purposes of this book we will define a model variable as a quantity that is calculated as part of the modelling solution and a model parameter as a quantity that represents the intrinsic character-istics of the system and is specified external to the model by the user. The **boundary** and **initial conditions** will also be specified external to the model. These are generally externally prescribed values for variables in the model. The boundary conditions apply only to the boundaries of the domain of interest; the initial conditions prescribe the values of all variables in the model at the start of a time-stepping run. Parameter values, boundary conditions and initial conditions are sometimes also known as the **auxiliary conditions** for a particular model application.

Confusion between these terms can arise because quantities with the same name that are parameters in one model might be variables in another, calculated on the basis of other externally specified parameters, while in some systems the boundary and initial conditions might also be treated as parameters of a model. In addition, par-ameter values are not necessarily constants but might be specified as varying in space and/or time by the user to reflect changing characteristics of the system. The basic definitions used here, however, should still hold.

Parameters are generally of two types. There are those that are intended to reflect the specific characteristics of the dynamics of a process; and there are those that are intended to reflect the specific characteristics of a location where the model is being applied. This distinction is often blurred and parameters that are given names that indicate that they are process-related (hydraulic conductivity, dispersion coefficient, bed roughness, aerodynamic resistance, partition coefficients, . . .) are used in practice to adapt the model to a particular location. Put another way, these parameters have to be calibrated in some way for each application of the model.

The **calibration** process is a particular problem in environmental modelling. If a formal model could be defined such that all parameters were universal constants then this problem would be eliminated. There are some constants and near-constants that appear in many environmental models (e.g. the gravitational acceleration constant, the latent heat of vaporisation of water) but achieving such universality is unlikely to happen more generally in environmental modelling given the need to represent specific locations with their unique characteristics (see Beven, 2000). Even if it were possible to *measure* all parameters when making a model application then this problem would be minimised but this is also not generally possible with current measurement tech-nologies. As it is, parameters are usually calibrated on the basis of very limited meas-urements; by extrapolation from applications at other sites, or by inference from a comparison of model outputs and observed responses at the site of interest.

All these three means of calibration have problems. Measurements may be them-selves subject to error and may be at scales different to that required in the model. Heterogeneity in space and non-stationarity in time may mean that directly measured values mean that the measured parameter and the parameter values required in the model to get good predictions may have the same name but they may be different quantities. For example, measurements of soil hydraulic conductivity are often made on small soil samples. Such measurements are known to exhibit order of magnitude variability over short distances (e.g. Nielsen et al., 1973), but a model requires values of hydraulic conductivity that will represent the response of a much larger spatial unit (sometimes the whole catchment scale). In fact, the model requires *effective values* of such a parameter that will provide good values of predicted states and outfluxes in the model that might have to compensate for things that are not in the model (Cushman, 1986; Beven, 1989). Such effective values might be difficult to compare with the measured values. The same may be true of state variables, such as representations of mass or energy. For example, measurements of soil moisture in hydrology (a state variable) are generally made at point scales. Such measurements often exhibit significant variability in space (e.g. Hills and Reynolds, 1969) but a model will predict only changes in mass averaged over some spatial unit of the catchment. Thus, these quantities might also be difficult to compare in model calibration.

Extrapolation from other sites runs into the problem that each site has its own unique characteristics that may affect the effective parameters required. In addition, if the values determined from other sites have been calibrated for a different model structure then the values may not be independent of the model structure used or even for different numerical implementations of the same equations with different discretisations or algorithms. Again, they may have the same name but they may be different quantities. This is often called the **commensurabilty** problem, though even where it is recognised as an issue, it is often ignored as a problem (see Section 1.5 below).

Inference of parameter values by comparison of observed and predicted responses is most generally carried out within an optimisation framework. The values of the parameters are changed until some "best fit" is found in representing the responses of the model. The parameters have to be calibrated within a **closed system** having made assumptions about the nature of the boundary and initial conditions. Both boundary and initial conditions will normally be subject to some uncertainty. The difficulty of finding an optimum model in general increases as the complexity of the problem increases relative to the information content of the observations. As computer power has increased it has been possible to add complexity to models. More and more process understanding is usually built into models with the aim of improving the science underlying the predictions at the cost of adding more and more parameters. The information in the observations has not necessarily increased at the same rate. These tendencies have, in very many environmental models, resulted in what is called **overparameterisation**. Pragmatically, this means that the model may have sufficient degrees of freedom in the parameter values to be calibrated to be able to give a good fit to the observations after optimisation, but it does not follow that the parameters are robustly estimated, or that the apparent "optimal" model is the *only* model that will give a good fit to the observations, or that it will give equally good predictions in the future with different boundary conditions.

These parameter identifiability problems are an important issue when considering the aim of modelling as science. As scientists we want to improve our representation of real world complexity but this will usually result in models with *more* parameters that cannot easily be specified for a specific application. This would not be a problem if it also meant that these parameters were more fundamental in the sense of being easily measured or estimated in any application. Unfortunately, this is usually not the case. They are more often parameters which still have the function of representing the physical, chemical or biological characteristics of particular locations and sometimes also particular time periods. It also remains the case that, although those parameters may be given *names* intended to reflect their physical, chemical or biological significance, the *values* of those parameters needed to get good model performance will depend on the particular model structure used. Despite their intended significance they are then acting as effective parameter values that are not then easily transferred from one model structure to another or one application to another. As we will see in the later chapters (especially Chapter 4), the availability of data with which to estimate effective parameter values then becomes of crucial significance to the value of the model results. Again, the modeller as pragmatic realist will suggest that this may be only a transitory phase: both descriptions and ways of estimating parameter values will improve in the future. Later chapters will demonstrate, however, that this is a problem that is unlikely to go away in the foreseeable future and that environmental modellers will need to grasp this particular problem in a more explicit way.

1.5 The scale problem and the concept of incommensurability

There is an exception to this general picture. This is particularly seen in atmospheric and oceanographic modelling which, until very recently, have been constrained by the available computing resources to use very coarse spatial discretisations of the atmosphere and oceans particularly at global scales. They have developed two methodologies for dealing with scale problems: to use *nested grids* to refine predictions where more detail is required; and to use *sub-grid parameterisations* to represent the effects of smaller scale processes. Treating the scale problem in terms of sub-grid parameterisations (a top-down approach to the scale problem) seems to be a much more justifiable approach than trying to aggregate representations of small-scale processes in the face of local heterogeneities (a bottom-up approach). The difficulty, of course, is deriving adequate sub-grid parameterisations. In atmospheric models parameterisations are required for momentum losses associated with sub-grid scales of turbulence; convection and cloud formation; snow and rain production; heterogeneous surface vegetation, albedo and soil moisture effects; and many other factors. Very often bottom-up arguments have been used to justify the required parameterisations, sometimes ignoring heterogeneities in the system (as in many past parameterisations of the land surface). All the parameterisations are simplifications of the perceptual model; the question is how far they can be useful in reflecting the complexity of the sub-grid scale processes. In most cases, we do not know the answer to this because we can check the grid-scale predictions of the processes only in very indirect ways because measured and predicted variables are incommensurate. This top-down way of looking at the scaling problem is, however, very useful (see discussion in Beven, 2006b).

If, as has been suggested above, the small-scale equations used in many environmental models are not easily scaled up in heterogeneous and structured flow domains, is it possible that improved parameterisations might be found? It is important to remember that there are (at least) two problems here. One is that associated with the problem of change of scale and heterogeneity of parameters, even if the small-scale equations were correct at the local scale. If this were the only problem then it might be possible to derive a theory of scaling that would allow scale-dependent parameters to be developed, either empirically or based on knowledge of a statistical model of the heterogeneity (see for example Dagan, 1986, and Neuman, 1990, for examples from groundwater flow and transport). However, it is not the only problem. The second problem is that the small-scale equations may not be correct at the local scale.

One example is the use of Darcy's law to describe flow in unsaturated soil. Darcy's law is a relationship that says that flow rate of water in a porous medium is linearly proportional to the gradient of potential with a constant of proportionality that is called the hydraulic conductivity of the soil (it is therefore analogous to Ohm's law for flows of electricity or Fick's law for diffusion processes). The relationship was first derived by Henri Philibert Gaspard Darcy (1803–1858) from the results of experiments on saturated sand samples (Davis et al., 1992). It is still the basis for most models of groundwater in saturated porous media. Darcian theory was later extended to unsaturated soils by Richards (1931), allowing for the highly nonlinearity change in hydraulic conductivity with moisture content. It has been shown to work well for unsaturated conditions in laboratory experiments for uniform porous media. The problem arises in taking this relationship out in the real world where soils may be heterogeneous (spatially variable characteristics), non-stationary (temporally variable characteristics), hysteretic (characteristics that vary depending on whether the soil is wetting or drying), and structured (with continuous larger voids due to cracks, root channels, earthworm channels etc that bypass parts of the soil matrix). In a structured soil, flow may be responding to quite different potential gradients for different parts of the pore space and may not have a simple linear relationship between flux rate and gradient (i.e. a problem of the second type). In hydrology, more realistic descriptions are also needed for runoff on irregular and vegetated surfaces; the controls of soil water on water use by plants (or of water use by plants on soil moisture when the roots grow faster than water can move by Darcian flow towards them and when a fundamental control on evapotranspiration appears to be levels of abcissic acid in the plant); and of the effects of soil layering on downslope flows. Similar issues will affect the representation of geochemical and transport processes.

Thus, it is clear that these complexities mean that local-scale equations, such as Darcy's law, will not easily scale up to the larger scales needed for real applications. Different representations that should ideally reflect the effects of local complexity will generally be required. In many modelling fields local-scale equations have been used as if they apply at larger scales because we just do not know what conceptual representations might be better at larger scales. This sort of works if effective parameter values can be calibrated for each application but it is not really very satisfactory.

Similar examples of the difficulties of finding an appropriate conceptual representation at the scales of practical interest could be taken from many different fields, where processes in the perceptual model are neglected in the conceptual model. The use of increasingly complex process descriptions will share some common features, regardless of the process to which they refer. Being more complex, they take more computer time and almost certainly require more parameters and state variables. Those parameter values may not be readily estimated, except by calibration against some data when observations on a particular process may not be directly available. Calibration of these models is already a problem; adding more parameters to be calibrated will only make the problem worse. Hence it is sometimes easier to just leave some of the complexity out (at least for the moment).

Thus, it would appear that there is a fundamental dilemma in "process-based" environmental modelling. The formal model will always be a simplification of the qualitative perceptual model. But adding more process understanding will introduce more complex equations with more parameters. As noted already, these parameters may not be easily measured and may also suffer from the problem of heterogeneity, which would therefore necessitate a large number of measurements to assess the degree and importance of spatial variability. And yet, for many environmental systems, the response to an external forcing is not that complex. In hydrology, a rainfall occurs, surface and subsurface flow rates increase, a hydrograph results. Reproducing the dominant modes of the integrated response at the catchment scale is not that difficult and requires a relatively simple model with few parameters (see for example, Young, 1998, 2003; Young and Parkinson, 2002). How can this lesson be adapted for the general case of environmental models representing the space–time responses at different scales? What would be the minimal model to describe the space–time responses given all the uncertainties in the modelling process? Would it actually be simpler than some of today's models? Would such a model be useful in predicting changed conditions?

These types of questions arise in most, if not all, fields of environmental modelling. In all applications there has to be a compromise between model complexity, techniques and resources available for data collection, and what needs to be predicted for a particular purpose. Absolutely central to the discussion of such a compromise and the model complexity dilemma are the issues of commensurability and uncertainty.

1.6 The model space

The modelling problem can be viewed as a form of *mapping* of the environmental system of interest into a model space. This conceptual framework will be useful throughout the discussions of this book. We have already noted how formal models mediate between the real system (or at least our perceptual model of the real system) and the predictions of practical interest. We can think of the formal models in terms of their parameters. For any defined set of boundary conditions, the predictions of that model will be different for different sets of parameter values. Thus, the parameter values are defining the functionality of the formal model, and this functionality can be mapped in a multi-dimensional space defined by axes of the parameter values.

This is demonstrated in just two parameter dimensions in Figure 1.1. For each combination of parameter values, we can calculate a model response. For a model producing time series of predictions, we can map the predictions at each time step, or map a summary measure over all time steps. The map can be extended to multiple parameter axes (although it clearly rapidly becomes difficult to visualise). Effectively, the complete map represents the potential range of outputs of that particular model structure, within the chosen parameter ranges, and for a particular set of inputs.

The map can be extended further by allowing for additional potential model structures (that might have their own parameter axes), by allowing for potential variability in the initial and boundary conditions used to drive the model, and by allowing for

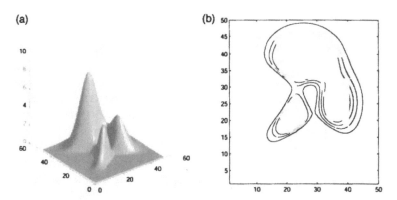

Figure 1.1 A two-parameter model space and response surface represented (a) in three dimensions and (b) as a contour plot

stochastic rather than deterministic outputs from the model. The important point is that, at least in principle, the model space can be filled with the model outputs. In other words, the model space can be considered known in principle (in practice, there may be practical difficulties in filling the model space due to computer costs, chaotic sensitivities to small errors in nonlinear models, and instabilities of solutions for particular sets of parameters).

So, let us for the moment consider a case where the model space can be filled. We then know the functionality of the model (at least for the ranges of parameters and boundary conditions considered). There is no uncertainty in the model predictions (even for stochastic and chaotic models). The uncertainty comes from mapping the environmental system into that model space (with one or more potential model structures) for a particular place or application, given the limited knowledge that we will have about the system.

This picture of the modelling process is quite useful in that it can reveal the limitations of the single optimal model of the system of interest. As well as the output variables of the model, we can also map some summary performance measures (or goodness-of-fit measures, or objective functions, or likelihood measures) for the model. The surface described by such a measure in the model space is then generally called the response surface of the model. It is then readily seen that choosing an optimal parameter set is equivalent to mapping the system to a single point in the model space, at a peak in the response surface (for maximising a performance or likelihood measure). Optimisation methods are designed to try to find the parameter set at the global peak for the performance measure (or lowest point for a minimisation problem) on what might be a very complex surface. A good optimisation method will find the optimum as efficiently as possible, without being distracted by local peaks or valleys.

The search for the global peak is clearly much easier if the response surface is relatively simple, as shown in Figure 1.1 (though this has some local peaks that might divert an optimisation algorithm). Adding dimensions to the model space, and uncertainty to the model structure and input data, however, often means that in environmental modelling problems the response surface is very complex in shape, with multiple local peaks and troughs and many different models with more or less equivalent levels of performance (even if they differ in the detail of their predictions). This is partly the result of different values of parameters in the model interacting in complex nonlinear ways to give similar predictions and partly the result of the predictions not actually being sensitive to changes in some parameter values.

Searching the model space will be considered in more detail in Chapter 4. For now, it is sufficient to note that, while it might be possible to find a global optimum in the model space, there is also a real possibility that the resulting parameters may not be invariant to either the calibration dataset used, the model structure used, the goodness-of-fit criterion or performance measure used, or even the initial set of parameter values used in the optimisation. In particular, we may often wish to evaluate model performance using more than one criterion. Then, because of the different sources of error in the modelling process, it will not generally be the case that the optimal model on one measure of performance will be the optimal model on a criterion measuring some other aspect of performance. If there is no single optimum on all criteria then there will inevitably be a compromise to be achieved in which improvements on one criterion are offset by deterioration on another. The result is a set of **Pareto optimal** models, all of which lie along the **Pareto front** which is the surface in the space of performance criteria on which none of the performance measures can be further improved without loss of performance on another (e.g. Figure 1.2, for the simple case of two criteria). Some methods for finding Pareto optimal models are presented in Chapter 4. As with the optimum for a single criterion, however, those models that are found to be on the Pareto front might change with calibration period or particular realisation of the inputs.

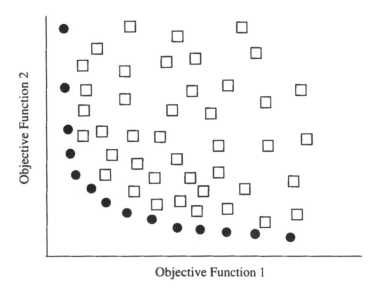

Objective Function 1

Figure 1.2 The concept of a Pareto optimum represented on two objective function axes. Each dot represents a model run with different parameter values. The circles are on the Pareto front, the squares are behind the Pareto front

As a result, it will be suggested in much of what follows in this book that environmental modellers will increasingly need to recognise the potential for many different models appearing to be acceptable descriptions of the environmental system of interest. In the past this has been mostly treated as a problem of identifying parameter values of a particular model structure given limited information. It was considered either as a problem of **non-identifiability** (difficulties of identification due to the lack of a clear optimum on the parameter response surface) or of **non-uniqueness** (difficulties of identification due to multiple optima on the parameter response surface or different optimal parameter sets resulting from different goodness-of-fit criteria). Both non-identifiability and non-uniqueness could arise from errors in the model structure, errors in observational data, inadequate observational data in identification, and mismatches in scale between observables and predicted variables. Such problems will be worse with models that are **over-parameterised,** and models that show interaction and co-variation between parameters. The result is that in calibration, the "optimal" model may vary for different calibration periods, variables or fitting criteria.

Such problems are now widely recognised but have not mitigated the continuing search for "optimal" models in very many studies. Duan et al. (1992), for example, demonstrated clearly the lack of a clear optimum in their hydrological modelling study (Figure 1.3), but suggested that the conclusion to be drawn was that better global optimisation techniques were needed. They provided a stochastic complex evolution algorithm that has since been very widely used in the calibration of hydrological and

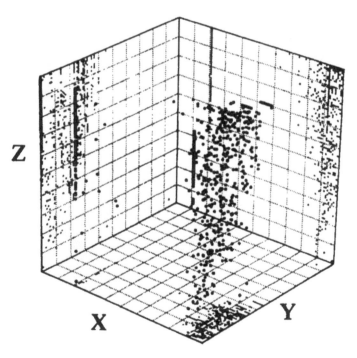

Figure 1.3 Three-parameter sub-space of a six-parameter model space for a rainfall runoff model. Each dot represents a local optimum for a sum of least square objective function

Source: Duan et al., 1992

other models. Others have suggested different types of genetic algorithm, simulated annealing, or multiple search criteria as means of refining the optimisation in the face of identifiability problems (see Chapter 4).

There is, however, another approach. That is to accept that it is very unlikely that our current model structures are truly realistic descriptions of the environmental system of interest so that there may indeed be many different models that can be shown to provide predictions that are acceptably consistent with whatever observed data are available. This is to treat the problem of identifiability as one of **equifinality** of model structures and parameter sets in reproducing the known behaviour of the system. This is a term originally used by Ludwig von Bertalanffy (1901–1972) in the context of General Systems Theory. It is not a view that has been widely taken but underlies the Generalised Likelihood Uncertainty Estimation methodology that is described in Section 4.5 (e.g. Beven, 1993, 2006a). In this view, equifinality is a result of the difficulty of deciding between competing models as hypotheses of how the system is working, given the limitations of the available information.

In prediction, identifiability problems, whether due to non-identifiability, non-uniqueness or equifinality, imply uncertainty. It is a prime motivation for this book that this uncertainty should be addressed in a much more explicit way in the future. Prediction is not the only aim of environmental modelling, however, and such uncertainty also has implications for explanation and understanding of the systems we are interested in.

1.7 Ensembles of models

As computer power has become more readily and more cheaply available there has been a change away from trying to find the optimal model in calibration (and sometimes estimating the uncertainty of the predictions in the region of the model space close to that optimum) to carrying out *ensemble* experiments. An ensemble is a collection of models, all producing different outcomes. Different types of ensemble experiment are used in different types of environmental modelling. Three essentially different uses may be distinguished.

The first is the use of ensembles to represent different scenarios of future conditions. An example of this is the use of the outputs from economic models to provide different emissions scenarios for use in modelling global climate change. The Intergovernmental Panel on Climate Change (IPCC[2]) provides outputs from a variety of global climate models running a small number of potential emissions scenarios. In this type of scenario modelling, an obvious question is which of the outcomes is the most likely? The point of modelling scenarios, however, is that it is impossible to say whether one scenario might be more likely than another. It is not even a good idea to suggest that they might be equally likely since we do not know that either. Scenario modelling is one way of dealing with uncertainty that cannot be dealt with in terms of chances, odds or probabilities (see Section 1.11 below).

A second use of ensembles is to explore the propagation of input uncertainties through a nonlinear modelling system. In this case the input uncertainties are defined

2 See http://www.ipcc.ch/.

a priori, very often as a form of probability distribution. For a linear system, such propagation can often be performed analytically, but for nonlinear systems it is often better to sample the range of input conditions and see what effect this has on the results. In certain cases, such as the simplified three equation nonlinear atmosphere model of Lorenz (1963, 1993), variability of the inputs can lead to chaotic behaviour in the system (that is the solutions for initial conditions differing by arbitrarily small amounts will diverge exponentially). We could expect similar behaviour for the effects of misspecification of initial conditions in the numerical weather prediction models that are used operationally in many countries of the world to provide the publicly issued weather forecasts. It is known that the predictions of such models are useful only up to a few days ahead before the predictions diverge from what actually happens. Thus, in many forecasting organisations (the UK MetOffice, MétéoFrance, the US National Weather Service) an ensemble of forecasts is being produced to try and span the range of possible developments of the atmosphere over the forecasting period. At the European Centre for Medium-Range Weather Forecasts (ECMWF) at Reading in the UK, ensemble forecasts up to ten days ahead are being produced (see, for example, Figure 5.1 in Chapter 5), and being used for a variety of purposes, including an operational system for identifying flood alert conditions for all the major river basins of Europe at the European Joint Research Centre (JRC) at Ispra in Italy.[3] Weather forecasting involves an updating or data assimilation step, each time a new set of observations becomes available and a new set of model runs is started. In ensemble weather forecasting, the different initial conditions for the ensemble are defined in a way that reflects how far the real atmosphere has departed from the past predictions. There are several different ways of updating ensembles in data assimilation (see Chapter 5).

The third use of ensemble simulations is to characterise the response surface in a model space. By running a sample of all possible models in the model space, the quantity of interest can be assessed (this might be a likelihood value, or the value of a particular output variable in the model) and, effectively, mapped in the model space. For simple models and low-dimensional model spaces this is a relatively simple problem and how the sampling is done will not matter too much. For high-dimensional spaces a large number of model runs would be required to characterise the response surface, and if each model would take a long time to run, then computing constraints would become an important limitation, even with modern high-performance parallel computer technology. Thus, efficiency of sampling is an issue. Two strategies can be distinguished. The first is to sample the whole space and represent the response surface associated with a value associated with each model run. The second is to make the density of sampling in the model space a direct reflection of the height of the response surface. If it is a likelihood measure surface, for example, then model runs are sought to make the density of sampling proportional to likelihood, with the highest density in the region of the highest likelihood. This strategy tries to ensure that model runs are only made where they are most important so as not to waste computer time. It has its limitations, however, where the surface is complex and areas of high likelihood are scattered through the model space. An additional step can be used with both

3 See the European Flood Alert System at http://ies.jrc.cec.eu.int/98.html.

strategies, which is to use a simpler model as a model emulator to interpolate between a smaller number of runs of the full model. All these strategies are considered in more detail in the later chapters.

1.8 Modelling for formulating understanding

A primary use of models in environmental research is as a way of formalising scientific explanation of environmental systems. Indeed, one aim of the research scientist is often to develop a model that will properly reflect his/her understanding of the environment in a way that will also be useful in practical prediction. Ultimately we would like to get the right predictions for the right reasons (see Beven, 2001d; Kirchner, 2006, and Chapter 2). The previous sections have outlined the main problems that we will meet in any environmental modelling application: problems of representing our understanding of the system in the perceptual model; problems of scale and commensurability; and problems of equifinality in reproducing the available observations.

These problems effectively provide limits to the extent to which environmental models can be used as explanatory constructs. Indeed, models used with the aim of formulating understanding are often used in a purely deductive sense, i.e. using a defined model to explore what the behaviour of a system *might be* if it did have the same characteristics as the model (for example, in exploring the impacts of climate change on water resources and terrestrial ecosystems).

An interesting discussion of the use of models for formulating understanding has been provided by Stan Schumm (1998) in the context of explaining landforms in geomorphology. Schumm has a different starting point from the modeller in wishing to understand the landscape on the basis of empirical evidence and theoretical reasoning, even if in only a qualitative way. In essence, therefore, his aim is to illuminate the perceptual model of the geomorphology of a landscape. Discussions of the application of scientific methodologies in explaining the earth have a distinguished history in geomorphology, and Schumm describes the contributions of Grove Karl Gilbert (1843–1918) (1886, 1896) and Thomas Chrowder Chamberlin (1843–1928) (1890) who each tried to formalise a method of multiple working hypotheses (see also the discussions in Haines-Young and Petch, 1986). This is clearly the methodology of explanation preferred by Schumm, but he recognises that in applying it to understanding landform development and form there are *"ten ways to be wrong"*. These ten problems relate directly to the types of modelling problems already noted here. They are grouped by Schumm into three classes as follows:

Problems of scale and space
- Time (observations are available only over a particular period and time span)
- Space (observations are available only at particular scales)
- Location (observations are available only at particular places)

Problems of cause and process
- Convergence (similar effects from different sets of causes, geomorphological equifinality)
- Divergence (different effects from similar causes)
- Multiplicity (effects due to multiple processes operating simultaneously)

- Efficiency (response to energy expended may not be a simple function of energy inputs)

Problems of system response
- Singularity (local differences due to unexplained variation or indeterminacy)
- Sensitivity (different responses to different initial conditions, especially close to thresholds)
- Complexity (complex responses of systems with multiple interconnected parts).

Schumm's discussion therefore serves to illustrate the limitations of even the perceptual model in describing the real system. His conclusion, however, is optimistic:

> The discussion of ten problems faced during explanation and extrapolation is not an attempt to discourage scientific investigation. Rather it is an attempt to emphasize the complexity of natural systems and to explain why explanation and extrapolation can often be suspect. The identification of these problems does not by a process of name magic solve anything, but it does help to develop a rational scientific approach to complex systems. Furthermore, recognition of the problems will lead to more thorough research plans. Consideration of the problems may, therefore, be difficult, time-consuming and expensive, but never as expensive as failure.
>
> (Schumm, 1998, p119)

In addressing the application of environmental models in what follows, we shall adopt a very similar attitude. The concept of models as multiple working hypotheses of how the system is working has much in common with an approach to modelling that recognises the equifinality of multiple models consistent with the available observations and it has been suggested that model evaluation be treated as a form of hypothesis testing rather than a matter of finding the optimal model (Beven, 2002a,b, 2006a). However, we must also then consider the possibility, suggested for example by Haines-Young and Petch (1986), that having multiple feasible models simply means that we are using poor models if we have not been able to properly distinguish between them (we will return to this question in Section 7.3). It may still be, of course, that our models (as hypotheses) are fine, but we simply do not have adequate observational data to really distinguish between them as explanations or predictions of the system of interest.

1.9 Modelling for practical applications

1.9.1 Simulation with no historical data available

Using models to simulate the environmental system of interest without any historical data available implies using models in a form of deductive reasoning. We are often forced into this situation. There are many applications where data are non-existent, or there is a lack of resources to collect data, or predictions are only made of future conditions (where it is impossible to collect data on what those conditions might be). Deductive reasoning has a long and prestigious history in science. It allows the consequences of a given theory and set of assumptions to be enumerated and, in many cases, later tested experimentally. There have been a number of well-documented cases of deductive predictions in science that have later been confirmed by observation (or at least by selected observations, see Press and Tanur, 2001). The implication is then that the assumptions of the theory are a good approximation to reality and from a strong realist viewpoint that the variables embodied in the theory are real variables. Confirmation in this way may be very important in persuading scientists to accept a theory, despite the fact that there may be numerous anomalies between theory and observation that still need to be explained. As noted earlier, the assumptions of the theory may be known to be wrong (or approximate), but anomalies will often be shelved pending further developments. One thing that the history of science teaches us is that even in 20 years time we will not be using the same methods and theories as now.

It is not necessary to make such strong realist claims for quantitative theorising about environmental systems, which will be necessarily incomplete, often at least partially based on empirical expressions and recognised as approximate. The important point is that precise deductions can be drawn from precisely defined assumptions and premises, regardless of whether those assumptions actually apply to any real system. Common sense suggests, of course, that it is more valuable to explore sets of assumptions that have some relationship to real environmental systems rather than those that do not!

Deductive models of this type define a virtual reality, or what Cartwright (1999) calls a **nomological** (law-like) system (see Chapter 2). The predictions of such models are valid only within the context of the model structure itself. As such, any reasoning about the nature of the real system that follows from deductive model predictions does so by analogy. A typical example would be the use of such models to examine the sensitivity of the outputs to different types of change in boundary conditions or parameter values. A deductive model might give an (approximate) indication of how the real system might operate conditional on the specific assumptions made about the processes, boundary conditions and parameter values but will not necessarily be very useful in predicting a specific instance of a particular application where the boundary conditions and parameter values might not be well known. The use of such deductive models will be considered in more detail in Chapter 3 in the context of exploring the model space.

The difficulty of such an approach, of course, is that the results depend precisely on the assumptions made in setting up the analysis. Whether these assumptions are realistic or not will depend very much on the value of the expert judgements made in deciding which models to use, which parameters to vary, what distributions should be used for each parameter, whether interaction between parameters can be taken into account, how much uncertainty should be allowed in initial and boundary conditions, and so on. Experience has shown that such expert judgements are not necessarily reliable in applications to real systems. It is probably best, therefore, if this approach is treated as the first stage in a learning process such that the initial expert judgements will be refined as some data become available for the model predictions to be evaluated (see Beven, 2007).

1.9.2 Simulation with historical data available

In many modelling studies, however, it is possible to gather together some data on how the system has responded in the past, subject to some limitations and uncertainties. Data will have been collected at specific points in space and time (which may not match the time and space discretisation of the model) and for only a restricted number of variables. Such data are commonly now stored and provided in the form of electronic files of numbers, without any record of how they have been collected or processed. A common problem in hydrology, for example, is to model discharges from a catchment given rainfall records. There are very many sites where discharge data are available to calibrate or evaluate model predictions. These can be obtained from the relevant agencies or, in some cases, directly on the Internet from public databases. The discharge in a stream is not normally, however, measured directly. It is inferred from another measurement, normally water level, by applying a "rating curve" developed for the discharge measurement site. The rating curve is often based on measurements of velocities in the cross-section taken at different water levels. As the level increases, however, it becomes more difficult and expensive to take such measurements, so that the discharges at flood water levels tend to be associated with a significant uncertainty that is rarely reported. Thus, modellers obtaining a file of such data will, lacking any further information about the uncertainties, generally take the numbers as "true". They should, however, be wary of doing so. Many other similar examples of different measured environmental variables could be added here, but there is still an expectation that such historical data should be useful in constraining the feasible models for a particular site.

Any reasonably complex environmental model is difficult to apply in a specific instance. Unlike a deductive model where the dependence of simulations on model assumptions, boundary conditions and parameter values is simply part of the deductive process, in any application to a *specific* place assumptions, boundary conditions and parameter values must be chosen to represent that particular part of reality in space and time. One of the implicit assumptions of a pragmatic realist approach to environmental modelling is that this should be possible; that, in principle, if we have a correct model structure then specific cases can be treated in terms of their specific boundary conditions and parameter values. In fact, the assumption made in practice is often stronger than this: that it should be possible to represent the unique characteristics of a specific case with specified boundary conditions by a unique set of parameter values. This ideal may not be reached because of data or measurement technique limitations but the principle underlies many modelling studies.

We are perhaps unlikely ever to reach a stage where attaining such an ideal is actually feasible but in any application it is necessary to use the available observations to decide on what the boundary conditions and parameter values appropriate to that case might be. In most cases this will entail a calibration process and in what follows in this book that calibration process will be represented as a form of inference from the observations about appropriate models for the specific case under study. Note that the *form* of model structure used may be exactly the same as a model chosen for deductive experimentation. The aim in inference will always be to use the available observations to obtain the best model or models for the application, or at least constrain the set of plausible models (as hypotheses) and reduce the prediction uncertainties. Induction as model calibration is an important tool in environmental modelling, albeit fraught with difficulties (see Chapter 2). It is an essential tool in applications to particular locations.

There is also the possibility that inference from model observations might be used as a more direct way of deciding on model structures by a process of hypothesis rejection. In situations where the model predictions do not appear to be consistent with the available observations then we might be able to learn about the usefulness of a model structure (or the data that are being used to drive it). There are certainly some published cases where this type of model evaluation has led to the rejection of all the models tried.

The role of induction and deduction in science has been the subject of extensive and continuing debate in the philosophy of science. We cannot really fully discuss the application of environmental models without at least some appreciation of the nature of the debate and its relevance to modelling. The next chapter therefore offers a short summary of the issues involved in the form of a philosophical diversion, while Chapter 4 deals with the techniques used in practice for making use of historical data in conditioning model predictions and constraining uncertainties.

1.9.3 Forecasting the near future

The last type of model application that we will consider is in making forecasts of the near future (sometimes called real-time forecasting or nowcasting). Examples include weather forecasting using atmospheric models and flood forecasting for decisions about warnings using hydrological models. In some cases coupled models might be required, for example in forecasting the transport of a pollutant wave down a river system when stream discharge is changing or for the effects of atmospheric pressure, tide and river flows on the forecast of a tidal surge in a river estuary.

In each of these cases, the issue is less about having a model that is correct in simulating the detail of the processes than having one that gives accurate forecasts with minimum uncertainties at the required lead time into the future. The concept of lead time is important in real-time forecasting, since in a decision-making context the lead time must be sufficiently long for any reaction, for example, by the emergency services and the general public, to be effective. We might be able to get very accurate predictions of flood discharges for a particular site at risk of flooding one hour ahead of time, but they would not be as useful as a much more uncertain prediction 6 hours or 12 hours ahead of time that would allow flood warnings to be issued to the public. The difficulty of providing warnings in systems with short response times is discussed in Section 5.1 with reference to floods in Boscastle, UK and Vaison-la-Romaine, France.

A second issue in forecasting the near future is that the speed with which model predictions can be made is important. This is currently an issue with numerical weather prediction models. The forecasters would like to do two things to improve their predictions. They would like to refine the grid scale of the models and they would like to increase the number of ensemble runs of the model for each forecast. The ensemble is made up of a number of different runs of the model, each with different patterns of initial conditions, reflecting the uncertainty in the knowledge of the atmosphere at the start of the run. Both decreasing the grid scale and increasing the number of runs would increase the computer run time.

Perhaps the most important issue, however, is the need for data assimilation in real-time forecasting. We expect, in any forecasting situation, that the model predictions will drift away from what actually happens. If we are making forecasts and it is possible to have some information about what is actually happening in the system, we can use that information to improve the forecasts. This is known as **data assimilation** or **adaptive forecasting** such that as the observations about the real system are received at the forecasting centre, any model bias or drift can be corrected. This requires that the observations are transmitted sufficiently rapidly to be useful within the lead time of the system but communication systems are now becoming cheaper and much more reliable, even in remote areas. There are now a number of different techniques of implementing adaptive forecasts that will be described in more detail in Chapter 5.

1.10 Guidelines for effective modelling

There have been a number of practical discussions about effective practice in environmental modelling. Reichert and Omlin (1997) discuss the use of process models in ecology in terms of the guiding principles of physicality, characterisability and identifiability, together with decision criteria of quality of fit, parsimony and "balanced accuracy" for model selection. Hill and Tiedeman (2007) discuss 14 guidelines for effective modelling in their particular context of groundwater modelling; Refsgaard et al. (2007) summarise work on the European Union HarmoniQuA project[4] in terms of five major steps and 48 tasks for good practice in integrated water management modelling. There is a lot of good common sense advice involved in these suggestions that is transferable to many other modelling problems and methodologies. There is also much more recognition that the modelling process should be expected to be imperfect and that the modeller should be aware of these limitations when interacting with the stakeholders who will use the model predictions in making decision (e.g. Olsson and Andersson, 2007; Stainforth et al., 2007b). The process of uncertainty estimation can provide a useful framework with which to structure such a discourse. Pappenberger and Beven (2006) and Faulkner et al. (2007) suggest the need to develop, with the input of stakeholders, codes of practice in different application areas (see also Section 6.11 later in this book).

My own experience suggests that the sets of questions below might be useful in structuring an approach to modelling applications that takes account of the potential uncertainties. The questions are addressed both to the modeller and to the relevant stakeholders.

1   Define the context of the problem, in discussion with the appropriate stakeholders. What type of predictions are required and at what degree of accuracy for effective decision making or hypothesis testing? What data are available to constrain the model predictions (for input data, boundary conditions and model parameters)? Are there qualitative data available that might constrain the model processes to be considered in the application? What additional data *could* be collected? What sorts of uncertainties are associated with the data?

4 See http://www.harmoniqua.org.

2   Define the modelling approach to be used. What model concepts are consistent
    with the context of the problem and the stakeholders' understanding of the sys-
    tem? Will the available model(s) provide the predictions required? Should more
    than one competing model structure be considered? (Would a simpler model do
    the job?)

3   Set up the model(s) carefully, including making basic consistency checks on the
    available data. Are there any obvious deficiencies in the data revealed by mass or
    energy balance checks? Are there space/time resolution issues in making accurate
    predictions? Are there discretisation issues in representing heterogeneities in the
    system?

4   Evaluate the performance of the model, including prediction uncertainties, against
    the available quantitative and qualitative data (where this is possible, of course).
    What is an appropriate uncertainty estimation method (see Section 1.9)? Where
    the model needs to be calibrated against observations of the system response,
    make sure that the predictions are also evaluated against additional data not used
    in the calibration exercise. Are there obvious deficiencies in the model predictions?
    Could these be the result of data errors/uncertainties? Could they be the result of
    model deficiencies (poor estimation of effective parameter values; poor represen-
    tation of heterogeneities; poor representation of processes; missing processes,
    ...)?

5   Consider whether the uncertainties associated with the model predictions can be
    constrained. Uncertainty estimation is not the end point of the modelling process
    but can be a guide to further work that might better define the response of the
    system. Would a particular set of additional measurements allow model param-
    eters to be better identified in a cost-effective way? Would a particular experiment
    allow different model structures to be differentiated as hypotheses about how the
    system is responding?

Most readers will see straight away that many of these questions have implications for
the resources available to the project. This will consequently have an effect on what is
possible or necessary to achieve the aims of a project and needs to be the subject of a
discussion between modeller and stakeholder in the light of the importance of the
application. In the UK, some £800m per annum is spent by the government and
Environment Agency on flood risk management, including the design and maintenance of flood defences. About 1% of that sum is spent on modelling studies but the
point has been made in formulating a modelling strategy for flood risk management
that if that 1% is not spent wisely then there is the potential that the remaining 99%
will not be used to greatest effect.

1.11 The meanings of uncertainty

Uncertainty means different things to different people. It is also sometimes used syn-
onymously with other terms such as ambiguity, vagueness, imprecision and
indeterminacy. There have been a number of attempts to define it in a more restricted
way so as to avoid uncertainty in its meaning, but since such constrained definitions
conflict more or less with everyday usage, it is probably more realistic to try and
enumerate the various meanings of uncertainty.

There are two classical technical formulations of uncertainty, one derived from set theory and one from statistics. In set theory uncertainty arises when there is a set of possible alternatives when only one is required. In statistics, probability theory allows the expression of uncertainty in terms of some measure, in the range 0 to 1, on a universal set of alternatives. The probability expresses the likelihood that any given alternative is the one that is required.

Both classical definitions of uncertainty have their limitations. In the case of set theory the limitation is primarily that of assuming that the set has a crisp boundary, i.e. any given element is either in the set or outside it (even if degree of membership for elements within the set might be described by a fuzzy measure). In the case of probability theory it is the assumption that the primary cause of uncertainty is randomness so that the probability measures can be interpreted as the long term (asymptotic) averages of a random process (these are called **aleatory uncertainties,** or sometimes just *noise*). The asymptotic assumption can be relaxed in a **Bayesian** interpretation of statistics (see Section 4.3 and Box 4.1) but there is still an expectation that for a stationary statistical process, the probabilities will converge to their long-term averages given sufficient information.

Both sets of assumptions can be questioned in practical applications and this has led to the development of some different representations of non-random uncertainty in terms of fuzzy sets, rough sets, fuzzy measures (e.g. Klir, 2006) and the Info-Gap approach of Ben-Haim (2006). Klir (2006) points out that uncertainty arises because of a lack of information or because of conflicting information. Thus, in order to constrain uncertainty it is necessary to take action to collect additional information (which may involve a cost) where this is possible. It is not always possible, of course, either because the measurement technologies do not exist or because we are unsure of the future. There may be many uncertainties about the future (in climate change, political and social change, technological change, and natural catastrophes) that cannot be assessed in any probabilistic way. Such non-random factors (**epistemic uncertainties**) will arise in environmental modelling in a variety of ways, but most particularly because of model structure and future boundary condition deficiencies that cannot be considered as simply random (though they can sometimes be treated *as if* they were random components where we have no better knowledge of how they might be non-random; see the discussion of the total error equation later in Chapter 4). Thus, alternative views of uncertainties will be worth considering, particularly where decision making in a practical application can tolerate a degree of imprecision in predictions. Fuzzy set methods aim to estimate the *possibilities* of potential outcomes rather than probabilities. The relationship between fuzzy set theory and probability theory is dealt with from a practical engineering viewpoint quite nicely by Ross (1995, Chapter 15) and more extensively by Klir (2006).

Fuzzy set theory derives from work in the 1960s by Lotfi Zadeh, though some of the ideas are recognised as being older. Klir (2006) notes that the theory of possibility has origins in the work of Ralph Hartley (1928) and the concept of potential surprise developed by the economist George Shackle (Shackle, 1949; 1955). In his 1965 paper Zadeh introduces the concept of a set with imprecise boundaries. One of the most important aspect of fuzzy set theory is that it allows descriptions of quantities in terms of imprecise *linguistic* variables (large, small, . . .) rather than simply crisp numerical values. Fuzzy sets are also capable of expressing *vagueness* and *non-specificity*. Both may be considered to be components of uncertainty within a fuzzy framework. Vagueness results from imprecision of definitions, as in the case of linguistic variables. Non-specificity results from lack of information about a quantity; as more information is gained, so our reasoning about that quantity should become more specific.

A closely related concept is that of the rough set (Pawlak, 1991). A rough set is an imprecise representation of a crisp set in terms of a lower and upper approximation. The lower approximation includes all elements from the universal set that are included in the set. The upper approximation includes all elements whose intersection with the set is not empty. An extension of the concept of rough sets is to make the approximations themselves fuzzy (fuzzy rough sets), or to represent a fuzzy set in terms of a rough set approximation (rough fuzzy sets) (Dubois, 1990; Dubois and Prade, 1992).

Fuzzy measure theory derives from work by Michio Sugeno (1977). It relaxes the assumption of additivity associated with classical measures in favour of weaker assumptions of monotonicity and continuity. There are several different classes of fuzzy measures based on different forms of monotonicity and continuity including plausibility measures, belief measures, possibility measures, necessity measures and, indeed, classical probability (Wang and Klir, 1992). Each allows the representation of forms of uncertainty in different types of applications.

The starting point for the Info-Gap theory of Yakov Ben-Haim (2006) is that there are many sources of uncertainty that cannot easily be represented by any form of distribution, either probabilistic or fuzzy. The Info-Gap theory is designed to cope with cases of extreme uncertainty by formulating nested sets of increasingly uncertain possibilities. He refers back to the work of Frank Knight in the 1920s. Knight (1921) differentiated between risks that people would be prepared to insure against (i.e. uncertainties that can be represented as statistical odds or probabilities) and what he called "true" uncertainties that cannot be assessed in this way. The latter, essentially unquantifiable epistemic uncertainties are now sometimes known as *Knightian* uncertainties. Many of the uncertainties in predictions of future economic change and its effect on climate change are of this type (e.g. Stainforth et al., 2007a).

The fact that Knightian uncertainties are difficult to quantify does not mean that they are unimportant or can be ignored. Many future scenarios used as boundary conditions in modelling change in environmental systems are of this type. We can make some guesses (create scenarios) about what the boundary conditions might be in the future, but we cannot very easily assess whether one scenario is more likely than another, or whether we have missed some likely potential futures because of lack of knowledge or understanding (Type III errors or Donald Rumsfeld's unknown unknowns).

There is one particular type of epistemic uncertainty that is often overlooked in the application of environmental models, that Linkov and Burmistrov (2003) call "modeller uncertainty". Their study of the uncertainty in model predictions in an International Atomic Energy Agency Biosphere Modelling and Assessment project demonstrates how the interpretation of scenarios and the use of different assumptions was greater in controlling the range of predicted outcomes than the effects of allowing for parameter uncertainty in the models used. In part, this was explained by the tendency for modellers to use approximate reasoning in making decisions in the face of uncertainty in ways that can lead to biased outcomes (as do other decision makers, see Morgan and Henrion, 1990). Linkov and Burmistrov (2003) report how the differences between different modelling groups decreased following meetings to develop a consensus interpretation of the scenarios to be modelled, and when data were available for (partial) model calibration. Whether such a consensus will reduce the potential for bias in the predictions, of course, is another question that can only be answered when there is the possibility of a post-audit analysis. Other model inter-comparisons have revealed large differences between different modelling groups, especially when different model structures are used. Refsgaard et al. (2006) report an interesting example from a groundwater water modelling exercise, when different consultancy companies were asked to evaluate the available geological data for an area west of Copenhagen in Denmark to define a conceptual model of the aquifer structure. The different groups produced quite remarkably different interpretations.

Regan et al. (2002) also discuss the issue of **linguistic uncertainties** in the context of ecology and conservation management which arises because "much of our natural language, including a great deal of our scientific vocabulary, is underspecific, ambiguous, vague, context dependent, or exhibits theoretical indeterminacies" (p.618). They discuss each of these types of linguistic uncertainty in turn and, from the discussion, it is clear that there is significant linguistic uncertainty about the classification of different types of uncertainty in applications to environmental problems. When, for example, is something that is vague (they use examples such as "endangered species" or "viable population") simply a result of epistemic uncertainty? Certainly some vague variables can be addressed using the fuzzy approaches briefly outlined above, while the indeterminacy of theoretical terms might also be considered as a form of epistemic uncertainty as outlined above. Their advice is to try to be as precise as possible in dealing with linguistic uncertainties, to accept that not all uncertainties can be dealt with in probabilistic terms, and to seek to represent vagueness using fuzzy, rough set, or similar measures.

This brief overview of different meanings of uncertainty is by no means complete but it is clear that these different ways of representing different sources of uncertainty in the modelling process will result in different estimates of the uncertainty of a model predicted outcome. Environmental modelling problems are complex enough that there is no clear consensus about which form of representation should be used for different types of application (except perhaps among those who think that probability is the only framework for dealing with uncertainty and who are prepared to gloss over the real difficulties in real applications of forcing some sources of uncertainty into a probabilistic representation; see Chapter 4). In one sense it does not really matter how the estimates of uncertainty are made as long as the assumptions of any analysis are set out clearly and openly. Scientists, decision makers and stakeholders can then assess those assumptions in an open way. Within those assumptions, each form of analysis can be carried out objectively. In another sense, however, it does matter a lot, because the possibility of multiple representations of uncertainty exacerbates the communication problem from scientists to decision makers and stakeholders. That communication problem remains an important hindrance to the more widespread use of uncertainty estimation in environmental policy and decision making.

1.12 Deciding on an uncertainty estimation method

It will be clear from the sections above and from simply reading the table of contents of this book that there are a wide variety of uncertainty estimation methods available to the environmental modeller. There are also some important controversies about what methods should be used for different applications. To try and guide potential users through the choice of an uncertainty estimation method a decision tree has been developed and included as part of the Wiki pages on risk and uncertainty developed within the UK Flood Risk Management Research Consortium[5] (Figure 1.4). The site includes descriptions of different methods, including illustrative case studies. Wiki pages are Internet sites where users can edit and add material themselves. In the case of the Risk and Uncertainty pages it is hoped that users will report additional case studies and report experience of using different methods (see also Pappenberger et al., 2006a).

It is immediately evident that Figure 1.4 assumes that a suitable model is already available. The approach taken in this book is similar. It does not attempt to deal with fitting statistical distributions or relationships to data. There are many books on statistics and geostatistics already available that cover that material well (see, for example, Clarke, 1994, and Hipel, 1995, in hydrology, or MacGarigal et al., 2000, in ecology, or the geostatistics texts of Clark, 1979, Cressie, 1993, and Kitanidis, 1997). The classical methods of linear statistics still have value in environmental modelling, although the user sometimes needs reminding that those methods can be used to estimate the uncertainties in, for example, multivariate regression relationships or geostatistical interpolations, as well as the best estimate values. Good recent examples are the regionalisation studies of Jones and Kay (2007) and Yadav et al. (2007) in which purely statistical models of catchment hydrological variability are used to extrapolate to ungauged catchments *with* estimates of uncertainty. Here, our starting point is to assume that a formal model structure has already been constructed and that the user wishes to investigate the uncertainties associated with its use (Chapter 3) and constrain those uncertainties by taking new observations or different types of observations into account (Chapter 4), perhaps for real-time forecasting (Chapter 5) or as an input into risk-based decision making (Chapter 6).

1.13 Uncertainty in model predictions and decision making

Models are often applied in the framework of a decision-making process. Indeed it is generally the case that an agency or company with responsibility for making decisions compatible with current legislation will commission a modelling study to help fulfil that obligation in a way that demonstrates that science has been brought to bear on the decision. Decisions made frequently, such as the definition of flood risk zones for planning purposes, might be associated with recognised modelling methods and standards for data collection (in that case for estimating flood frequencies and predicting the inundation associated with a design flood). In other cases, important decisions, such as where to site a nuclear waste repository, might involve a multi-million dollar

---

5 See http://www.floodrisknet.org.uk/methods/Introduction.

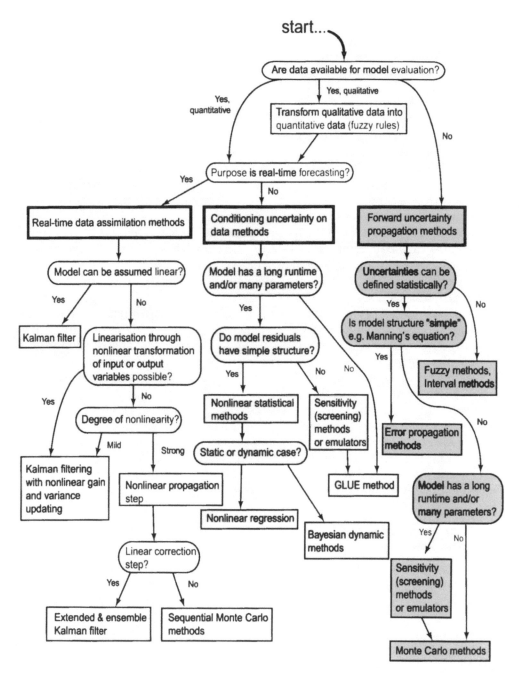

Figure 1.4 A decision tree for choosing an uncertainty estimation method

Source: http://www.floodrisknet.org.uk/methods/Introduction

data collection and model development programme to inform the decision-making process. The proposed radionuclide waste disposal sites at Yucca Mountain in the US and Sellafield in the UK are both good examples of this where, despite a huge expenditure and modelling effort, the uncertainties of what might happen to radionuclides stored at depth were not adequately resolved because of different interpretations of the data and model implementations by different scientists.

It is important to recognise that most such decisions have to do with individual sites, and that there will always be some uncertainty about the model representation and characterisation of the site given the limitations of the data that might be available. This suggests that the decision-making process should take account of such uncertainties (at least for any non-trivial decision). This has rarely been the case in the past. Environmental modellers have not been very good at assessing the uncertainties in their predictions, and decision makers have not been very good at knowing how to deal with scientific uncertainties when the modeller supplies them. There is a myth that decision makers, in government, environmental agencies and local authorities do not want to know about scientific uncertainty. They only want to be given the *best* possible scientific estimate on which to base their decision.

And yet, decision and policy makers have always had to deal with uncertainties in making decisions, whether formally or informally, and it is surely a duty of the scientist to make the best estimate of uncertainties possible in any application as an input to that decision-making process. There are a variety of methods available for decision making under uncertainty (see Chapter 6) and perhaps both modellers and decision makers just need to be better trained about what is possible and how to discuss and communicate the nature and meaning of prediction uncertainties from scientist to decision maker to public and vice versa (van der Sluijs et al., 2005a; Faulkner et al., 2007).

However, it is also true that any decision will not depend simply on the science; other political and societal considerations will also come into play with their own uncertainties (and in fact those factors can already be seen in shaping some model prediction uncertainties provided to policy makers, such as the range of potential climate changes produced by the IPCC which provide a censored consensus range of outcomes without any attempt yet – for good computational reasons – to assess the real range of uncertainty in the climate dynamics). Environmental decisions involve a complex interplay between scientific, societal, and political questions and uncertainties. Indeed, social scientists have often argued that the framing of the scientific question itself is not without a political and social context that often does not include the input from all stakeholders in the problem. They advocate a much more participatory decision-making process. The way in which modelling can inform decision-making processes will be considered in more detail in Chapter 6, including the use of Info-Gap theory to deal with the non-probabilistic Knightian uncertainties.

1.14 Summary of Chapter 1

This chapter has been a general introduction to some of the issues and problems that arise in trying to apply environmental models for real applications, and the sources and types of uncertainty that are involved. These may be summarised as follows:

Problems of representation of understanding
- The expected limitations of our perceptual models as representations of the real system
- The simplification inherent in moving from a perceptual model to a formal model
- The additional approximation that might arise in solving the formal model on a digital computer

Problems of scale and space
- The need for system closure by defining appropriate boundary and initial conditions
- The fact that the procedural model will require scale-dependent effective boundary conditions, initial conditions and parameter values that may be different for different model implementations
- The problem of calibrating the model boundary conditions, initial conditions and parameter values in any application to particular places
- The problem of incommensurability of measurements (and theory) with what is required at scales of modelling
- The problem of transferring parameter values from one model application to another

Uncertainty and practical applications
- Different methods of uncertainty estimation are required for different types of problem
- In particular, uncertainty estimation without historical observations available, uncertainty estimation with conditioning on historical observations, and real-time forecasting are distinguished

Problems of representation of uncertainty
- The problem that not all uncertainties are quantifiable
- The problem that models of modelling errors may be difficult to construct and may involve non-stationarities arising as a result of input and model structural error
- The problem that different model structures and parameter sets may do equally well in fitting the available observational data
- The problem of using different representations of uncertainty in decision making

Uncertainty and decision making
- In real application, uncertainty estimation is only a means to an end: the end of making better decisions
- Methods for decision making in the face of uncertainty exist but how to best represent uncertainties, and the basis for the estimation of uncertainties, to decision makers is still an unresolved issue.

Chapter 2

A philosophical diversion

When looking for positive guidance from philosophy we must rest content with some vague generalisations about the need to be specific.

Alan Chalmers, 1989

All models are wrong, but some models are useful.

George Box, 1979

2.1 Why worry about philosophy?

It is, of course, quite possible to develop and use environmental models without any explicit underlying philosophy. Many practitioners do, although most might have the aim of developing and using models that are as "realistic as possible", given the constraints of current knowledge, computing capabilities and observational technologies. This type of implicit or pragmatic realism seems quite natural and appears to be intrinsic to modelling efforts such as the coupled Global Circulation Models being used to predict future climate change. The philosophical subtleties are not really necessary to the practising environmental modeller who only needs to know that achieving realism is still difficult in the practical prediction of complex environmental systems. Current implementations of such models are known to have their limitations but it is implicitly accepted that they will continue to evolve towards more realistic representations of the earth–ocean–atmosphere system.

While this pragmatic realism has served environmental science well it has long been demonstrated that it has major flaws. It should be replaced, but experience suggests that philosophical arguments alone have had little impact on the way in which modelling is actually done. This is, perhaps, beginning to change and there is the start of a recognition of the need for a more scientifically robust assessment of the foundations, capabilities and limitations of model predictions. This view stems not from any deep notions of environmentalism, nor from any higher-level holistic principles of nature, nor directly from philosophical arguments about the concept(s) of realism itself, but from practical experience of the difficulties of modelling in different areas of environmental science. In this chapter, a philosophical framework for environmental modelling is outlined that might satisfy this need while allowing for model structural error and the incorporation of improved knowledge over time.

What I am calling pragmatic realism here naturally combines elements of foundationalism,[1] actualism, empiricism, idealism, instrumentalism, Bayesianism, relativism and hermeneutics; of multiple working hypotheses, falsification, and critical rationalism (but allowing adjustment of auxiliary conditions); of confirmation and limits of validity; of competing scientific research programmes while maintaining an open mind to paradigm shifts; and of the use of "scientific method" within the sociological context of competing research groups and the politics of grant-awarding programmes. Within the philosophic tradition, refined and represented in terms of ideals rather than practice, it probably comes closest to the transcendental realism of Roy Bhaskar (see Section 2.3). However, in environmental modelling, at least, the *practice* often appears to have more in common with the entertaining relativism of Paul Feyerabend (1975), not least because theories are applied to open systems, artificially treated as closed by the specification of approximate boundary and initial conditions (Morton, 1993).

As Nancy Cartwright (1999) has pointed out, in open systems, even the fundamental equation *force=mass*acceleration* may be difficult to verify or apply in practice. Hydrologists also know only too well the difficulties of verifying or applying the mass and energy balance equations in open systems (Beven, 2001a,b, 2006b). Other areas of environmental science will have similar examples of fundamental principles that cannot be verified or easily applied because of lack of knowledge of all the factors influencing the system of interest. This does not, of course, mean that such principles or laws should not be applied in practice, only that we should be careful about the limitations of their domain of validity (as indeed are engineers in the application of the force and force balance equations when they will often be careful by applying a factor of safety to design calculations).

Cartwright (1999) suggests that open systems might best be represented in terms of the *capacities* of real entities to respond in a particular way to external influences. She argues that representation of capacities is only possible within a nomological system (law-based formal system) with its own defined constraints and limitations. Thus, the problem of defining a formal model of the system is, within this framework, a matter of defining a nomological system that may not in itself be realist in terms of being totally consistent with the perceptual model but which can be used to produce quantitative predictions within the limits of its own definition. We should also note here that even having designed a consistent nomological representation of an environmental system then there may be further approximations necessary in implementing that system on a digital computer, such as in the approximate discrete numerical solution of the continuum partial differential equations that are the basis of many environmental models (what was called the procedural model in Chapter 1).

It is in the critical rationalist idea that the description of reality will continue to improve that many of the problems of environmental modelling have been buried for a long time. This apparent progress is clearly the case in many areas of environmental science such as weather forecasting and numerical models of the ocean. It is not nearly so clear in areas such as hydrological modelling even though many people feel that, by analogy, it should be. It has led to a continuing but totally unjustified determinism in

---

1 Definitions of all the highlighted terms will be found in the Glossary at the end of the book.

many applications of modelling and a lack of recognition of the limitations of the available predictive models.

There may be a fundamental difference between different areas of environmental modelling here. It seems that some areas, such as hydrology, are no longer primarily limited by computational requirements but by other *informational* constraints, whereas areas that apparently continue to advance are still primarily computationally limited. If this is the case then the scope for improvement in areas where information limitations are more important than computational limitations may be severely constrained (at least until new measurement techniques become available). On the other hand, areas where computational constraints are still a controlling limitation (such as in atmospheric modelling) may not yet have reached their limit of improvement, but can be expected to do so some time in the future. That will be the point at which knowledge of system characteristics, boundary and initial conditions, for everywhere in the domain of interest and limited by the available measurement technologies, starts to become a dominant controlling factor on prediction uncertainties.

2.2 Pragmatic realism

Computer simulation has become a common research and practical methodology in the environmental sciences. It can be used as a framework for formulating and testing theories. It can be used to make predictions for practical applications in response to demands from policy and decision makers. As noted in Chapter 1, particularly interesting examples are the predictions of global climate change and its consequences, and predictions of the impacts of underground repositories for nuclear waste. Very large research programmes have been funded with a view to improving the accuracy of these predictions. There is an implicit belief underlying these efforts that such improvements are possible, despite the nonlinear and open nature of the systems being studied and the approximations necessary to implement our qualitative understanding of the systems under study in a working computing program.

The foundation of this belief is the form of pragmatic realism noted above. I am a hydrologist. As a hydrologist I intend that the computer simulation models I develop should represent real water; in modelling contaminant transport in flowing water I intend that the models should represent the real contaminant. This is in spite of the fact that, as a hydrologist, I would recognise the varied nomological status of many of the concepts that are used in my models (Beven, 1989; 1996b; Shrader-Frechette, 1989; Davis et al., 1992; Hofmann and Hofmann, 1992; Section 3.1). In the same way, the atmospheric modeller will intend that the variables in his computer should represent the real atmosphere and the aquatic geochemist will intend that the (approximate) computer solutions to systems of stoichiometric equations should represent real aqueous solutions. In time, new understanding and knowledge gained from experiment about reality will be incorporated into improved simulations. In this way the research programme of environmental science should progress.

This common working version of realism is, of course, rather naïve. It is naïve not only from a philosophical point of view (a more sophisticated realist position is outlined in the next section) but also from a scientific point of view, since it requires that the systems under study be knowable. Clearly, for many environmental systems the complexities that can be included in a purely qualitative perceptual model are such that all the boundary conditions, auxiliary conditions and system characteristics cannot be knowable given current measurement technologies. Thus, in applying this pragmatic realism to any particular open environmental system, it is clear that we can recognise much more complexity than it is possible to represent in a mathematical model that, implemented as a computer simulation, will make quantitative predictions. In these cases the need for approximating the perceptual model into a set of mathematical concepts at a certain scale is obvious (Beven, 2006b). A good example is the parameterisations of sub-grid scale processes in global circulation models that are as much a result of computational constraints as lack of understanding of smaller-scale processes. In other cases the process is not so self-evident, such as when equilibrium geochemical codes for solutions are applied to mixtures of waters (and mineral surfaces and organic colloids) from multiple sources at field scales. Even though the principles of the perceptual model may be well understood, the implementation of those principles in practice may be difficult both for reasons of finding realistic approximations of complex open systems (within the limitations of computational constraints) and also because of lack of knowledge of the local characteristics of the system and its boundary conditions.

This may be in part because we still have some surprises to learn about the nature of the processes but it would seem that even if we had a *perfect* theoretical description, the real problem is that the detailed characteristics of the system, that have an important control on how the system works, may be essentially unknowable. Keeping to the sphere of water, there is an analogy here with the study of turbulence. Homogeneous turbulence has received much study and simplified representations of the energy dissipation down to viscosity-dominated scales are used widely in computational fluid dynamics (CFD). It is true that there are still debates about the best closure schemes to use at modelling scales above what is computationally possible with direct numerical simulation, but for systems that are largely self-organising, even approximate CFD models appear to produce reasonable results.

However, there are many fluid dynamic systems in both laminar and turbulent flow regimes in which the interactions with the boundaries dominate the energy dissipation. This is true for shallow turbulent water flows in streams and rivers (where form roughness and the aquatic and bankside vegetation may also play an important role in energy dissipation depending on the discharge in the channel), requiring the specification of effective roughness parameters to reflect these small-scale effects; it is true for (mostly) laminar porous media flows in soils and groundwaters. Again, it is not so much that the principles of the interactions are not understood in the perceptual model (though we may still have much to learn) but that it is necessary to characterise the energy dissipation at every point in the flow domain or, more correctly, at the scale of the elements of the approximate representation of that flow domain in the implementation of a formal model to produce quantitative predictions.

It could be argued from a pragmatic realistic perspective that the problems of defining the parameterisations of a formal model and defining effective parameter values in system characterisation are not, in themselves, sufficient to require a new philosophy of environmental modelling but merely reflect the technological constraints of today (computing limitations, measurement limitations, theoretical limitations etc). The expectation is that new technological developments will reduce the significance of today's problems. There is always the hope of unforeseen technological innovation but this argument does not survive an examination of the effects of trying to apply models to specific locations with their own unique characteristics.

2.3 Other philosophical concepts of realism

The pragmatic realism of the previous section does not correspond directly to any of the concepts of realism expounded by philosophers of science, of which there are many. In fact, there are almost as many concepts or modifications of concepts of realism as there are philosophers of science who have written on the subject. A good (and relevant) example is provided by the article by Alan Chalmers (1989) entitled "Is Bhaskar's concept of realism realistic?" Philosophers, of course, tend to be concerned primarily with clarifying the principles underlying a realistic approach to science, as opposed to the rather messy practice sketched in the view of pragmatic realism outlined above.

One of the more interesting realist accounts of science produced in recent years is that due to Roy Bhaskar (1980; 1989). This is his transcendental realism, sometimes referred to by others (and accepted by Bhaskar) as critical realism (see, for example, Collier, 1994). Bhaskar's realism is transcendental in the sense that it uses the form of transcendental argument made popular in philosophy by Immanuel Kant (1724–1804) as, *if X is observed what are the conditions necessary for the observation of X*. This is the type of argument used all the time in applications of models when the model is being used to formalise the conditions necessary to reproduce an observable (at least approximately).

It is worth noting, however, that this argument involves three stages. The first is the *capacities* necessary in the system for the occurrence of X; the second is the set of *conditions* necessary for the occurrence of X (which may involve a specific trajectory of the system); and the third is the translation of the real quantity X into an observed quantity, X'. Within this framework, however, X and the conditions and capacities that caused it to happen can be considered to be real. Indeed, any conditions and capacities that are currently *non-observable* can also be given a realist interpretation. Many practising environmental modellers would, I am sure, be quite happy to incorporate this type of position into their pragmatic realism, however difficult it is in practice to represent those real capacities and conditions in their current models.

Chalmers (1989) provides a useful critique of Bhaskar's approach in respect of the use of models. Although he states quite clearly that he considers transcendental realism to be the best account of a philosophy of science yet available, he suggests that it can still be criticised on some points of detail.

2.4 Models as instrumentalist tools

There is another philosophical position that, given the many practical problems of environmental modelling, might often seem attractive. This is instrumentalism, in which models are seen only as tools, without making any realist claims. In the philosophical literature, instrumentalism is therefore often presented as an anti-realist position, though this is not necessarily the case. There will be many practitioners and environmental modellers who will be pragmatic realists in intention, but instrumentalists in practice. Instrumentalism is, if you like, the engineering view of environmental modelling. Practical engineering requires models for prediction in practical applications. Model limitations may be recognised and factors of safety introduced in the use of the predictions, but one suspects that predicted pressures are still usually understood as "best estimates" of the real pressure distribution; predicted velocities are still usually understood as "best estimates" of the real velocity distribution; predicted contaminant concentrations are still usually understood as "best estimates" of real concentrations. The model is a tool, but the best tools should have a basis in reality.

Recent expressions of instrumentalism are due to Bas van Fraassen (1980), Nancy Cartwright (1983, 1999) and Susan Haack (2003). At the centre of the instrumentalist view is that all scientific theories of the past have proven to be false so that it is logical to expect that all current theories will turn out to be false. This need not, however, prevent us from making statements about the nature of things, including quantities that are not directly observable. Such statements, including models, can allow us to make useful predictions in practice, it is just that we should not believe that they are real. This is the sense in which instrumentalism is anti-realist, especially since it also allows that some statements about the nature of things may be subjective. It follows that the only justification for scientific theorising is empirical adequacy, a concept that is also intrinsic to the pragmatic realism outlined above (and in much of what follows in this book).

2.5 The model validation issue

The concept of model validation appears widely in the environmental modelling literature. It tends to be used in the context of testing a model that has been previously calibrated for a particular purpose. The test might comprise prediction of a different period of observations to that used in calibration, or a different set of variables (see, for example, Klemeš, 1986, and Refsgaard, 1996, in the context of hydrological models; and Rykiel, 1996, in the context of ecological models). Models that survive such a test are deemed validated or verified.

This terminology is not acceptable from a philosophical point of view since the terms "validation" and "verification" imply a truth value or degree of realist belief that we are aware that models, as approximations, do not have. The issues of validation and verification in respect of environmental models have been discussed in detail by Oreskes et al. (1994) in a contribution from professional philosophers as a response to the evaluation of the predictions of groundwater models by practising environmental scientists (Konikow and Bredehoeft, 1992). Similar issues arise in other areas of the application of simulation models (e.g. Herskovitz, 1991; Kleindorfer et al., 1998). Oreskes et al. suggest that verification and validation of models of open natural systems is impossible, despite their widespread use in the modelling literature. They point out that few, if any, models are entirely confirmed by the available data, but equally few are entirely refuted. Models of such systems may be non-unique and only a conditional confirmation is possible. It is conditional because it may depend on the calibration of parameters or other auxiliary conditions and may also depend on the period of data used in evaluation.

If models cannot be verified or validated but only conditionally confirmed, can we have any belief in their predictions? This is not just a philosophical issue but also a very practical issue, for example when there are competing model predictions involved in a court case or public inquiry. Bair (1994), for example, describes his experiences as an expert witness in the $500m lawsuit that concerned the injection of hazardous wastes into a deep groundwater system. Two models produced by the plaintiffs and defendants produced very different predictions. That submitted by the plaintiff predicted waste migration three times greater than that submitted by the defendant. In addition, the lawyer for the plaintiffs produced a copy of the Konikow and Bredehoeft (1992) article in court with a view to casting doubt on all groundwater model predictions.

Bair notes that, although two different codes, both quality-assured, were used, they were not both used to the same standards of good practice. In this case, it was possible to demonstrate more belief in one set of predictions than the other or that one model was more valid as a description of the real system than the other. The jury apparently agreed, despite the impossibility of validation in principle.

Another interesting example, also involving two groundwater modelling exercises, occurred during an official public inquiry into a proposed deep-rock characterisation facility at Sellafield in the UK. The application for the facility was submitted by NIREX Ltd., a company set up by the UK government, which has responsibility for planning for the disposal of intermediate-level radioactive waste in the UK. Having originally failed to get approval for a proposed near-surface disposal site (which happened to be in a marginal constituency held by the government of the day), they had developed a plan for a deep-rock disposal site at 650 m below sea level beneath Sellafield (the site in West Cumbria of the nuclear fuel reprocessing plant run by British Nuclear Fuels plc.). This site has the advantage that waste from the Sellafield nuclear fuel reprocessing plant would only have to be transported a short distance. The site has the disadvantage that it lies beneath the water table within the Borrowdale Volcanic Group geological series which, while not highly porous, is subject to significant fracturing. There would therefore be potential for radionuclides to eventually move from the depository back to the surface. It turned out that the safety case for the depository depended on the nature of the assumptions made about the geology and boundary conditions over the next 10,000 years. The deep-rock characterisation facility was proposed, at huge cost, to reduce the uncertainties inherent in predicting the groundwater flows.

The amount of information about the geology at depth was limited. Some 13 boreholes had been drilled for NIREX (at a cost of some £1 m each) but this still allowed a significant degree of flexibility in interpretation of the geological structure of the Borrowdale Volcanic Group. AEA Harwell produced a model, on behalf of NIREX, that suggested than the groundwater flows would only reach the surface below the Irish Sea. Haszeldine and McKeown (1995) produced a model, on behalf of Greenpeace who were opposing the planned facility, that suggested that the proposed repository was in a zone of flow towards the surface and that the range of experimentally measured conductivities in the boreholes actually spanned the range over which the repository would fail the safety requirements. Both models were based on quality-assured groundwater modelling codes; both made assumptions about the nature of the system that were consistent with the geological evidence and future scenarios; but the resulting predictions were quite different. In this case, neither model was considered to be more valid than the other; both appeared to be feasible representations of the groundwater flow system. However, even though the public inquiry was convened only to consider the establishment of the deep-rock characterisation experimental facility and not the waste repository itself, there was enough doubt about the possible safety implications arising from the modelling that the NIREX planning application was rejected. The UK was left with no plans for an intermediate-level waste disposal site.

2.6 The model falsification issue

One of the most influential 20th-century philosophers of science, Karl Popper (1902–1994), argued that the most that could be hoped for from a scientific theory was a degree of *verisimilitude* with respect to the real phenomena (Popper, 1963). This neatly avoids (by a simple change of words) any debate about verification or validation in terms of truth value, but raises questions about whether there are measures of verisimilitude that would allow theories or models to be ranked or preferences to be expressed. Popper never really addressed this issue in that way but put much more focus on the possibility of falsifying models as a means of avoiding the problems of induction in developing theory (Popper, 1979). Indeed, the primary requirement that Popper demands for a theory or model to be scientific is that it should be falsifiable by empirical observation. He did, however, admit that there may be degrees of falsification or testability. He suggests that theories that survive testing by a wider range of potential falsifying statements are to be preferred, while those that cannot be falsified (the set of potential falsifying statements is empty) should be considered as unscientific.

This does not really allow for the practical application of environmental models which mediate between some theory and some practical application with all its complexities of process and place. As has already been pointed out, mediating models are necessarily approximate and require additional auxiliary conditions (effective parameter values, boundary conditions) that may be specific to that application but can only be determined by some process of calibration against observed variables. This calibration process can save a model from being falsified. If the model fails some test, it may be possible to avoid rejection by changing the auxiliary conditions or simplifying assumptions (Beven, 1993; Morton, 1993). This may be the case even if the observations really do call the assumptions of the model into question. Models are, in consequence, purpose-specific. The process can become even more circular if the *interpretation* of observational data depends on the theory that underlies the model, further reducing the possibilities for falsification.

In practice, it would appear that falsification is not really an issue in most environmental modelling. This is partly due to the fact that all models can be considered false if examined in sufficient detail, and partly due to the fact that, in any practical problem that demands some predictions, at least one model must be retained to carry out the simulations. As long as a set of parameters can be found that give some *acceptable* fit to the available observations, then predictions can be made. We will address the issue of when a model is acceptable in later chapters. For the moment it is sufficient to note that in the same way that there will be degrees of confirmation or degrees of falsification of a scientific theory, the environmental modeller will have to deal with degrees of acceptability of one or more models for a particular purpose.

2.7 The model confirmation issue: Bayesian approaches

The fact that all environmental models can be falsified does not mean that all models are equally poor; some might be useful as the eminent statistician George Box suggested in the quotation at the head of this chapter. Any modeller knows from practical experience that some models (or parameter sets, or sets of boundary conditions) will do a better job at reproducing the limited observed data that is normally available than others. We should therefore expect to have a greater belief in the predictions of some models than others; we should also expect that there will be an infinite class of models of a particular system in which we should have no belief at all. The class of models that are, in some sense, acceptable is likely to be much smaller. Those models can be considered, to some degree, confirmed by the available data as hypotheses of how the system being predicted actually functions. A probabilistic approach to model confirmation can be formalised in terms of Bayes theory. The philosophical background to a Bayesian approach to science is discussed more extensively in Earman (1992) and Howson and Urbach (1993).

The Rev. Thomas Bayes was a Sussex clergyman who also studied mathematics. After his death a paper titled *An essay towards solving a problem in the doctrine of chances* was discovered amongst his effects and communicated posthumously to the Royal Society of London by his friend Richard Price in 1763. The paper contained the first expression of what is now called Bayes theorem. For our purposes here we can define Bayes theorem in a form that, given a set of feasible models or hypotheses M and evidence E, then the probability of any M given E is given by

$$P(M|E) = P(M) \, L(E|M) \, / \, C \qquad [2.1]$$

where $P(M)$ is some prior probability defined for all feasible models, $L(E|M)$ is the likelihood of simulating the evidence given the model, and C is a scaling constant to ensure that the cumulative of the posterior probability density $P(M|E)$ is unity. Bayes himself allowed only for a uniform prior probability distribution.

Bayes theorem can be applied sequentially as new sets of evidence become available. The posterior probability density at the end of one step becomes the prior density at the next step. It therefore provides a rigorous mathematical basis for the expression of degrees of confirmation of different models as long as the different components of [2.1] can be defined adequately (see, for example, Bernado and Smith, 1994).

This sequential process must of course start somewhere, so that some initial or prior probability density is required. This is often defined subjectively by the modeller on the basis of past experience or subjective judgement. After repeated application of [2.1], however, the effect of different subjective assumptions about the initial prior should be reduced by the information introduced by the evidence E (at least providing that the available evidence does add information about how far one model or hypothesis is better than another). For this reason, when there is no strong prior evidence to express greater belief in one model than another, a *non-informative* prior is often used.

The second requirement is a definition of the likelihood $P(E|M)$. Bayesian statistics is predicated on the assumption that a likelihood function can be defined that formally represents the probability of predicting the evidence E given a model M. A large body of theory is available for likelihood functions appropriate for different assumptions about the nature of the errors in modelling the available observations (the evidence). As will be seen later in Section 4.5, it is also possible to use subjective likelihood assessments, though then, clearly, the resulting posterior will be conditional on the assessment criterion chosen. If, however, the posterior is interpreted as a measure of belief (or disposition to accept risk on the basis of the correctness of the predictions for a particular model), and the subjective likelihood assessment reflects the effect of the evidence on that belief, then Bayes theorem would appear to be still useful to assess the changing degree of confirmation over a set of feasible models. This may include the rejection of some models that are no longer consistent with the new evidence.

2.8 The information content of observations as evidence for the confirmation of models

The use of Bayes theorem in estimating predictive uncertainties of environmental models depends on the use of a measure of likelihood, L(E|M) in [2.1], determined from how well the model fits any available observations. The likelihood measure should therefore represent the information content of the observations in conditioning the feasible models of the system.

In the theory of Bayesian statistics, the specification of the likelihood function has generally been based on assuming that all the different sources of error in the modelling process can be represented in the summary additive form

$$O(x, t) = M(\underline{\Theta}, \underline{I}, x, t) + \varepsilon(x, t) \tag{2.2}$$

where $O(x, t)$ is an observation made at position X and time t; $M(\underline{\Theta}, \underline{I}, x, t)$ is the output of a model at x, t given vectors of parameter values $\underline{\Theta}$, and inputs $\underline{I}$; and $\varepsilon(x, t)$ is the model error for that observation. Multiplicative errors can also be handled simply in this approach by simply using the logarithms of all three terms. The simple form of [2.2] assumes that the effect of all sources of error in predicting any observation $O(x, t)$ can be subsumed into the total error series *as if* the model was correct and that the input and boundary condition data and observations were known precisely.

Furthermore, if the total error $\varepsilon(x, t)$ can be assumed to have a relatively simple form (or can be suitably transformed to a simple form, as in the log transformation for multiplicative errors, though other transformations are often used in attempts to stabilise the variance of the error series, see Box 4.1) then a formal statistical likelihood function can be defined, dependent on the assumed error structure. As an example, for an evaluation made for observations at a single site for total model errors that can be assumed to have zero mean, constant variance, independence in time and a Gaussian distribution, the likelihood function takes the form:

$$L(\varepsilon \mid M(\underline{\Theta}, \underline{I}, x, t)) = (2\pi\sigma^2)^{-T/2} \exp\left[-\frac{1}{2\sigma^2} \sum_{t=1}^{T} \varepsilon_t^2\right] \tag{2.3}$$

where $\varepsilon_t = O(x, t) - M(\underline{\Theta}, \underline{I}, x, t)$ at time t, T is the total number of time steps and σ^2 is the residual error variance. For total model errors that can be assumed to have a constant bias, constant variance, autocorrelation in time and a Gaussian distribution, the likelihood function takes the form:

$$L(\varepsilon \mid M(\underline{\Theta}, \underline{I}, x, t)) = (2\pi\sigma^2)^{-T/2} (1 - a^2)^{1/2}$$

$$\exp\left[-\frac{1}{2\sigma^2}\left\{(1 - a^2)(\varepsilon_1 - \mu)^2 + \sum_{t=2}^{T}[\varepsilon_t - \mu - a(\varepsilon_{t-1} - \mu)]^2\right\}\right] \qquad [2.4]$$

where μ is the mean residual error (bias) and a is the lag 1 correlation coefficient of the total model residuals in time (the background to these equations is given in Box 4.1).

A significant advantage of this formal statistical approach is that, when the assumptions are satisfied, the theory allows the estimation of the *probability* with which an observation will be predicted, conditional on the model and parameter values, and the probability density functions of the parameter estimates (which under these assumptions will be multivariate normal). As more data are made available, the use of these likelihood functions will also lead to reduced uncertainty in the estimated parameter values (even if the total error variance is not reduced). The amount of information added by new observations can also be formally evaluated within this framework. Lindley (2006) and O'Hagan and Oakley (2004) suggest that this probabilistic framework is the *only* satisfactory way of addressing the issue of model uncertainty. They suggest that without a formal theory for making probability estimates, statements about modelling uncertainty will have no meaning.

Others (e.g. Klir, 2006; Ben-Haim, 2006; Beven, 2006a) would disagree. There is an important issue about when probability estimates based on additive (transformed) error structures are meaningful. From a purely empirical point of view, a test of the actual model residuals $\varepsilon(x, t)$ for validity relative to the assumptions made in formulating the likelihood function might be considered sufficient to justify probability statements of uncertainty. From a theoretical point of view, however, there has to be some concern about treating the full sources of error in Equation [2] in this type of aggregated form. Model structural errors will, in the general case, be nonlinear, non-stationary and non-additive. Input and boundary condition errors, and any parameter errors, will also be processed through the model structure in nonlinear and non-stationary and non-additive ways.

Kennedy and O'Hagan (2001) have tried to address this problem by showing that all sources of error might be represented within a hierarchical Bayesian framework. In particular, where any model structural error is simple in form, it might be possible to estimate this as a "model inadequacy function". In principle, this could take any (nonlinear) form (although the most complex in the cases they considered was a constant bias, which can, in any case, be included as a parameter in [2.4]). The aim is to extract as much structural information from the total error series as possible, hopefully leaving a Gaussian iid (independent and identically distributed) residual error term. The model discrepancy function can then also be used in prediction, under the assumption that the nature of the structural errors in calibration will be "similar" in prediction.

It should be noted, however, that the model discrepancy function is not a direct representation of model structural error. It is a compensatory term for all the unknown sources of error in the modelling process, conditional on any particular realisation of the model (including specified parameter values and input data) and which might well change in different ways between different simulation periods. These sources of error could, in principle, be considered explicitly in the Bayesian hierarchy if good information were available as to their nature. This will be rarely the case in environmental modelling applications. In applications of hydrological models, for example, rainfall inputs to the system may be poorly known for all events in some catchments, and even the most fundamental equation – the water balance – cannot be closed by measurement (Beven, 2001c, 2002a,b, 2006b). Thus, disaggregation of the different error components will be necessarily poorly posed and ignoring potential sources of error, including model structural error, may result in an overestimation of the information content of additional data, leading to an unjustified overconfidence in estimated parameter values. In representing the modelling process by the simplified form of [2.2], the error model is required to *compensate* for all sources of deficiency.

This can be expressed in a more complete form of the modelling error equation as follows (Beven, 2005, 2006a):

$$O(x, t) + \varepsilon_O(x, t) + \varepsilon_C(\Delta x, \Delta t, x, t) = M(\underline{\Theta}, \underline{\varepsilon}_\theta, \underline{I}, \underline{\varepsilon}_I, x, t) +$$
$$\varepsilon_M(\underline{\Theta}, \underline{\varepsilon}_\theta, \underline{I}, \underline{\varepsilon}_I, x, t) + \varepsilon_r \quad [2.5]$$

The error terms on the left-hand side of [2.5] represent the measurement error, $\varepsilon_O(x, t)$, and the commensurability error between observed and predicted variables, $\varepsilon_C(\Delta x, \Delta t, x, t)$ for the model discretisation scale defined by $(\Delta x, \Delta t)$ (see section 1.5). The model prediction, $M(\underline{\Theta}, \underline{\varepsilon}_\theta, \underline{I}, \underline{\varepsilon}_I, x, t)$, will depend on the error in input and boundary conditions $\underline{\varepsilon}_I$, model parameters $\underline{\varepsilon}_\theta$, and model structure. The model structure error term, $\varepsilon_M(\underline{\Theta}, \underline{\varepsilon}_\theta, \underline{I}, \underline{\varepsilon}_I, x, t)$, can now be interpreted as a compensatory error term for model structural deficiencies, analogous to an inadequacy or discrepancy function in a Bayesian statistical approach, but which must also reflect error in input and boundary conditions, model parameters, and model structure. Finally there may be a random error term, ε_r.

Equation [2.5] has been written in this form both to highlight the importance of observation measurement errors and the commensurability error issue and to reflect the real difficulty of separating input and boundary condition errors, parameter errors and model structural error in nonlinear cases. There is no general theory available for doing this in nonlinear dynamic cases. The one simplification that can be made in [2.5] is that, if it is applied on a model-by-model basis, model parameter error has no real meaning. It is the model structure and set of *effective* parameter values together that process the (non-error free) input data and determine total model error in space and time. Thus [2.5] could be rewritten, for any model structure, as:

$$O(x, t) + \varepsilon_O(x, t) + \varepsilon_C(\Delta x, \Delta t, x, t) = M(\underline{\Theta}, \underline{I}, \underline{\varepsilon}_I, x, t) + \varepsilon_M(\underline{\Theta}, \underline{I}, \underline{\varepsilon}_I, x, t) + \varepsilon_r \quad [2.6]$$

and $\varepsilon_M(\underline{\Theta}, \underline{I}, \underline{\varepsilon}_I, x, t)$ is a model-specific error term. However, in any model application, all that can actually be determined is:

$$\varepsilon(x, t) = O(x, t) - M(\underline{\Theta}, \underline{I}, \underline{\varepsilon}_I, x, t) \quad [2.7]$$

so that:

$$\varepsilon(x, t) = \varepsilon_O(x, t) + \varepsilon_C(\Delta x, \Delta t, x, t) - \varepsilon_M(\underline{\Theta}, \underline{I}, \underline{\varepsilon_I}, x, t) + \varepsilon_r \qquad [2.8]$$

for which only the random error term ε_r is likely to have a simple form (and that by assumption). The other terms may exhibit non-stationarities and complex nonlinearities that will be difficult to determine in any particular application. It is then clear how, in fact, it is impossible to separate all the different sources of error in [2.6], *unless* some very strong assumptions are made about each of the terms, assumptions that would generally be difficult to justify.

The issue that then arises is just what is the real information content of an observation or the associated total model residual $\varepsilon(x, t)$ in conditioning the choice of an appropriate model? The value of $\varepsilon(x, t)$ will certainly reflect the model error, but it will also reflect the other, impossible to separate, sources of error. The extended model inadequacy function of Kennedy and O'Hagan (2001) is one way of trying to allow for these more complex error terms in the formal Bayes approach, but one that introduces additional statistical parameters to be estimated. All the terms on the right-hand side of [2.6] could also be included, if we were prepared to make strong assumptions about their form, although this would add even more statistical parameters to be estimated. Yet, by neglecting these terms, the formal Bayes approach will, in general, overestimate the information content of new observations in conditioning the feasible models. This can lead to over-conditioning of the model space, which can be visualised as a likelihood response surface that is much too peaky, giving the impression that the best models are far better defined than is really justified by the data (see for example the discussion of the paper by Thiemann et al., 2001, in Beven and Young, 2003, although in that study the assumptions of the Bayesian analysis neglected very obvious correlations in the residuals, so some over-conditioning could have been avoided within a formal Bayesian approach by a better set of assumptions about the nature of the errors).

The conclusion therefore is that Dennis Lindley and Tony O'Hagan are right when they say that the best approach to estimating model uncertainties is a Bayesian statistical approach, but that this will only be the case *if* all the assumptions associated with the error model can be justified. Given all the different terms in [2.6], the form of [2.1] with simple assumptions about the error term may be difficult to justify as more than a convenient approximation to the real nature of the errors.

So what is the real information content of observations in model conditioning or choosing between different models as hypotheses of how the system is working? The fact is that we do not actually know. But we do know that oversimplifying the problem by making convenient formal Bayesian assumptions may certainly result in overestimating the real information content of the data in conditioning the model space (see Beven et al., 2007). We will return to this issue later in Chapter 7, after more detail has been given of the Bayesian approach and of some alternatives in Chapter 4.

2.9 Explanatory depth and expecting the unexpected

The discussion in the previous section about the information content of observations in conditioning the model space is also relevant to the explanatory power of models. I hope the reader will, by now, be willing to accept that environmental models, in Adam Morton's phrase, have assumptions that are wrong and are known to be wrong. This then poses two interesting questions. The first is how approximate can a model be and still retain some utility or an element of realism in explaining and predicting the quantities and fluxes of interest for some practical purpose? The user of the model predictions (and the modeller as instrumentalist) is not necessarily interested in the explanatory *depth* of the model as in its explanatory *power* for the phenomena being predicted; that is the power to predict states and fluxes of real quantities. Is it therefore possible that a model that is useful in prediction (purely instrumentalist) need not be based on any depth of explanatory realism?

The answer to this question is clearly yes, as there are empirical modelling techniques that are based purely on deriving some useful predictive relationships directly from the available observations without any claims to explanatory depth. There are many examples in environmental modelling, including simple statistical regression, regression trees, support vector machines, transfer functions, Bayesian belief networks and artificial neural networks. These modelling methods are purely inductive in nature, in which the inferences drawn from the observations do not necessarily imply any theoretical or causal linkages.

What about deductive models based on a theoretical description of the relevant processes as a formal model? Here it is not necessarily the case that adding greater complexity implies greater explanatory depth (or indeed accuracy) in prediction. The degree of explanation will be necessarily linked to the problems of formal model complexity and knowability of system characteristics at different scales. It may therefore be possible to recognise greater explanatory depth in the perceptual model than in the formal model that is used to run simulations. That does not imply that the perceptual model will always be a correct representation of the real system, nor that the formal model chosen is the correct approximation to the perceptual model for a particular application. Explanatory depth, however complex, does not necessarily imply explanatory power if there is a conflict between the formal model and the processes or characteristics of the system of interest. In such a case, approximate models for which the parameters may be more easily identified may be advantageous.

The second question is how far can models that are necessarily approximate be expected to predict the unexpected, i.e. modes of response of the system that have not yet been observed? A fully realistic model should do this; a model based purely on inductive inference from observations made in the past may not be able to predict the unexpected, either because some change occurs or because conditions outside the range of past observations involve an unexpected process or nonlinearity. An argument is therefore often used in model development that additional processes and complexity perceived as having an effect in the real system should be represented in the model, even if only as phenomenological relationships of no great explanatory depth, because those processes *might* be important in predicting future responses.

This has sometimes led to misguided addition of complexity in the name of realism. One example is the great vertical complexity of land surface parameterisations used in GCMs, for example, at the expense of neglecting the spatial heterogeneity that is often important in controlling the fluxes of latent and sensible heat to the atmosphere integrated to the GCM grid scale. The explanatory depth of the vertically complex description may be greater for predictions at a point but the explanatory power at the scale of the GCM grid (even now, tens of kilometres), the scale at which predictions are required, was significantly compromised. Spatial heterogeneity is finally being introduced into the land surface representations of new versions of some GCMs but, from a hydrological point of view, it has been a long time in coming.

Similar issues arise in a variety of other fields. There is a description of mixing in river channels based on turbulent shear in the vertical velocity profile that leads to an advection dispersion equation (ADE). The approach has been shown to give accurate predictions in laboratory flume experiments and has been applied very widely in natural channels (Rutherford, 1994). In many natural channels, however, the ADE will give incorrect predictions. While it has depth of interpretation in terms of fundamental mixing processes, it completely neglects the transverse mixing and internal shear zones that are important in controlling mixing in natural channels of irregular three-dimensional geometries and boundary conditions (Young and Wallis, 1993).

The ADE has also been applied to the problem of predicting transport of solutes in porous media. Again it has been shown to give good predictions of solute break-through curves in laboratory columns of well-defined media. Adding terms to the ADE allows processes such as adsorption/desorption to be included so that a wider range of solute behaviours can be predicted. It has been used very widely to predict transport of contaminants in real soil and groundwater systems, after some calibration of the parameters involved. An interesting example of its application has been in the prediction of the movement of highly sorptive contaminants (pesticides, herbicides, phosphorus) applied at the surface of the soil. The predictions are that such contaminants will be retained in the near-surface soil. Sampled vertical concentration profiles at study sites confirm this impression, with a rapid decline in measured concentrations down from the soil surface. However, gradually other types of measurements revealed that this was not the whole story. Unexpectedly, pesticides, herbicides, phosphorus and other contaminants were being found widely in field drainage waters, stream waters and at depth in groundwaters. The ADE does not treat adequately the possibility of flow in preferential flow pathways in the natural soil structure that allows at least small amounts of contaminant, possibly sorbed to colloidal material, to move rapidly to depth during rainfalls. Without highly frequent sampling in the field during storm periods this process was not evident from the measurements.

All these examples show that when taking formal models into the environment and applying them to unique places, theoretical rigour of the formal model should not be confused with explanatory depth of the real phenomena, nor with explanatory power in prediction. All that can be demanded is consistency of explanation at the scale of application, including consistency with perceptual understanding and observations at larger and smaller scales (Beven, 2006b). The importance of preferential flow in contaminant transport should not have been a surprise, for example. It had already been recognised in experimental work by Lawes et al. (1882) at the Rothamsted Experimental Station in England, and the translocation of clay particles in large pores had been recorded from soil-thin section work for over 100 years (see the review of preferential flow soils in Beven and Germann, 1982). The unexpected may not turn out to be so unexpected when we are forced to review a model because of a failure in prediction. This is one situation in which we might learn more from a model failure than from a model continuing to work (sort of) reasonably well.

The unexpected may also occur as a result of change to the system, often the result of human interference, either inadvertent or as a result of policy and management.

Anthropogenic effects may change either the characteristics of the system or its boundary conditions at the scale of closure. Often these impacts are not very well quantified even in hindcasting and history matching (the construction of records of inputs of greenhouse gases into the atmosphere since pre-industrial times is a good example of uncertainty in estimating such effects). Thus, it may only be possible to treat such impacts as potential scenarios, giving potential outcomes but without any objective quantification of the associated uncertainty. Most predictions of the impacts of climate change are of this type.

The unexpected may be difficult to predict but might prove useful in model confirmation or in discriminating between alternative model structures as hypotheses of how the system is working. Where direct testing of models is not possible then it may be just a question of continuing to monitor the system, waiting for the unexpected. All the time that the chosen models remain consistent with new observations then the predictions will appear to be acceptable. If, however, a model can be rejected on the basis of new, perhaps unexpected, responses then perhaps the model can be refined. When the truly unexpected occurs and all the available models must be rejected, then the science will truly progress (even if only by considering more realistic auxiliary conditions). Such rejections or model falsifications will be an important part of the learning process about how to model environmental systems (see also the discussion of Tarantola, 2006).

2.10 Uncertainty, ignorance and factors of safety

In Section 1.11, a variety of meanings of uncertainty were introduced. In particular, a distinction was made between uncertainties that can be dealt with in terms of random variation or probabilities (**aleatory uncertainties**), those that result from a lack of knowledge (**epistemic/Knightian uncertainties**) and those due to the vagaries of language (**linguistic uncertainties**). Both epistemic and linguistic uncertainties can be the result of ignorance, a concept that has been the subject of significant discussion in both philosophy and law (see, for example, Smithson, 1989; Furmston, 1992; Wynne, 1992; Ferson and Ginzburg, 1996). Clearly, one of the reasons why we might be concerned about uncertainty is because we are often in a situation of a degree of ignorance: ignorance about the past factors affecting a system; ignorance about how a system works now; ignorance about future boundary conditions; ignorance of the expertise and experience of other scientists; ignorance of factors that are not yet recognised as important (the unknown unknowns).

Ignorance raises a number of interesting philosophical issues, particularly in relation to the unexpected, that have a long history, although the tendency in philosophy has been to discuss knowledge (or epistemology) rather than ignorance. How far is ignorance excusable as innocence? When does ignorance become incompetence? What is an appropriate response to a recognition of irreducible ignorance? These are not only questions of philosophy, however, they have real practical implications in, for example, engineering design; the assessment of responsibility in law when failures occur; and the assessment of future climate change. A moment's thought will reveal that these issues are not just matters of science. They also have a sociological context. Ignorance can only be assessed in the context of a particular social group or individual. It will, necessarily, change over time with the dynamics of the group, in interaction with other groups. Ignorance, as the complement of knowledge, is, like knowledge, a social construct.

Smithson (1989) provides a taxonomy and useful discussion of different types of ignorance, from ambiguities to taboos. Some can be treated by the uncertainty estimation methods presented in this book (probabilistic, interval, fuzzy, Info-Gap and other methods). Others are beyond the scope of any quantitative analysis. The interesting aspect of ignorance in respect of environmental modelling is when we are happy about some quantitative framework of dealing with uncertainties but are rather ignorant when it comes to the specific assumptions about the nature of the uncertainties that might be appropriate for a particular application in making an analysis based on that framework.

This is not, of course, a new problem. It is as old as the history of the design of tools, machines and structures when there might be some past experience of success and failure in use, but when it was not known exactly what forces they might be subjected to in future. This is not so different to the application of environmental models. We have some past experience of success and failure in calibration, but we are not sure exactly what boundary conditions and changes might occur in future.

In the context of engineering design a formulation of handling ignorance of future conditions has been developed based on the *factor of safety* concept. For a structure, the factor of safety is defined as the ratio of the expected working stress on the structure to the critical stress at failure for a particular design. It is then a question of negotiated standards as to what factor of safety is acceptable for a given type of application, given the risk to society associated with the application and past experience of failures. There may, of course, be multiple sources of stress and uncertainties associated with the specification of the stresses and the design that can, in complex design problems, make the assessment of appropriate factors of safety an enormous task in what is now called *reliability analysis* (e.g. Blockley, 1992). These uncertainties mean that designs need a certain degree of robustness, as implied by applying a factor of safety. While there are cost savings to be made by optimising designs, there are also risks of failure as a result of neglecting uncertainties (as is evidenced by the history of box-girder bridge design where a number of failures in the 1970s led to improved design criteria). An interesting example of the use of freeboard as a factor of safety in the design of flood defences is discussed in Hine (2007) using the Info-Gap methodology for uncertainty decision making of Box 6.2.

In the application of environmental models we are similarly trying to make predictions in the face of uncertainty and real ignorance, but the factor of safety concept has not been used. Any reliability analysis has generally been a matter of assessing the probabilities of model predictions given some probabilistic assessment of the uncertainties in model inputs. We know, however, that model uncertainties are not only a matter of probabilistic uncertainties and that this might be important in making robust decisions (see Chapter 6). This might be a question of the experience of failure. In engineering design, failures occur and that experience conditions future practice. In environmental modelling there are few examples in the literature of reported failures (though see, for example, Konikow and Bredehoft, 1992; Parkin et al., 1996; Freer et al., 1996; Choi and Beven, 2007). There are many more examples of the presentation of the optimal model results after calibration against some historical data. Negotiated standards therefore have tended more towards consensus about best practice (as in the various IPCC reports on predicting climate change) than taking account of ignorance in avoiding failure in prediction. But, as noted in the previous section, we will always learn more from model failures than from models that sort of work. Failures will result from ignorance about model structures and parameter values, errors in the data used to drive the model, and errors in the data used to evaluate the model. We can use the term ignorance here, rather than uncertainty, because each of these potential sources of failure might involve unknown unknowns as well as quantifiable uncertainties. If this is the case, then quantitative uncertainty analysis will have its limitations, and some methodology analogous to the factor of safety might be useful. This should be borne in mind when reading the rest of this book.

2.11 Summary of Chapter 2

This chapter has been a summary of some of the philosophical issues that arise in making models of environmental systems. It is, of course, perfectly possible to do modelling without worrying about philosophy at all, and, in fact, the vast majority of environmental modellers do just that. The following points are, however, important to what follows in the rest of this book:

- Most environmental modellers (even if they have not thought about it) have a form of *pragmatic realism* as an underlying philosophy. They generally want to use models that are, in some sense, realistic in their descriptions of the controlling processes and that have some predictive power for practical problems, while recognising the approximations inherent in using any form of model.
- There are some severe criticisms of this type of philosophical position since it is difficult to provide empirical evidence to support even the most fundamental mass and energy balance equations in open environmental systems. The best basis for the utility of models is a form of instrumentalism in which models that might have utility in prediction are not falsified.
- It is shown how the continuing process of model confirmation by evaluating predictions against observations is a form of learning process that can be formalised by the use of Bayes equation (although other forms of evaluation are also possible).
- The learning process depends on the real information content of the available observations in learning about those models that might be useful in prediction. There is no realistic theory of information content in the face of multiple sources of uncertainty, when the assumptions made in a formal Bayesian analysis might be too strong.
- It is necessary to expect the unexpected in environmental systems, with more extreme conditions than have been seen before. In this case, however, more severe tests of model performance are possible. Learning from model failures/falsification in such circumstances may be an important part of the learning process.
- Ignorance cannot always be assessed using quantitative uncertainty assessment. In that case, a concept analogous to factor of safety might be useful in assessing predictions. This, however, implies negotiated standards based on experiences of failure. Until environmental models are subject to failure tests, learning and the possibility of defining standards will be limited.

Chapter 3

Simulation with no historical data available

One does not choose to doubt: doubt is an unavoidable attribute of the human circumstance. That chunk of the wider world which one experiences and knows, that projection of the greater reality which comes one's way, is insufficient to characterise the entirety which must remain unknown. Uncertainty, in this view of things, is the conception of alternative possible worlds that we cannot distinguish.

Yakov Ben-Haim, 2006

3.1 Sensitivity, scenarios and forward uncertainty analysis

As noted in Chapter 1, in many cases there is a requirement to make predictions using environmental models without any historical data available for model calibration or conditioning. There are also many applications when it is necessary to make predictions without any information available about the boundary conditions that might occur, in particular in considering possible futures. We do not know what the future holds and can only speculate about future boundary conditions as scenarios of what *might* happen.

In all of these types of situation, the results of any modelling study are totally dependent on the assumptions made by the modeller, fed forward through the model predictions. This is what is known as a *forward uncertainty analysis*. To obtain predictions the modeller must decide on one or more feasible model structures, feasible values for model parameters, and feasible boundary conditions for the simulation period required. In such cases, what is considered feasible is generally a subjective decision by the modeller, or the result of some consensus agreed between modellers and stakeholders. It may not be possible, for example, to define probabilities for the different model structures, parameters and boundary condition scenarios, so that any estimation of uncertainty associated with the predictions will be itself necessarily uncertain. It will be a preliminary estimate based on and *conditional* on the prior assumptions.

In such cases it is very useful to have an idea of what the predictions are most sensitive to. Such information would allow the modeller to concentrate effort on making the choices that will have the most impact on the predictions. This requires a sensitivity analysis. There are many different forms of sensitivity analysis and, for any realistically complex modelling problem, no definitive answer as to what the model predictions are most sensitive to. The relative sensitivity of different model structures, parameters and boundary conditions will generally depend on the type of analysis, the form of sensitivity measure chosen, and the output variable for which sensitivity is being examined. Some examples are given later in this chapter. However, there will often be some general agreement between the different types of analysis, so that the most important sensitivities in any modelling problem for the output variable or variables of interest can generally be recognised.

There is a straightforward link between sensitivity analysis and forward uncertainty analysis. In effect, both are types of exploration of the model space (see Section 1.6) based only on the prior assumptions about the nature of that space made by the modeller. We do not necessarily know *a priori* which parts of that space might be the most important or relevant in representing the real system, we do not even know whether we are searching the most relevant model space in terms of choice of models, parameter ranges and boundary conditions, but the space is usually defined on the basis of past experience and reasonable common sense or consensus prior assumptions that would not be disputed by other modellers. So we can assume, at least, that if we make sensible choices the results should have some relevance to the response of the real system.

There is, however, an immediate problem in carrying out either a sensitivity analysis or forward uncertainty analysis for any reasonably complex model application. That is that the dimensionality of the model space is generally high. There may be more than one potential model structure, there may be many parameters associated with each model, there may be many alternative boundary conditions. In addition, the computing cost of even a single model run may be high. Thus, even if a full exploration of the model space is possible in principle, in practice it may be computationally infeasible. This is the case, for example, for the global circulation models used in climate change prediction. The computational time of a single run of a coupled ocean–atmosphere model is high. The number of parameters in these models is high. The exploration of the uncertainty of a full global model is then generally limited to simple sensitivity analyses and small numbers of ensemble runs. If the model is simplified, then many more runs and a more detailed search of the model space are possible (see the results of the *climateprediction.net* study in Allen et al., 2000 and Piani et al., 2005), but then there is always the doubt about how relevant the results obtained with a simplified model are to inferences about the full model (see Section 3.6 below on model emulation techniques). Other than for quite simple problems, there is generally a need for efficient search algorithms for exploring the space of feasible models. Some of the possibilities are outlined in Section 3.3.

A common feature of this type of sensitivity and forward uncertainty analysis is the use of scenarios of boundary conditions (and sometimes of changing system character-istics) to drive the model. For predictions about the future (and often for hindcasting of the past) there can be no direct measurements of the variables required. We cannot make measurements of future conditions and we cannot go back and make additional measurements of what happened in the past. We can only construct feasible scenarios of what might be/have been (see section 3.7). It is clear that it will be very difficult to associate any probabilities with such scenarios, even if they are based on model predictions, where the model has been evaluated for its performance when data are available. Thus, the use of such scenarios will result in results (for both sensitivity and uncertainty estimates) that are conditional on the choice of scenarios. The construction of the scenarios may be more subjective in some applications, less subjective in others, but in all cases some subjectivity will be involved and the conditionality on the assumptions made should be recognised.

Sometimes, particularly when only a small number of feasible scenarios is considered, this conditionality is made clear. This is the case, for example, in the latest assessments of climate change by the Intergovernmental Panel on Climate Change (IPCC) where a limited range of future scenarios for output of greenhouse gases, each based on different assumptions in a socio-economic model of future change, are considered (IPCC, 2007). Different climate models are applied to the same scenarios, and the conditionality of the results on the scenarios is clear (e.g. Figure 3.1). Other conditionalities in applying these models, however, are much less clear (such as effects of the choice of parameters and sub-grid representation of atmospheric and ocean processes in each model).

Multi-model Averages and Assessed Ranges for Surface Warming

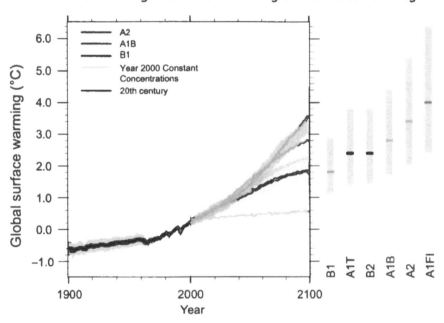

Figure 3.1 Predictions of change in global temperature under different emissions scenarios, including range of predictions from an ensemble of different models. Shading represents +/− 1 standard deviation around multiple model annual averages. The grey bars at right indicate the best estimate (solid line within each bar) and the range assessed for six different emission scenarios using a wider range of evidence

Source: *Climate Change 2007: The Physical Science Basis*, Summary for Policymakers, Intergovernmental Panel on Climate Change

There is a general rule here to bear in mind in what follows. It is very difficult (except for the simplest possible problems) to be totally objective about any sensitivity or uncertainty analysis. The results will always be conditional on the assumptions made, and therefore it will always be worth examining the assumptions made for plausibility, in so far as that is possible, since many studies do not explicitly list all the assumptions on which they are based, even in the refereed scientific literature. Many papers can be found that purport to be objective in their analysis but where the assumptions of the analysis are not consistent with a more complete perceptual model of the real system.

In some cases, of course, this may be just science being done badly. In other cases, however, it is more interesting in that we may not know how to represent the behaviour of the system properly and the best models today are still an exploration of how to do so. Such an exploration is essential in understanding how to do better. This is the case for global climate models. The current generation of coupled ocean–atmosphere models is greatly improved over past models, and can now avoid the oceanic energy flux corrections that were necessary when the atmosphere circulation was modelled alone. Such advances are (as yet) as much to do with increasing computer power as scientific understanding, and with each incremental improvement more understanding is gained, even though the current generation of models still leaves much to be desired in representing the global climate. As noted above, a full sensitivity and uncertainty analysis of such global models is still computationally infeasible, but the same principle of conditionality applies to simpler models. The results will depend on the decisions made in setting up the analysis.

3.2 Making decisions about prior information

Thus, making decisions about prior information is critical in any sensitivity and uncertainty analysis. Such decisions define the space of feasible models to be considered and all results of the analysis will be conditional on those decisions. It is therefore a good idea to make the best possible choices but this might often be difficult. We may not be sure which model structure will best represent the system; we may not be sure which values of the parameters best represent the effective parameter values required by the model at the scale of application; we may not be sure which boundary conditions might be appropriate for the period of simulation. In some cases, estimation of parameters and boundary conditions may depend on the predictions of other models produced by another set of modellers and with, often hidden, choices and assumptions in their own domain of expertise that we may not be competent to evaluate (as in the case of all the different inputs that feed into the IPCC global change scenarios). Thus, the problem of making decisions about prior information is essentially a question of belief. Given a variety of alternatives of models, parameters and boundary conditions, which do we have more belief in?

3.2.1 Prior distributions of parameters

It is easiest first to consider the estimation of prior distributions of parameter values in a particular model structure chosen for a particular application. We should know something about how such parameters might vary with different characteristics of the system (soil, vegetation type and structure, geology, human influences etc.). We might have some information on values from text books, or past applications of the model, or guidance in the user manual, but the estimation of the *effective* values required by the model might still be rather uncertain. Ideally we would want to give prior estimates for some distribution of each parameter and the co-variation of the parameters in providing good simulations whether in a probabilistic, fuzzy or even qualitative way. We would certainly want to avoid wasting computer time by running the model with combinations of parameters that would not be expected to give good results, but it might be very difficult to decide which combinations of values are feasible and which are not. It might even be difficult to decide on what is a feasible *range* for each individual parameter, let alone the full joint distribution of all the parameters. And the more parameters that need to be specified to run the model, the worse this problem will be.

Consider the following example. We wish to carry out a forward uncertainty analysis of a model to predict the phosphorus concentrations of a river draining a rural catchment area of mixed land use, farms and settlements. We have a model available (we will use the example of the INCA-P model of Wade et al., 2007, but there are a variety of other models to predict phosphorus in streams that could be used, each with different representations of the processes involved). This model attempts to reproduce, in a relatively simple way, the way that flow is generated on the catchment area and the processes by which phosphorus is transported to the stream and transformed within the stream channel. The catchment can be subdivided down into small sub-catchment areas where it is thought that there might be significant differences in the runoff and inputs or transport of P. Each sub-catchment area can have different values of the parameters to allow those differences to be reflected in the predictions. In principle, each reach of the channel network could also be allowed to have different parameter values, although these are usually specified as values for the whole network.

The user manual for INCA-P gives a list of the parameters required and some expected ranges for the parameters based on the experience of the model developers. There are 47 parameters that could be allowed to vary for each land use category (up to six land use types are allowed in each sub-catchment), and 20 parameters that could be allowed to vary for each reach of the channel network, and 25 other parameters. Even a small number of reach and land use subdivisions therefore results in a very large number of parameters that must be specified before the model can be run. There is no information on how the parameters might show co-variation in producing a good simulation. This is not untypical for many environmental modelling problems, especially models that are distributed in space and that attempt to simulate complex interacting processes. The parameter estimation problem has to be simplified to be manageable.

Sensitivity analysis can be useful in doing so. Those parameters to which the predictions are less sensitive might be able to be fixed at constant values, reducing the number of parameters that needs to be varied. In general: the co-variation of parameters is also often ignored at this stage (unless evidence is available to suggest that some parameters interact strongly) so that the parameter distributions can be estimated independently. Lacking better knowledge of which values of a parameter might be more likely to give good results, a uniform prior distribution is often assumed between some upper and lower limits for the feasible range. This is not a truly non-informative prior distribution in a Bayesian sense, since an assumption is then being made that outside the specified range there is no expectation of a good simulation (this issue will arise again in considering model calibration in Chapter 4). Assuming a uniform prior does not, however, give undue weight to any values within the range when there is little evidence to support this and, in fact, the need to consider values only within a certain range is really a computational issue, to avoid making runs where we do not expect to find useful model predictions.

Sometimes it will be possible to define prior distributions more closely, such as by providing location and scale parameters for each parameter (such as the mean and standard deviation for a normal or Gaussian distribution). There are many different possible distributions that can be used and some examples are given in Figure 3.2. Many of the candidate distributions in such cases (e.g. normal, log normal, exponential, gamma distributions in probability) have infinite tails and, again for computational reasons, need to be truncated to limit the sampling of values once the prior probability for a value gets very small. Having chosen to use a Gaussian distribution, for example, we might truncate the distribution at +/– three standard deviations away from the mean value so that we would be sampling 99% of the values for that distribution, neglecting the more extreme values. Other distributions (such as the triangular, trapezoidal or beta distributions often used in the representation of fuzzy variables) can be specified with upper and lower limits.

The question again, however, is how to decide on how much belief to give different values of a parameter, especially in a complex nonlinear model with many parameters that might co-vary in a consistent way. The basis for representing variability as a probability distribution in this way is outlined in Box 3.1. Essential assumptions are that all possible values are included in the distribution. The probability density function (PDF) can then be scaled such that the integral under the curve is unity. The integral of the pdf up to any particular value of the variable or parameter is the cumulative density function (CDF). It will sometimes be difficult to specify such distribution functions precisely. Box 3.1 also shows how probabilities can be represented as intervals or imprecise probabilities.

3.2.2 Belief networks

Belief, of course, is something personal (and essentially subjective). Different modellers can be expected to have different beliefs. In some cases, we may be able to evaluate those beliefs for different sets of circumstances to those for which predictions are required, which might allow us to condition our beliefs as to how well a model (and parameters and boundary conditions) performs and therefore its feasibility as a predictor for the circumstances we are interested in. The techniques of Chapter 4 are relevant when such conditioning data are available.

There is no guarantee, however, that the knowledge gained for one set of circumstances will carry over to the circumstances we are interested in. This will be true if we have evaluations of a model for a place of interest under current conditions for which we have some measurements and we want to make predictions for future conditions. It will be even more true if we need to use information from modelling one place in predicting responses at another place, even if those places have apparently similar characteristics. This may then cause us not to have too strong a belief in knowledge extrapolated from elsewhere but, on the other hand, we do not want to make our choices too wide; we only want to consider models, parameters and boundary conditions that really are feasible as simulators for the application of interest (in so far as it is possible to do so).

	Probability Density Function (pdf)	Cumulative Density Function (cdf)
Uniform Distribution $p(x) = \alpha$; xmin<x<xmax		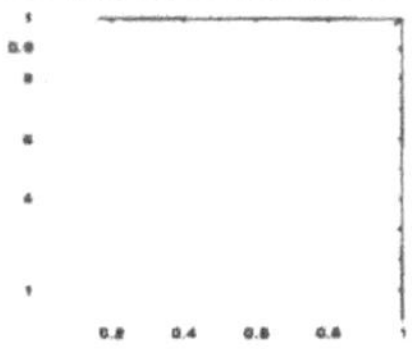
Triangular Distribution $p(x) = \alpha x/x_\alpha$; $x < x_\alpha$ $p(x) = \alpha(1-x)/(1-x_\alpha)$; $x > x_\alpha$		
Trapezoidal Distribution $p(x) = \alpha x/x_1$; $x < x_1$ $p(x) = \alpha$; $x_1 < x < x_2$ $p(x) = \alpha(1-x)/(1-x_2)$; $x > x_2$		
Exponential Distribution [λ=0.2] $p(x) = \lambda \exp(-\lambda x)$; $x \geq 0$ $p(x) = 0$; $x \leq 0$		
Gamma Distribution [a=5, k=8] $p(x) = \dfrac{1}{a\Gamma k}\left(\dfrac{x}{a}\right)^{k-1}\exp(-x/a)$		
Gaussian (Normal) Distribution [μ=0.5, σ=0.15] $p(x) = \dfrac{1}{\sqrt{2\pi}\sigma_x}\exp\left(-(x-\mu_x/\sigma_x)\right)$		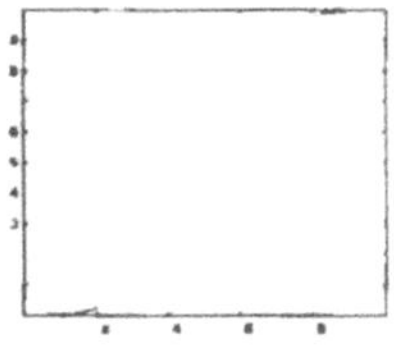
Log Normal Distribution [μ=-1, σ=0.4] $p(y) = \dfrac{1}{\sqrt{2\pi}\sigma}\exp\left(-(x-\mu/\sigma)\right)$; $y = \ln(x)$		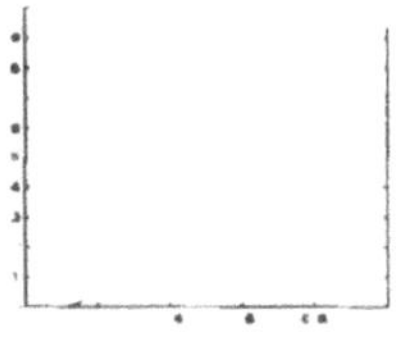
Beta Distribution [b=3, c=2] $p(x) = \dfrac{\Gamma(b+c+2)}{\Gamma(b+1)\Gamma(c+1)}x^b(1-x)$; $0 < x < 1$ (can be either left- or right-skewed depending on the parameters b,c)		

Figure 3.2 Definition of Selected Probability Distributions with scaled probability density function (pdf) and cumulative density function (cdf)

Even for experts in any domain, however, it is not easy to make such choices. So how should you decide about your prior beliefs? One technique is by eliciting the advice from multiple experts. There is always a possibility that experts with similar training might be similarly biased in their opinions, and experts with different backgrounds may conflict in their opinions; they may not actually have the relevant experience, or they might be simply wrong, but the aim is to average out some of these uncertainties across the range of opinions offered. This is actually how science works. Over long periods of time, a consensus of opinion arises about the best theory or representation of a process or system. Sometimes that consensus is shown to have been wrong as new evidence arises or a new theory is shown to give a more convincing explanation of the available data. As discussed in Chapter 2, such changes of paradigm are often resisted by established experts in the field but, when they occur, often occur rapidly.

In making predictions, however, we have to work within an existing paradigm. The predictions are needed now within some decision-making context and we cannot necessarily wait for new evidence. Thus, we may depend on eliciting the current expertise of the scientists in the field, even if we expect their opinions to change over time. Expert elicitation has been the subject of much research (see, for example, O'Hagan et al., 2006). Here we mention briefly a technique for deciding about beliefs called Belief Networks.

Belief Networks allow that it may not be possible to achieve a complete consensus amongst experts about choices of model structures, parameters and boundary conditions. It might be possible, by means of a scoping study, to outline what the range of feasible choices of models, parameters, or scenarios of boundary conditions will be for an application. The choices can then be structured as one or more networks, with each pathway through the network given an initial estimate of relative belief. This then provides a structure for eliciting the opinions of different experts about relative degrees of belief in the options. As each new set of opinions is added to the information available, the set of belief measures can be updated, often using a form of Bayes equation (see Box 4.1), giving then a Bayesian Belief Network (e.g. Varis, 1997). Since this technique is primarily used in decision analysis we will return to it in Section 6.3. Later in this chapter we will consider fuzzy set representations of uncertainty. There is a fuzzy analogy to Bayesian Belief Networks in the use of systems of fuzzy rules (see Box 3.4).

3.3 Sampling the model space

To define the model space for a given environmental modelling problem, we need information about the prior possibilities of both model structures and model parameter values. We are then interested in quantities within that space, produced when a model (or set of models) is run with certain boundary and initial conditions. A model here is intended to represent a particular model structure with a particular set of parameter values. It might be deterministic, it might be stochastic, it might be a set of fuzzy rules: the important point here is that it will produce a set of outputs when driven by a set of inputs (which might also be deterministic, stochastic or fuzzy).

In fact, something that is often forgotten is that, *at least in principle*, the model space can be filled completely with results from such runs (see Section 1.6). In practice, of course, there is a limit to how far this is possible because there are still computational limits on making such calculations in large model spaces and where run times for a single model run are long. However, that does not change the principle. We shall return to this issue in Chapter 4 for the case where some evaluation data are available and the response surface of interest is a performance measure or likelihood, but for the moment we will concentrate on ways of sampling all the possible behaviours in the model space.

For low-dimensional model spaces it may be possible to make an exhaustive search of all the possible model structures and parameter values, but this rapidly becomes more difficult with larger model spaces. A sampling strategy is needed, and an exhaustive search may not be the most appropriate strategy where a lot of the space may not actually be of very great interest to the modeller. This will be the case where there are parts of the space that are not of great interest because the outputs are all the same ("plateau" or "insensitive" areas of the response surface), where the prior beliefs or likelihoods of sets of parameter values are very low (whether represented in terms of either probabilities or fuzzy measures), or where parameter sets are giving outputs that are assessed as being unrealistic compared with the behaviour of the real system. Thus, we want to differentiate the parts of the model space that are of interest and use the available computer time to explore the response surface in more detail or sample more densely in those regions.

But how do we know where the interesting regions of the model space might be? This requires information, and there are two basic ways of acquiring this information, often used in combination. The first is to use some prior information about what model structures might be interesting to explore and what values of the parameters might be interesting to explore. It is common, for example, to choose just one model structure (because of past experience on performance, or because it has been widely used in similar applications, or because it is freely available rather than having to pay for it, or just because it is a model you wrote yourself and therefore, by definition, best in some sense or another). As noted above, it usually possible to define prior probability distributions or fuzzy membership values for the potential values of the parameters. This provides information to structure an initial search.

The second method is to learn about the surface as the sampling proceeds. This can be particularly useful where there are strong interactions between parameter values in producing an output or performance measure. These methods are often called guided search algorithms and are closely related to optimisation algorithms (since finding an optimum model is also a matter of searching a complex response surface in the model space for a peak in the relevant performance measure). However, because of the issues discussed in Chapters 1 and 2 about whether the concept of the optimal model is ultimately useful in environmental modelling we will say little about optimisation algorithms in what follows, concentrating instead on the more general search for regions of interest. There is more about searching and conditioning the response surface as data are made available in Chapter 4.

3.3.1 Analytical propagation of probabilistic uncertainty

For simple linear models in which the outputs are directly proportional to the inputs and where there are only a small number of uncertain variables it may be possible to propagate uncertainties analytically. This is much easier if the uncertain variables are independent (see Box 3.1). Where co-variation of the variables is an important consideration then, even for a linear model, the results will be very dependent on a proper specification of the probability of occurrence of a value of one uncertain variable, conditional on the values of all the other uncertain variables. This is normally specified as an $N \times N$ covariance matrix, where N is the number of uncertain variables that must have entries in each of the N^2 elements, although other methods are possible (see Kurowicka and Cooke, 2006 and Section 3.3.5 below). Clearly, as N gets larger, this will become more and more difficult and assuming at least some of the variables as independent (with the relevant off-diagonal elements of the covariance matrix set to zero) increasingly attractive. This will not always be a good approximation to the co-variation in reality, however, and the effect of co-variation is generally to decrease the uncertainty in the resulting outputs. Thus neglecting co-variation might result in a pessimistic assessment of the uncertainty in model outputs. This also applies where it is not possible to propagate uncertainties analytically, because of nonlinearities in the model formulation and resort must be made to some approximate numerical evaluation of the shape of the response surface in the model space, e.g. using random Monte Carlo sampling.

3.3.2 Discrete samples or random Monte Carlo search?

In exploring the model space numerically, the first choice to be made is whether to sample by taking values at discrete intervals for each parameter, or take random values across the range being considered (see Box 3.2), or to combine the two strategies by sampling randomly within discrete intervals in the model space. These simple search strategies are illustrated in Figure 3.3, for the case where the prior assumption of a normal distribution of the parameter to be sampled has been made. Figure 3.3a shows the probability density function (pdf) and Figure 3.3b the cumulative density function (cdf) produced by discrete interval sampling across the range of the parameter.

Random sampling strategies can also be used. With a specified distribution, there are two ways of representing the distribution. The first is to use uniform random sampling and represent the prior distribution as weights associated with each sample (Figure 3.3c); the second is to sample randomly but in a way that reflects the prior probability density function (Figure 3.3d). In the latter form of density-dependent sampling the samples can be treated as having equal weight in propagating the uncertainty associated with that parameter. These sampling strategies are also shown for the case of a skewed beta distribution in Figure 3.4.

The last two sampling strategies are also illustrated in two dimensions in Figure 3.5 for the case of two parameters varying independently, one with the normal distribution of Figure 3.3a and one with the beta distribution of Figure 3.4a. In Figure 3.5a the use of random uniform sampling with weights is shown, with the weight associated with each sample indicated by the size of the symbol. The weights are greater where there is greater joint probability for the two sampled parameters. Figure 3.5b shows the density-dependent sampling strategy. Here higher joint probabilities are represented as a higher sampling density so that each sample can be treated as having equal weight as in the single parameter case of Figures 3.3d and 3.4d. These different sampling schemes can all be extended to model spaces of higher dimensions. We will return to them again when we are looking at evaluating the way in which models can represent observations in the next chapter.

First, however, we must emphasise the difficulty of sampling complex model spaces by any of these methods. Consider doing a search of a two-dimensional model space for only one model structure (as in Figure 1.1 in Chapter 1). Splitting the space into discrete interval samples of 100-parameter values for the first parameter, and 100 for the second will give a good idea of the response surface of the model in this two-dimensional space and will usually be computationally feasible except for models with very long runs times. Only 10,000 runs are needed. Adding a third parameter makes it a little more difficult (1,000,000 runs). For six parameters it might be better to increase the width of the sampling intervals for each parameter (ten values for each parameter requires 1,000,000 runs for six parameters) but, for larger dimensional spaces, even a coarse resolution of ten values per parameter can still require an infeasible number of runs. Iorgulescu et al. (2005), for example, sampled 2,000,000,000 (two thousand million or two American billion) parameter sets of a 17-parameter model randomly (finding 216 acceptable model runs), but the equivalent discrete

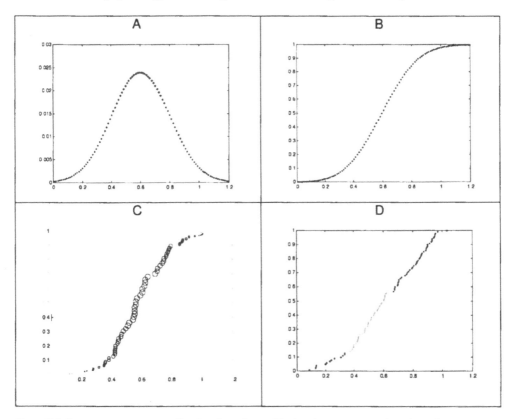

Figure 3.3 Representing a normally distributed variable by discrete interval and random sampling strategies. A. Probability density function defined by discrete increment sampling for values of the variable. B. Cumulative density function defined by discrete increment sampling of the variable. C. CDF defined by uniform random sampling with weights representing probability density associated with each sample (size of symbols proportional to weights). D. CDF defined by density-dependent sampling. All samples of size 100

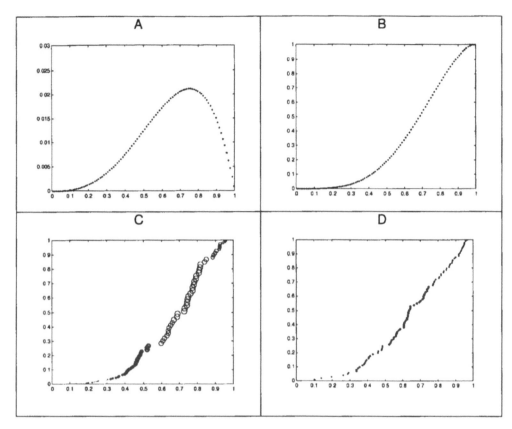

Figure 3.4 Representing a skewed beta-distributed variable (with parameters [4,2] in the range [0–1]) by discrete interval and random sampling strategies. A. Probability density function defined by discrete increment sampling for values of the variable. B. Cumulative density function defined by discrete increment sampling of the variable. C. CDF defined by uniform random sampling with weights representing probability density associated with each sample (size of symbols proportional to weights). D. CDF defined by density-dependent sampling. All samples of size 100

search over all parameters with just ten values on each axis is 100,000,000,000,000,000 (100 million American billion) samples. Their two billion runs could therefore be considered as rather a small sample in this case. Many environmental models, of course, have many more uncertain parameters and inputs and will be subject to the problem of making enough runs to characterise the whole model space. Even where some Monte Carlo runs are made to assess the potential uncertainty in the outputs of such models, it is more likely that tens of thousands of runs have been made rather than billions. The issue then is whether there are import-ant parts of the model space that remain unsampled in terms of representing the dynamics of the system.

One way of making this type of search more computationally feasible is to use Latin Hypercube sampling (LHS). In LHS, each parameter dimension of the model space is split into a number of discrete values (x_i, $i=1....N$ for the ith parameter). Where a

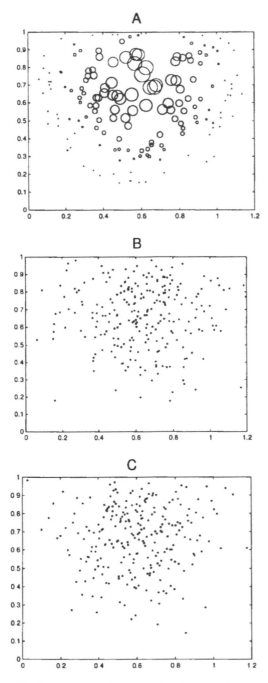

Figure 3.5 Representing the joint distribution of two parameters, one normally distributed (x axis) and one beta-distributed (y axis). A. by uniform sampling with weights (with circles proportional to weights associated with each sample). B. by density dependent sampling. C. by Latin Hypercube sampling using 250 sampled values on each parameter axis combined randomly without replacement

prior distribution of the parameter is assumed, discrete intervals of probability are usually used. However, unlike a pure discrete sampling strategy, not all combinations of all the discrete values are sampled. Instead, it can be shown to be much more efficient (at least for simple shapes of response surface) if parameter sets are chosen based on sampling an interval randomly for each parameter without replacement. This means that only N sets of parameter values will be generated and run, but that the sample will be scattered through the model space. An example of Latin Hypercube sampling is shown in Figure 3.5c. More flexibility can be added to the LHS method by sampling the parameter values randomly *within* each discrete interval. Further variations on the LHS method are presented in the next section.

However, the modeller faces a dilemma in applying a more structured sampling strategy where there is not strong prior information about the important parts of the model space. Ideally, we would then want to use an adaptive sampling strategy or learning strategy so that we can refine the search of the regions of interest as they become identified within the model space. But, we have to start with some (small) initial sample generated either by discrete or random sampling. There is always a possibility of *not* sampling a region of interest, especially if the region or regions of interest might be small compared with the initially coarse sample. We cannot learn about a region of interest (for example where the output variables are very sensitive to small changes in parameter values or, as considered in Chapter 4, where a good fit to data is found) if that region has never been sampled. The higher the dimension of the model space, of course, the greater the possibility that an initial sample might not identify some of the regions of interest. Thus, it is often a useful strategy, even where the guided search strategies considered below are used to try and increase efficiency of search, to continue a low level of background sampling in areas that have not been previously sampled.

Whatever sampling strategy is chosen it is important to be careful about the sampling properties of the random number generator used. Most generators are based on a deterministic algorithm and it is not possible to generate entirely random numbers from such an algorithm. They generate pseudo-random numbers that appear random when their statistical properties are analysed. Random number generation is the subject of a considerable body of literature. Different algorithms have different repeat patterns and sampling characteristics. This is discussed in Box 3.3, where a simple algorithm suitable for most purposes is outlined. It is worth noting that many algorithms require the user to specify a starting value or seed for the generator. For a particular algorithm, sequences of numbers starting from the same seed should have exactly the same sequence of numbers (at least using the same compiler on the same type of computer), i.e. they are not actually random. It follows, however, that if you do not change the seed between runs of the sampler you will get exactly the same set of realisations and model results. This can be avoided by randomising the seed for different runs, but it is also a useful characteristic in program development so that the results of different runs can be compared.

3.3.3 Pseudo-random numbers and the realisation effect

The statistics of variables that are derived from generated pseudo-random numbers will depend on the number of samples used. This realisation effect is demonstrated here for a single generated series (using the random number generators included in Matlab).

Figure 3.6 shows a sequence of uniform random numbers in the range [0–1], together with the mean of the values as more data are added. This is easily converted to any other range by adding a minimum value and scaling the unit range of the generated numbers to the difference between the maximum and minimum values of the range required (as in Eqn. [B3.3.3] in Box 3.3).

Figure 3.7 shows a sequence of random numbers chosen from a normal probability distribution with mean = 0.5 and standard deviation = 0.25. Note that with a coefficient of variation (the ratio standard deviation/mean) of 0.5, some of the numbers generated will be below zero. The normal distribution has infinite tails but 95% of the values will be within the range +/– 2 standard deviations from the mean and 99% of the values between +/–3 standard deviations from the mean. Thus, there is a 2.5% chance of drawing a negative value with this coefficient of variation.

Figure 3.8 shows the mean and standard deviation of the sequence from Figure 3.7. These settle to close to the parameters of the generating distribution after about 200 samples. We are often interested in the extreme values from a distribution, however, in particular in estimating the upper and lower quantiles that might serve to define uncertainty bounds for a predicted value. These extremes, or values in the tail of the distribution, would be expected to converge to the true value more slowly. This is shown in Figure 3.9 where the 5% and 95% quantiles of the data in Figure 3.7 are shown together with the theoretical estimates for a normal distribution derived from the calculated mean and variances as more data points are added. It can be seen that, although the two estimates gradually converge to common values, there is an

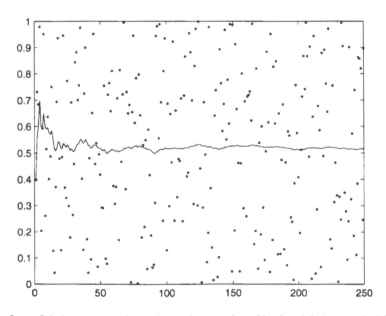

Figure 3.6 A sequence of pseudo-random numbers (dots) and their mean (solid line) drawn from a uniform distribution with the range [0–1]. X-axis is number of data points, Y-axis the value of the random number

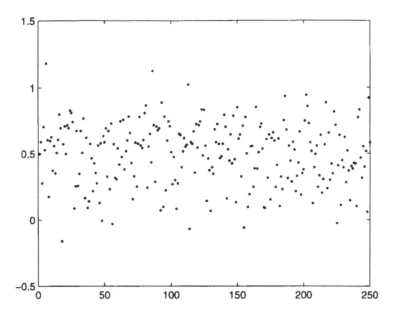

Figure 3.7 A sequence of pseudo-random numbers drawn from a Gauss (normal) distribution with mean of 0.5 and standard deviation of 0.25. X-axis is number of data points, Y-axis the value of the random number

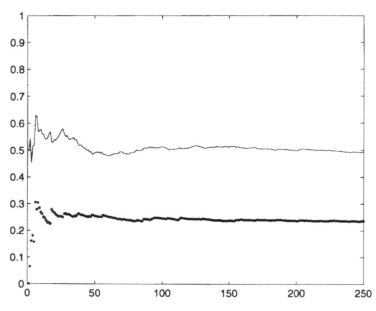

Figure 3.8 Mean (solid line) and variance (dots) of the data in Figure 3.7 calculated as more data points are added. X-axis is number of data points, Y-axis the value of the mean or variance

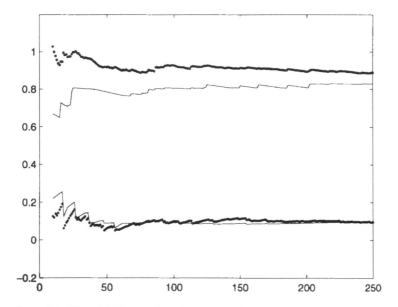

Figure 3.9 5% and 95% quantiles of the data in Figure 3.7 estimated using the mean and variances (solid lines) shown in Figure 3.8 (and directly from the cumulative distributions of the data points as new data are added (dots)). X-axis is number of data points, Y-axis the values of the 5% and 95% quantiles.

important realisation effect in the early stages of sampling when single new values can have a significant effect on the calculated statistic.

These example plots are only for a single variable. The realisation effect of sample statistics will be more important as more and more sampled variables are added, particularly in estimating the statistics of extreme values. Similar principles apply in sampling real environmental variables by measurement. Convergence to the final asymptotic values will also be slower if successive samples are correlated, which can be the case if sampling real variables. The general rule to be followed is always to take as many samples as feasible for a given problem and to check on whether convergence of the statistics of interest has been reached by doing the type of plots shown in Figure 3.9 as more realisations are added.

3.3.4 Guided Monte Carlo search

If we have some prior information about feasible parameter sets, or if we are learning about regions of interest from an initial discrete or random search of the model space, then we can start to use some form of guided search strategy. The idea behind all guided search strategies is to concentrate samples in the regions of interest once they have been identified.

In some cases, they are identified purely on the basis of the prior definition of distributions or fuzzy membership values for the parameters. Even where it is difficult to define prior information for all parameters and their expected co-variation or interactions, this immediately provides some information of what values to sample.

The efficiency gains of adopting a guided search method can be very large. Hahn and Meeker (1991), for example, point out that if a set of parameter values is assumed to have independent multivariate Gaussian distribution, then already when there are 6 parameter dimensions covering ± 3 standard deviations from the mean for each parameter, then 95% of the model space has very low prior likelihood. Nearly all the useful information is in the remaining 5% of the model space so that savings of 1 in 20 could be gained over a purely random search if the parameter sets are chosen based on the prior distribution.

This type of guided sampling can be achieved in a number of ways. The most commonly used when there is such prior information is a variation on the Latin Hypercube sampling (LHS) method in which the intervals to be sampled are chosen to represent equal prior probability (or fuzzy membership) intervals. This will then automatically concentrate samples in the area of interest, while still limiting the number of runs to the number of discrete intervals in each dimension. To give just two quite different environmental examples of a forward uncertainty analysis using LHS, Catelinois et al. (2005) used it in an epidemiological study to estimate the number of excess cases of thyroid cancer in France resulting from Chernobyl fallout, while Sykes et al. (1996) used it in a forward uncertainty assessment of aquifer remediation.

A further extension is possible to take account of correlation or interactions between the parameters by carrying out a principle components analysis of an initial sample to create new orthogonal axes for sampling (see Iman and Conover, 1982). This is, however, a linear analysis technique and might not prove to be useful in applications of highly nonlinear models. New developments in this field include the use of copula sampling (see Section 3.3.5 below).

The Latin Hypercube sampling methodology can also be applied as part of a learning strategy. An initial set of samples can be made to evaluate the model outputs of interest. These can then be used to redefine the sampling increments, and a further set of samples run to give greater detail in the regions of interest. Another way of learning about the model space in this way has been suggested by Spear et al. (1994) based on regression tree analysis. This is a classification methodology, here used to subdivide the model space into discrete volumes containing different types of behaviour (see Figure 3.10). In their application to a water quality model they were trying to define regions of the model space that provided good fit to the available observations and regions that did not give good fit and were therefore not worth sampling further. The same technique can be used, however, without observed data, where the interest is just in classifying the outputs from the model, some of which might be of more interest than others.

The general name for this type of iterative refinement of the search is importance sampling. The idea is to learn about the nature of the surface of interest in the model space and concentrate samples where they will provide the greatest amount of information. One form of importance sampling is to gradually modify the sampling distribution until the density of sampling is directly proportional to the magnitude of the response surface (i.e. the value of the model output of interest).

One method of guided Monte Carlo search strategies with this aim is called Markov Chain Monte Carlo (MC^2) methods. These provide a way of learning about the structure of a response surface and creating samples with a density proportional to the

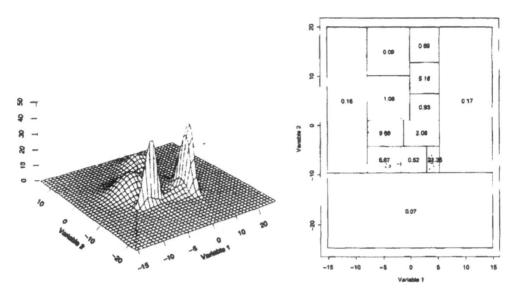

Figure 3.10 Result of regression tree analysis of a two-dimensional model space. Left: perspective plot of true density. Right: regression tree subdivision of space into areas identified as having different density with resulting samples of parameter combinations

Source: Spear et al., 1994, Copyright ©1995 American Geophysical Union, reproduced with permission

height of the surface. They can be used in forward uncertainty analysis where the response surface is defined by an output variable of the model but are most often used in cases where the height of the surface is a likelihood measure, and an integration of the full likelihood surface is required. If samples can be generated with density proportional to likelihood then all the samples will have equal weight in the integration. MC^2 therefore represents a learning process, but, when used for likelihoods, the learning process will require the availability of observed data with which to compare the model predictions. More details about MC^2 will therefore be left to the next chapter (and Box 4.3) where model conditioning based on observations is considered.

3.3.5 Copula sampling

In some applications, it is well known that there are strong dependencies or correlations between parameters in the model space. It is not too difficult to take account of such dependencies in low-dimensional spaces, e.g. by using the Latin Hypercube sampling technique, but the problem can be complex in high-dimensional spaces. Such dependencies are often revealed by soliciting expert opinions to provide prior information into a modelling problem and may deviate strongly from the multivariate Gaussian assumptions on which much of the rich statistical literature on correlated variables is dependent.

A recent technique for addressing this problem is the use of **copula** sampling (see, for example, Nelson, 1999, Kurowicka and Cooke, 2006) to generate samples of the required structure. Kurowicka and Cooke (2006) give a good introduction to the subject that includes a number of environmental examples. They also introduce a software package, UNICORN, specifically for high-dimensional dependence modelling (see Software Appendix).

The starting point for the UNICORN package is an assessment of the uncertainties and dependencies associated with all the inputs required to run a model, including the parameter values. Normally, the information about the marginal distribution of each input or parameter value is likely to be much stronger than information about their interactions. This is not surprising since, given N inputs required by the model, there is the possibility of $2^N - 1$ subsets of variables, each of which might interact with any other subset. Thus, N does not have to be very large before the number of potential interactions becomes very large, and it becomes very difficult to specify, even approximately, the potential interactions.

This is a problem within statistical representations of uncertainty based upon the multivariate Gaussian distribution (or the so-called meta-Gaussian methods where more complex distributions are transformed into a multivariate Gaussian structure, see Box 4.1). This is because, as already noted, the generation of samples from the multivariate Gaussian distribution requires the specification of a complete covariance matrix. This is already a demanding challenge in simple cases; when there is only partial information about the interactions in high-dimensional cases it becomes very difficult to use.

Copulae, coupled with Dependence Trees and Dependence Vines, were developed to cope with such cases. The essence of the copula concept is to transform the marginal distributions and partial information about dependencies to a unit space based on axes with the range [0–1]. Any inconsistencies in the information about dependencies can be resolved and random samples generated on the unit space in a way that conserves information about marginal distributions and any dependencies expressed as rank correlations. Because, in sampling on the unit space, rank correlations can be preserved regardless of the marginal distributions of the variables, they can be used as a constraint on the dependencies between variables. The samples in this space can then be transformed back to the actual values in the original model space in a more general way than the meta-Gaussian transform. The major attraction of the copula approach is that the definition of the marginal distributions can be separated from the specification of the dependency as a rank correlation. This gives great flexibility, including the possibility of using non-parametric marginal distributions directly from empirical data, rather than first fitting the data to a parametric distribution, such as those described earlier in this chapter.

Copulae can be represented as pdfs in the multivariate space. They have an apparently strange form (Figure 3.11a shows an example for two variables) because they effectively represent an inverse transform from the uniform sampling on the unit space to the original marginal distributions of the variable. A variety of copula structures are available to generate samples with different forms of dependency, including the Gauss copula, χ^2 (chi-squared) copula, diagonal band copula, elliptical copula, Frank's copula, Clayton copula, and minimum information copula (see Kurowicka and Cooke, 2006). The major problem in this approach is finding the right form of copula to represent the dependencies in a way that is consistent with the co-variation in the original variables. In the high dimensions often found in environmental problems, where there may be conflicting information about dependencies, this is a difficult (if

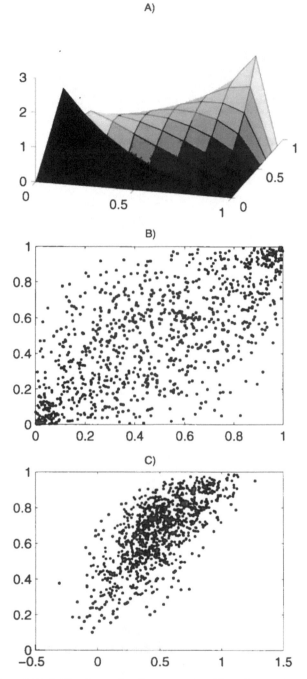

Figure 3.11 A. The form of a Gauss copula with rank correlation of 0.7. B. Samples generated using the Gauss copula with rank correlation 0.7. C. The same samples transformed to variables normally distributed (x axis) and beta-distributed (y axis) with same parameters as Figure 3.5

not impossible) problem that generally requires invoking some strong constraints. Figure 3.11b shows an example of samples generated using the Gauss copula with a rank correlation of 0.7; and Figure 3.11c shows the transformation back to samples in the space of the original variables with beta and normal marginal distributions, for comparison with Figure 3.5. The results of sampling from other copula forms are illustrated in Box 3.2. Some families (such as the Gauss and χ^2) are more easily extended to multivariate problems than others (Kurowicka and Cooke, 2006).

3.3.6 Case study: Copula sampling in mapping groundwater quality

An interesting application of copulae is provided by Bardossy (2006) in an application to groundwater modelling. The advantage of the copula approach in this case was that the data (groundwater quality parameters in a large number of wells in Baden-Württemberg in Germany) suggested that the variables of interest had a non-Gaussian distribution and a complex covariance structure. The large quantity of data in this case allowed an empirical copula to be developed, using a chi-square copula, representing the structure of the spatial variation of the data, in a way more flexible than can be handled by the Gaussian 2nd order stationary assumptions of classical geostatistics (see, for example, Clark, 1979, for a good clear introduction to geostatistics as a way of expressing uncertainties in spatial patterns). As Bardossy notes, spatial problems of this type will have a large number of dimensions representing the variability between pairs of points at different distances. This makes estimating the dependences difficult. In his study, rank correlations were used as an empirical constraint on the copula model. Testing the resulting spatial model showed that it successfully reproduced the spatial structure of the data more adequately than a multivariate normal geostatistical model. Simulations of the field using the copula showed that the technique was able to reflect the way that structure was linked to the magnitude of the variable of interest in a way that classical geostatistics cannot (e.g. Figure 3.12). Other applications of copula sampling to environmental problems include the rainfall simulations of De Michele and Salvadori (2003) and joint frequency analysis in Favre et al. (2004) and Renard and Lang (2007). Reichert and Borsuk (2005) use copulas to represent dependence between parameter values in assessing and deciding between different policy scenarios in a problem of managing phosphorus inputs to a lake.

3.4 Fuzzy representations of uncertainty

It has been noted earlier how not all uncertainties can easily be represented as probabilities. One attempt to provide an alternative view was the fuzzy set theory of Zadeh (1965) where uncertainty is expressed in terms of fuzzy measures as an expression of possibility of different potential outcomes. Klir (2006) and Zadeh (2005) show (in their different ways) how this is one form of a more general class of non-probabilistic ways of expressing uncertainty based on set theory. Here we are concerned with using fuzzy sets in the propagation of uncertainties of inputs and parameter values through an environmental model to define possibilities of outcomes rather than probabilities.

A fuzzy set is defined by the **degree of membership**, μ, of the members of that set, with the range 0–1 (see Box 3.4). Any fuzzy variable may be defined with respect to the

A)

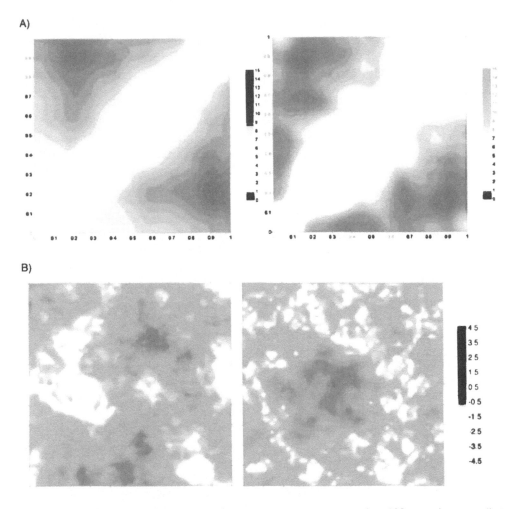

B)

Figure 3.12 A. Empirical copula densities from measurements at more than 600 groundwater wells in Baden-Württemberg, Germany for spacing of 3,000m. Left: chloride, right: pH. B. Two realisations of a groundwater quality variable. Left: using standard multivariate Gaussian distribution and right: using chi-square copula with normal marginals

Source: Bardossy, 2006, Copyright ©2006 American Geophysical Union, reproduced with permission

fuzzy set of its values, with particular values having particular degrees of membership. An example of a fuzzy variable that is commonly used in explaining fuzzy set theory is that of the height of people. Height can be measured on a length scale, but what does it mean to say that someone is "short", or "tall" or of "medium height"? The class of tall people is a fuzzy set, and might extend over different parts of the length scale in different circumstances (urban Europe or rural China for example).

A more relevant environmental example is that of temperature. Temperature can also be quantified on an agreed scientific scale, but we might also think of an examination of old diary entries to obtain information about historical weather patterns. A diary entry might say that a day was "exceptionally hot". What is "hot"? What is "exceptionally hot"? Clearly there is no precise definition of such descriptions and they might be better treated as fuzzy variables. In England or Sweden, a day with temperatures of over 30°C might be described as hot by one person, exceptionally hot by another. In the south of France or Texas, 30°C on a summer day would not be described as exceptionally hot.

Figure 3.13 shows another example of representing the hydraulic conductivity of different types of soil and aquifer materials as a fuzzy variable. What is the hydraulic conductivity of a sand in an unconfined valley bottom fluvio-glacial aquifer used for water supply purposes? In setting up a water resources assessment model of such an aquifer it would be useful to have some prior estimate of the possible range of hydraulic conductivity (as well as porosity, thickness, inter-bedding with other materials, unsaturated soil characteristics etc). Here the fuzziness in hydraulic conductivity ranges across orders of magnitude. If we are interested in pollutant transport in that aquifer, perhaps because of an old industrial site on the surface or agricultural pesticide applications, then there will be other parameters that could be considered as fuzzy variables, such as the longitudinal and transverse dispersion coefficients, reaction and degradation coefficients (e.g. Schulz et al., 1999; Zhang et al., 2006).

Joint possibilities of multiple variables can also be expressed in this way. Degree of membership is then defined with respect to the joint set. The range of values for which $\mu > 0$ is called the **support** of the fuzzy set. Note that if the support of the fuzzy set has a strictly limited range of values it constitutes a crisp set, but the membership of any value in that set will be more or less fuzzy, depending on the degree of membership.

Different degrees of uncertainty in a particular variable (or joint possibilities of several variables) can be expressed in terms of nested α-cuts of the fuzzy set. These are crisp sets defined by the range of values for which $\mu \geq \alpha$ ($0 < \alpha < 1$). Fuzzy variables defined in this way can be manipulated by standard operators of fuzzy arithmetic (including addition, subtraction and general linguistic operators; see, for example, Box 3.4 and Cox, 1994; Bardossy and Duckstein, 1995; Ross, 2004; Klir, 2006). Thus, for some simple cases of uncertainty propagation it would be possible to obtain analytical estimates of the possibilities of a model outcome given a fuzzy definition of the input variables and parameters.

More generally, Monte Carlo methods can again be used in uncertainty propagation of fuzzy inputs. By sampling values across the support of a fuzzy set (or the range of a chosen α-cut of that set), that value can be used in driving the model, carrying along a weight equivalent to its degree of membership. Sampling a full set of inputs and parameter values in this way will result in a simulation associated with a set of degrees of membership from each of the fuzzy inputs. It then remains to combine those degrees of membership to provide a possibility value for the outputs for that run (see the discussion of techniques in Box 3.4). This can be thought of as equivalent to defining a degree of membership for the output of the model throughout the feasible range of the model space (i.e. a possibility response surface associated with the actual outputs from the model). Defuzzification of the output fuzzy set is possible at this stage to provide a single crisp estimate, but in general will not be of interest here when we are actually interested in the uncertainty of the outcomes. Given the map of degrees of membership for a given output in the model space, α-cuts of the possibility surface can be made to define the nested sets of model output values at successive α levels.

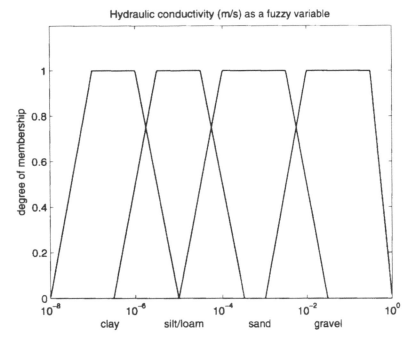

Figure 3.13 The hydraulic conductivity of different materials treated as fuzzy variables

3.4.1 Case study: Forward uncertainty analysis using fuzzy variables

An interesting example of a forward uncertainty analysis using fuzzy variables is provided by Schulz et al. (1999) in an application concerned with uncertainty in chemical equilibrium calculations for an aqueous cadmium sulphide system in the anaerobic sediment of the Silbersee near Nürnburg, Germany. The same authors have also been involved in studies on unsaturated flow in soils (Schulz and Huwe, 1997). The aim was to try to allow for the uncertainties inherent in the thermodynamic parameters in applying chemical equilibrium theory in the field. They note that some previous forward uncertainty analyses for chemical equilibrium calculations had used a probabilistic framework and Monte Carlo simulation (e.g. Schecher and Driscoll, 1988) but argued that a fuzzy approach might be more appropriate when it was difficult to define potential variability in terms of probabilities. Various forms of information might be vague rather than random, and therefore more appropriately represented possibilistically.

The cadmium sulphide system involves 11 different chemical species and eight different reaction coefficients. Information about the coefficients from the literature showed a significant range of values in each case. These were used to define the support for a fuzzy representation of each coefficient and the shape of the membership functions (e.g. Figure 3.14).

Schulz et al. provide results for a number of different levels of calculations. In the first, only the two coefficients in Figure 3.15 are varied, with calculations being made

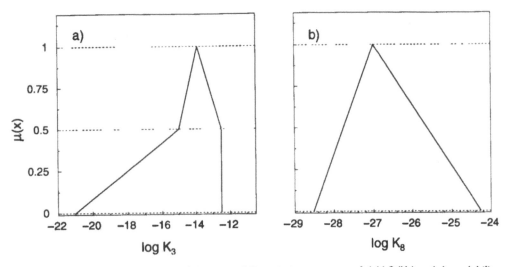

Figure 3.14 Membership functions of the second dissociation constant of a) H_2S (K_3) and the solubility product of b) CdS (K_8)

Source: Schulz et al., 1999, reproduced with permission from Elsevier

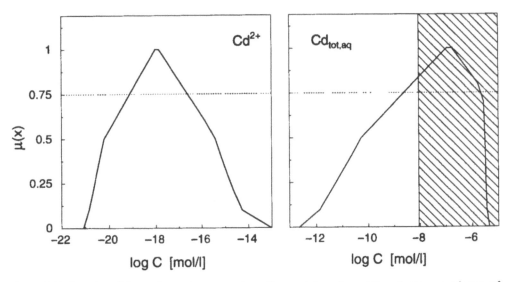

Figure 3.15 Results of fuzzy forward propagation of uncertainty for solid and aqueous phases of cadmium in equilibrium cadmium-sulphide aqueous geochemistry calculations treating parameters as fuzzy variables. The vertical axis is the degree of membership. The hatched area represents the concentrations above the WHO limit of Cd in drinking waters

Source: Schulz et al., 1999, reproduced with permission from Elsevier

at a number of different α-cuts. Then more coefficients are introduced and the uncertainties in the analytical data available are taken into account using membership functions chosen to be close to the Gaussian distributions normally used when analytical error is assumed to be random. Finally all the sources of variability are taken into account. The results of Figure 3.15 are from this last case, showing a high possibility that the levels of total cadmium might exceed the WHO admissible range for drinking waters.

3.5 Sensitivity analysis

3.5.1 Point sensitivity analysis

In considering the propagation of uncertainties through a model based only on prior information, it is useful to make an assessment of the sensitivity of the results to individual parameters or combinations of parameters (or parameters and other inputs to the model). We can relate the concept of sensitivity back again to the response surface for a predicted variable in the model space. At any point in this (generally high-dimensional space) the sensitivity to a particular factor can be thought of as related to the gradient of that surface and a traditional single factor sensitivity measure is obtained by evaluating the local gradient at a particular point in the model space, normalised by the value at that point. Thus, for parameter i taking values x_i in producing a predicted variable P, a sensitivity index, SI, can be calculated as:

$$SI_i = \frac{dP/dx_i}{x_i} \qquad [3.1]$$

The gradient term, dP/dx_i, is often difficult to evaluate analytically (which would require differentiating the equations of the model) and so is generally evaluated numerically by using runs of the model with slightly different values of x_i. A finite difference approximation to [3.1] then becomes:

$$SI_i = \frac{\{P(x_i - \Delta x_i) - (P(x_i + \Delta x_i)\} / 2\Delta x_i}{x_i} \qquad [3.2]$$

where Δx_i represents a small increment in the values of the parameter x_i.

Such sensitivity measures can be used to examine the relative sensitivity of different factors in the model space. Because they are point measures, the measure for a particular parameter or input variable might vary (sometimes rapidly and discontinuously) throughout the model space. They should also be expected to vary depending on what model output variable (P) is considered. Thus, while they might be a good preliminary guide to sensitivity of individual inputs and parameters, there are the same problems of exploring the way in which sensitivity might vary through the model space as there are in exploring the responses themselves. In fact, the problem is somewhat worse in that there may be sensitivities to the joint occurrences of certain values of parameters and inputs that are not revealed by the single factor measures above.

Thus, some more sophisticated form of sensitivity analysis is needed. A full treatment of this topic is beyond the scope of this book and the reader is referred to the book of Saltelli et al. (2004) and the review of Saltelli et al. (2006) for more details on some of the methods available. It is worth noting, however, that studies that have compared sensitivity analysis methods in applications to environmental models have shown that the apparent sensitivities of different factors vary with the type of sensitivity analysis used (e.g. Borgonovo, 2006; Pappenberger et al., 2006b; Tang et al., 2007b). There is no unique answer to the sensitivity problem, but the different methods will generally reveal the most important factors (if not always in the same rank order). We will look at just two contrasting sensitivity methods, both of which have been called Generalised Sensitivity Analysis (GSA). Both are *global* methods that attempt to derive a measure of sensitivity from model responses sampled throughout the model space.

3.5.2 Global sensitivity analysis: Sobol' generalised sensitivity analysis

The first is an extension of the simple point sensitivity methods described above to the case of a global sensitivity analysis of both single factors and joint combinations of factors. It is a linear variance decomposition method, originally developed by Sobol', that analyses the variation of a sample of points on the response surface for a variable as a result of variation in different factors (usually parameter values or other inputs that are uncertain). A good introduction to this form of variance-based sensitivity analysis is provided by Sobol' (2001) and Saltelli et al. (2004).

The Sobol' variance decomposition is defined by

$$V(Y) = \sum_i V_i + \sum_i \sum_{j>i} V_{ij} + \sum_i \sum_{j>i} \sum_{k>j} V_{i,j,k} + \ldots V_{1,2,\ldots,k} \qquad [3.3]$$

where Y is the model outputs of interest, X_i are the factors of interest, $V_i = V_{X_i}(E_{X_{\sim}}\{Y \mid X_i\})$, $V_{ij} = V_{X_iX_j}(E_{X_{\sim}}\{Y \mid X_i, X_j\}) - V_i - V_j$, and so on. The Sobol' sensitivity indices are then defined as :

$$SI_i = V_i \mid V$$

$$SI_{ij} = V_{ij} \mid V \qquad [3.4]$$

$$SI_{ij}^c = SI_i + SI_j + SI_{ij}$$

All the sensitivity indices are scaled to be in the range [0–1]. The SI_i values are known as the main effects, the SI_{ij} are the second-order interactions, and the SI_{ij}^c are the second-order closed effects representing the total effect of the two factors i and j. Higher-order interaction terms can be calculated to give an indication of the interaction of three or more factors, but become increasingly sensitive to the sample used and sources of error in the modelling process. The main and total effects are the easiest to interpret. Effectively those factors with the highest main effect sensitivities are those that, if determined more precisely, will have the greatest effect on reducing the variance in the model output. It is often found that, even for quite high-dimensional problems, only a small number of factors are important on the basis of the magnitude of the main effect sensitivities.

Relative sensitivities determined in this way might be quite different to the local gradient measures of sensitivity discussed in the previous section since the measures are based on all the samples in the space. If it is suspected that there might be distinct changes in sensitivity to different factors in different parts of the model space (which may well be the case with complex environmental models), it is possible to apply this form of global sensitivity analysis to different subsets of samples separately. For models which produce time series or spatial patterns of outputs, it is also possible to examine the changes in the calculated sensitivities in time or space (e.g. Ratto et al., 2005; Hall et al., 2005).

The calculated sensitivities might often be dependent on the number of samples over which the variances are calculated, particularly for complex models, and there have been a number of studies that have suggested ways of making the process more efficient. One way is to use a form of interpolation or filtering of the sample of responses in the model space (see Section 3.6 below).

Hall (2006) has extended this variance decomposition method of global sensitivity analysis to the case where the factors are known only as imprecise probabilities. Ratto et al. (2001) have applied the Sobol' sensitivity analysis approach to a model calibration problem within the GLUE Monte Carlo methodology described in Section 4.5, to determine which factors might be most important in influencing changes in a calibration measure.

3.5.3 Case study: Application of Sobol' GSA to a hydrologic model

In an application of this form of generalised sensitivity analysis to the sensitivity of the results of a distributed flood inundation model, Tang et al. (2007a) have used the Sobol' variance decomposition GSA method to calculate the first-order sensitivities of a distributed rainfall runoff model (the US National Weather Service Hydrology Laboratory Research Distributed Hydrological Model, Koren et al., 2004) for which 18 parameters were varied over each spatial cell in the distributed model. Predictions for the case shown in Figure 3.16a were made using an hourly time step and the total sensitivities for each parameter aggregated to monthly results. It is also possible, as shown in Figure 3.16b, to aggregate the results to show the sensitivities for each spatial element in the model and their interactions (in this case rather a coarse discretisation of the catchment area is used). This study also investigated the uncertainty in the calculated sensitivity indices by using a bootstrapping method to resample the simulation runs. The sensitivities are then recalculated for each sub-sample of the full set (see Efron and Tibshirani, 1993; Archer et al., 1997; or Saltelli et al., 2004).

3.5.4 Global sensitivity analysis: HSY generalised sensitivity analysis

The second form of global sensitivity analysis is quite different. It was first used by George Hornberger, Bob Spear and Peter Young (HSY) in the late 1970s in assessing a model of eutrophication and algal growth in the Peel Inlet near to Perth in Western

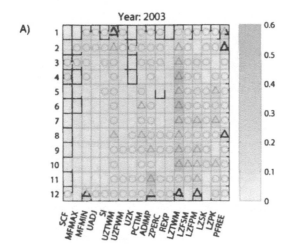

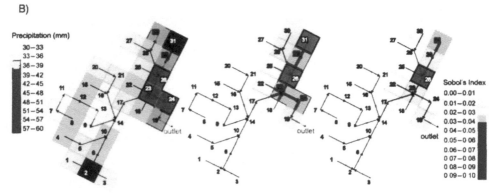

Figure 3.16 Sensitivity analysis for the parameters of a distributed rainfall-runoff model using the Sobol' GSA method. A. Total sensitivity indices for monthly periods of simulation for the year 2003. Columns represent the parameters varied; rows represent months. Triangles represent highly sensitive parameters that contribute at least 10% of the overall model output variance; circles represent sensitive parameters that contribute at least 1% of the overall model output variance; shading represents relative sensitivity. B. September 2003 event on the Spruce Creek catchment: rainfall totals (left), first-order Sobol' indices for each cell (middle), cell level interactions (right)

Source: Tang et al., 2007a, Copyright ©2007 American Geophysical Union, reproduced with permission

Australia (Hornberger and Spear, 1980) and rivers in the UK (Whitehead and Young, 1979). It also requires samples from throughout the model space, normally selected randomly. Their interest was in the model behaviour as a whole, not point sensitivities, so that they examined the way in which different types of behaviour responded to variations in parameters and inputs. Their analysis required that the different runs of the model be separated into two bins or classes with different types of behaviour. Then, in examining the sensitivity to individual factors they compared the cumulative distributions for those factors in each of the bins. Insensitive factors would show little difference between the distributions in the two classes; the most sensitive factors would show strong differences between the distributions in the two classes. The non-parametric Kolmogorov–Smirnoff d statistic for the difference between distributions was used as a relative measure of sensitivity (see Figure 3.17). The power of that test is not great when there are many samples contributing to the test, as there often will be in such model-generated samples, but the value of the statistic d still provides a relative measure of global sensitivity for the individual factors.

There are two problems with this form of global sensitivity analysis. The first is how to define the bins; the second is whether the values of an apparently insensitive parameter might still be important in contributing, as part of a parameter set, to whether a model output is classed in one bin or another, at least in locally sensitive regions of the model space. As with the linearity assumption of the Sobol' variance decomposition technique, the second is a limitation of using global analysis that will always filter local sensitivities and again, if local differences are expected, then this form of GSA can also be applied to sub-domains of the model space.

The first question is interesting as it is under the control of the user of the model in deciding what differences in model behaviour are important to consider. In the original application to the Peel Inlet model, the bins represented models that seemed to produce the right sort of behaviour when evaluated against chemical and biological observations (behavioural models) and models that did not (non-behavioural models). Making such distinctions in such applications can be difficult, however, particularly when there are no observations to compare against, or where there is a continuous range of outputs and no clear distinction between classes can be determined. In other applications, such as the rainfall-runoff modelling example of Hornberger et al. (1985), the latter case has been dealt with by taking the top 25% or 30% of models

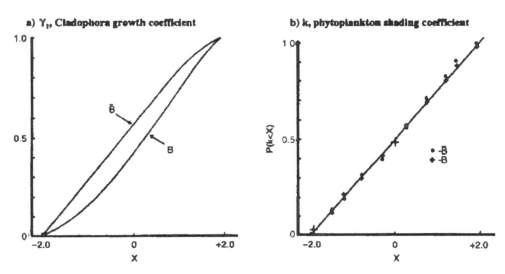

Figure 3.17 Example of HSY generalised sensitivity analysis results showing cumulative density functions for behavioural (B) and non-behavioural (B') models of phytoplankton growth in Peel Inlet, Australia. a) a sensitive parameter (Cladophora growth coefficient); b) an insensitive parameter (Phytoplankton shading coefficient)

Source: Spear, 1997, reproduced with permission from Elsevier

ranked by some predicted output variable as one class and the rest as the second class, with no implication as to whether one class or the other is "behavioural" with respect to the real system. In other cases it might be easier to define behavioural models. Guven and Howard (2007), for example, give an application of the method to the prediction of cyanobacterial blooms in the St. Johns River in Florida with the behavioural models being defined on the basis of characteristics of predicted chlorophyll-a concentrations during the months of the cyanobacterial blooms.

The HSY generalised sensitivity analysis is not, however, limited to using only two classes of model performance, and more detailed information can sometimes be revealed about differences in model performances by breaking the behavioural models down in small classes or subdividing the model space to look at different sensitivities. Figure 3.18 (a screen dump from the Lancaster GLUE demonstration program, see Software Appendix) shows the results of an analysis of sensitivities for four parameters in a rainfall-runoff model, where the simulations have been classified into ten subsets based on level of performance. The HSY sensitivity analysis was the starting point for the development of the GLUE methodology described in Section 4.5.

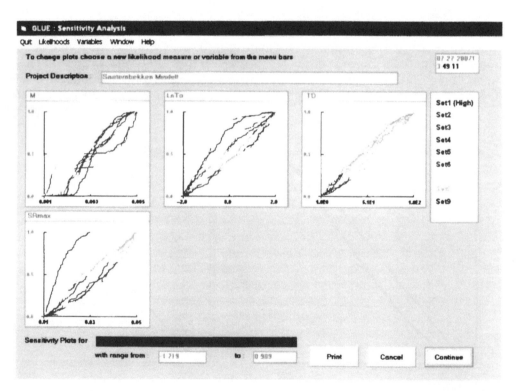

Figure 3.18 HSY generalised sensitivity analysis for four parameters of a rainfall-runoff model applied to the Saeternbekken Minifelt catchment in Norway. The simulations have been divided into subsets based on performance based on the Nash–Sutcliffe efficiency measure, with Set 1 being the highest performance

3.6 Model emulation techniques

We have already seen how computational limitations will limit the potential for a full exploration of the response surface in high-dimensional model spaces. Clearly, if the run time for a single model run is long then only a limited sample will be feasible (as, for example, with global climate models where multiple ensemble runs are only just becoming available, and then only in small numbers or with very coarse resolution). This has resulted in a variety of attempts at emulating the responses of complex models by simpler models so that the results of the limited number of runs of the full model can be approximated at many more points in the model space. This is called model emulation.

There are two types of model emulation. The first is where a simple model is used to mimic the behaviour of a much more complex model that takes a long time to run. The simpler emulator model is usually fit to representative sets of outputs from the more complex model (which may include variables of interest that are not observable directly by measurement). The emulator can then be used to run many more sets of input conditions quickly to expand the range of available results. Model emulation is not a replacement for making more runs of the full model (especially for highly non-linear models) but, in some cases, some remarkable reductions of dimensionality can be achieved. Young (1998), for example, shows how the mean global temperature output from a 26-parameter global carbon model can be modelled by a simple second-order linear transfer function emulator with a remarkable accuracy (see Figure 3.19). The transfer function model could then be used as an emulator to explore the responses to a much wider range of inputs within seconds, rather than waiting for further runs of the full model. This illustrates the problem of model emulation quite nicely. We could expect the results of the simple transfer function to be useful within the range of input conditions for which it will provide a good simulation of the full model. The issue is how wide is that range, or more generally, how far is the model emulation valid? We cannot know, of course, without running the full model more times to check the emulator results.

The second type of model emulation is essentially an interpolation technique. The aim is to use an emulator to interpolate the shape of the response surface in the model space from the limited number of runs of the full model. This is equivalent to interpolating the variation of an environmental variable in physical space (for which there are well-developed methods using geostatistics or other interpolators) but in the higher dimensions of a model space. In this case the variation in the output of the full model (the response surface of interest) is a function of the variations in parameter values along each dimension of the model space, not just of the inputs. Remember that the response surface could represent any model-predicted variable of interest or some performance or likelihood measure.

Many different model emulation techniques have been used. None, of course, will be perfect since they are all inferring a large number of unknown values from a much smaller number of known values on the response surface. The complexity of the surface (governed by the nonlinearity in the model responses) will essentially control how successful the interpolation model will be. One method is the Gaussian emulator (see, for example, Oakley and O'Hagan, 2005, and the review of O'Hagan, 2006a), where the nature of the interpolation can be varied throughout the space, depending

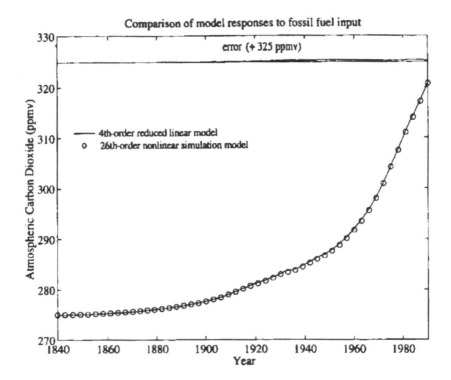

Figure 3.19 Example of model emulation: representation of the results of a high-dimensional
model of the global carbon budget by a 4th-order linear transfer function model

Source: Young, 1998, reproduced with permission from Elsevier

on the information available. Other methods for model emulation that have been used
include nearest neighbour methods (Beven and Binley, 1992; Klepper and Hendrix,
1994; Osidele et al., 2006); regression analysis (e.g. Iooss et al., 2006); artificial neural
network methods (e.g. Widrow et al., 1994; Broad et al., 2005); and State-Dependent
Parameter (SDP) methods (Ratto et al. 2005).

By definition, all these methods can only learn from the sample of runs of the full
model that are available. As with any interpolation technique, there will be some
uncertainty in estimating the shape of the response surface. For simple surfaces the
uncertainty should not be large. For complex surfaces, the uncertainty might be very
large. With some techniques the uncertainty of the interpolation can be assessed at an
interpolated point as well as a best estimate of the variable being interpolated.

3.7 Uncertain scenarios

A final type of uncertainty for which there is no historical data available that needs to
be considered is uncertainty in the form of discrete future scenarios for which it may
be impossible to make any quantitative estimate of probability or possibility. The
different scenarios used to drive global climate models are of this type, each the result
of the predictions of a model based on assumptions about the future. The modellers of

such scenarios have been required to make "realistic" estimates of what might happen to greenhouse gas emissions in the future as a result of different assumptions about global economic growth and social changes, but the uncertainties associated with those assumptions are unknown. Politicians and decision makers would, of course, be very interested in whether one scenario is more "likely" than another, but, in reporting the results of the climate change model runs based on different scenarios, the IPCC have been careful not to associate the outputs with other than a very general qualitative expression of "likelihood".

This is a particularly extreme example of assessing uncertainty in outputs by evaluating different scenarios because the computer run times of fully coupled atmosphere–ocean models are extremely long, even on the best high performance parallel computers. Thus, only a small number of runs is possible. It is also, of course, an extremely interesting example, in that the resulting outputs are of great practical importance and are being used to drive other simulation models to predict the impacts of climate change on hydrology, ecology, agriculture and other applications (without, it should be said, the possibility of coupling the future changes at the land surface back into the predictions of the climate model, although this is just beginning to be implemented in global models).

So if the predictions of environmental systems are uncertain for current conditions, even after calibration or conditioning against real data (see Chapter 4) what is the significance of using the outputs of climate change scenarios to predict future behaviour of a system of interest? Similar issues arise in other situations where only scenario predictions of future boundary conditions are possible. We do not know the probability or possibility of particular scenarios, so it therefore follows that any predictions dependent on those scenarios also cannot be associated with any objective assessment of probability or possibility, even if the uncertainty of representing the system under current conditions is used to make some estimate of the equivalent uncertainty under future scenario conditions (as, for example, in Cameron et al., 2000; Cameron, 2004; Wilby and Harris, 2006). The predictions remain simply scenarios, conditional on the assumptions about the inputs to the model running the scenarios and on the limitations of that model in predicting future responses. We may, or may not, wish to express some measure of belief in the different scenarios, but in general this will be possible only by subjective expert judgement or consensus. The way in which such scenarios might be considered in making decisions about the future is considered later in Chapter 6.

3.8 Summary of Chapter 3

This chapter has reviewed the issues and methods for the forward propagation of uncertainties in inputs and parameter values through a model for cases where only prior estimates of the relevant uncertainties are available. The following important points can be taken from the chapter.

* The results from a forward uncertainty analysis are totally dependent on the prior assumptions made about what is to be considered uncertain and how that uncertainty should be represented. Choices about prior uncertainties can be difficult, especially where there might be a large number of uncertain variables and where interactions or co-variation between the variables might be important.

- Ignoring co-variation between uncertain variables might give misleading results about model uncertainty, but lack of knowledge may often mean that it is difficult to avoid assuming that variables are independent. The assumption of independence makes sampling much simpler, but there are still choices about what ranges and distributions to sample over.
- Monte Carlo methods are useful in sampling the model space, but in high dimensional spaces very many realisations may be needed to adequately represent a model response surface unless it is very simple or unless strong prior information about the uncertainties in input variables is available so that guided sampling can be used.
- Given a sample of the response surface, both local and global sensitivity analyses can be used to decide which might be the most important variables in controlling uncertainty in model output.
- Depending on the complexity of the response surface, model emulation can sometimes be used to interpolate between the results of a more limited number of runs of the model in the model space.
- Some uncertainties that might be important in decision making, particularly about potential future change, can only be represented as scenarios. Results can then only be made conditional on the choice of scenarios considered.

Box 3.1 Simple operations with probability-distributed variables

We will consider two variables A and B that, based on some experimental data I, can be described as having probability distributions $p(A)$ and $p(B)$. We need not worry about the *form* of the distributions at this stage. We need only consider that the probabilities $p(A)$ describe the expected frequency of occurrence of values of A across the set of all possible values of A, and that $p(B)$ describes the expected frequency of occurrence of values of B across the set of all possible values of B. When dealing with probabilities, these are *crisp sets*, i.e. an occurrence of A is either contributing to the set or not (though the theory can be extended to imprecise probabilities; see, for instance, Klir, 2006).

B3.1.1 Axioms of probability
Probabilities must obey certain rules, the axioms of probability. These are that for any set of variables A in a set of all possible values of A:

1   $p(A) \geq 0$
2   $\int p(A) = 1$
3   If A and A' are mutually exclusive then $p(A + A') = p(A) + p(A')$

All of the classical theory of statistics can be shown to follow from these three simple rules (e.g. Papoulis, 1965).

B3.1.2. Combining probabilities
In environmental modelling we are often interested in the joint occurrence of values of A and B, making use of our expectations about $p(A)$ and $p(B)$. In fact,

when thinking about the use of uncertain variables in predicting a set of outputs of an environmental model, M(A,B), with values of A and B as inputs or parameter values, we are interested in the probability of an *output* given the joint occurrence of values of A, B.

We can easily calculate the output probabilities under different types of conditions based on the rules of probability. This is because, for most types of models, there will be a simple link between the probabilities of values of the inputs and the probability of the outputs. If the model is linear (or can be approximated locally as linear system) the calculations are relatively simple and often can be carried out analytically. If the model is nonlinear, calculation of the output probabilities may be onerous (and perhaps better approached using the Monte Carlo methods of Box 3.2) but it does not change the principles involved.

Thus, in this case we wish to calculate p(M(A,B)) given the joint occurrence p(A) and p(B). A consequence of the axioms of probability is the product rule:

$$p(AB) = p(A)p(A \mid B) \text{ or, equivalently, } p(AB) = p(B)p(B \mid A)$$

where p(AB) is the probability of the joint occurrence of A and B and the symbol | indicates conditionality (or, in words, p(A | B) is the probability of an occurrence of A conditional on the occurrence of a value of B). We need to consider the conditional probability p(A | B) (or p(B | A)) because we will not always be able to assume that the two variables occur quite independently, even if independence is commonly used as a simplifying assumption in modelling studies if no other knowledge is available. In fact the data $\underline{I}$ on which we are basing estimates of the probabilities might well indicate that A and B cannot be considered to be independent. If (and only if) they can be considered to occur independently then B3.1.1 reduces to

$$p(AB) = p(A)p(B) \tag{B3.1.1}$$

But this simplifying assumption should be made with care.

When we can assume a simple dependence of the outputs of the model on the joint occurrence of the uncertain inputs, then the probability of the model outputs p(M[A,B]) conditional on the joint occurrence of the inputs A and B is most easily expressed in terms of the cumulative density function (CDF) of the outputs. For any variable A the CDF is defined by

$$F(A < a) = \int_{-\infty}^{a} p(A)dA \tag{B3.1.2}$$

For variables that only vary over a finite range of values the lower limit of the integration can, of course, be replaced by the minimum value. These expressions are general, but to apply them in practice we have to give form to the distributions of A and B (see, for example, Figure 3.2 in the main text) and their conditional occurrence or covariance. This expression can be extended to take account of many different variables, but then we will have to allow for the conditionality of one variable on joint occurrences of all the other variables, that is their **covariance**. This becomes intractable for more than even a small number of variables except if (and only if) the variables can be considered to be independent (but see Kurowicka and Cooke, 2006, for ways of dealing with this in complex real cases with limited information using the copula approach discussed in Section 3.3.5 in the main text and B3.2.2 below).

B3.1.3 The normal and multivariate normal distributions

The simplest form of distribution commonly used in statistics because of its mathematical tractability (though not suitable for all environmental problems) is the normal or Gaussian distribution (Figure B3.1.1) with variance σ^2 which has the mathematical form

$$p(A) = \frac{1}{\sqrt{2\pi\sigma^2}}\, e^{-(A-\bar{A})^2/2\sigma^2} \qquad\qquad [B3.1.3]$$

where $\bar{A}$ is the mean value of variable A. This can be simplified by scaling the variable A to a standardised normal distribution with mean 0 and variance 1 using the transform $Z = (A - \bar{A}) / \sigma_A$ such that

$$p(Z) = \frac{1}{\sqrt{2\pi}}\, e^{-Z^2/2} \qquad\qquad [B3.1.4]$$

With CDF

$$F(Z < z) = \frac{1}{2}\left[1 + erf\left(\frac{z}{\sqrt{2}}\right)\right] \qquad\qquad [B3.1.5]$$

where erf represents the mathematic function called the *error function*.

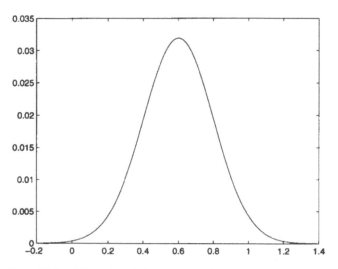

Figure B3.1.1 Normal probability density function for a single variable with mean 0.6 and standard deviation 0.2

For two normally distributed variables with covariance $Cov(A,B) = \sigma_{AB}^2$, the probability for occurrence of joint values of A and B, $p(AB)$, is a bi-normal distribution such that:

$$p(AB) = \frac{1}{2\pi\sigma_A\sigma_B\sqrt{1-\rho_{AB}^2}}$$

$$\exp\left\{\frac{1}{2(1-\rho_{AB}^2)}\left[\left(\frac{A-\bar{A}}{\sigma_A}\right)^2 - 2\rho_{AB}\left(\frac{A-\bar{A}}{\sigma_A}\right)\left(\frac{B-\bar{B}}{\sigma_B}\right) + \left(\frac{B-\bar{B}}{\sigma_B}\right)^2\right]\right\} \quad \text{[B3.1.6]}$$

where the correlation between occurrences of A and B is defined by $\rho_{AB} = \dfrac{Cov(A,B)}{\sigma_A\sigma_B}$.

In terms of two scaled normal variates, Z_A and Z_B, both with mean zero and unit variance and correlation ρ:

$$p(Z_AZ_B) = \frac{1}{2\pi\sqrt{1-\rho_{AB}^2}}\exp\left\{\frac{1}{2(1-\rho_{AB}^2)}[Z_A^2 - 2\rho Z_AZ_B + Z_B^2]\right\} \quad \text{[B3.1.7]}$$

A bivariate-normal distribution for two correlated variables is shown in Figure 3.1.2.

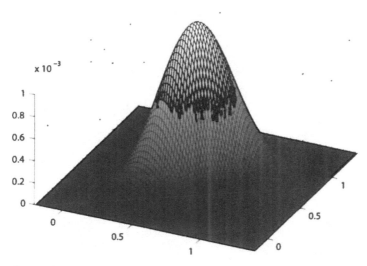

Figure B3.1.2 Three-dimensional plot of a bivariate-normal probability density function (z axis) for two correlated variables (x and y axes) with a correlation coefficient of 0.7

And for the general multivariate case for a set of n variables $\underline{A}$:

$$p(\mathbf{A}) = \frac{1}{2\pi^n \det[Cov(\mathbf{A})]} \exp\left\{ -\frac{1}{2} \sum_{j=1}^{n} \sum_{k=1}^{n} [Cov(A_j A_k)]^{-1} \right.$$
$$\left. (A_j - \bar{A}_j)(A_k - \bar{A}_k) \right\} \quad \text{[B3.1.8]}$$

B3.1.4 Means and variances for combinations for two random variables

Sometimes, we may wish to evaluate the joint occurrence of uncertain variables before they are used as inputs to the model. They may need to be added or subtracted. In these cases there are some simple rules for operating with variables of known means and variances. Consider again two variables A and B with means $\bar{A}$ and $\bar{B}$, and variances $\sigma_A{}^2$ and $\sigma_B{}^2$.

For the case where the variables are independent, adding the variables gives:

$$\overline{(A + B)} = \bar{A} + \bar{B} \,;\, \sigma_{A+B}^2 = \sigma_A^2 + \sigma_B^2 \quad \text{[B3.1.9]}$$

And subtraction:

$$\overline{(A - B)} = \bar{A} - \bar{B} \,;\, \sigma_{A-B}^2 = \sigma_A^2 + \sigma_B^2 \quad \text{[B3.1.10]}$$

For the case where the variables are co-varying with correlation ρ_{AB} then

$$\sigma_{AB} = \rho_{AB}\, \sigma_A \sigma_B \quad \text{[B3.1.11]}$$

And the relevant expressions are

$$\overline{(A + B)} = \bar{A} + \bar{B} \,;\, \sigma_{A+B}^2 = \sigma_A^2 + \sigma_B^2 + 2\rho_{AB}\sigma_A\sigma_B \quad \text{[B3.1.12]}$$

$$\overline{(A - B)} = \bar{A} - \bar{B} \,;\, \sigma_{A-B}^2 = \sigma_A^2 + \sigma_B^2 - 2\rho_{AB}\sigma_A\sigma_B \quad \text{[B3.1.13]}$$

B3.1.5 Gaussian error propagation

A general form of error propagation for under-assumptions of linearity and Gaussian distributed variables can be derived analytically. For a variable x that is dependent on uncertainty in some vector of uncertain but independent variables $y_1, y_2, \ldots y_n$, then, the variance of x can be derived as:

$$\sigma_x^2 = \left(\frac{\partial x}{\partial y_1}\sigma_{y_1}\right)^2 + \left(\frac{\partial x}{\partial y_2}\sigma_{y_2}\right)^2 + \ldots + \left(\frac{\partial x}{\partial y_n}\sigma_{y_n}\right)^2$$
$$= \sum_{i=1}^{n} \left(\frac{\partial x}{\partial y_i}\sigma_{y_i}\right)^2 \quad \text{[B3.1.14]}$$

The partial differential terms in [B3.1.16] (e.g. $\partial x/\partial y_1$) represent the gradient of the functional dependence of x on the different input variables, y. For a linear system these gradient terms will be constant, regardless of the values of y. For a nonlinear system they will not be constant which is why this form of error propagation is strictly applicable only to linear systems. The gradient terms can be directly related to the sensitivity coefficients of Section 3.5.1.

The input variables, y, cannot always be considered independent, of course. When the input variables are known to be correlated, then the effects of the correlation need to be taken into account. The equivalent equation is:

$$\sigma_x^2 = \sum_{i=1}^{n} \left(\frac{\partial x}{\partial y_i} \sigma_{y_i}\right)^2 + \sum_{i=1}^{n} \sum_{j=1, j \neq i}^{n} \left(z\rho_{y_i y_j} \frac{\partial x}{\partial y_i} \frac{\partial x}{\partial y_j} \sigma_{y_i} \sigma_{y_j}\right)$$ [B3.1.15]

Where $\rho_{y_i y_j}$ is the correlation coefficient between variables y_i and y_j, and $\sigma_{y_i y_j}^2$ is the covariance between variables y_i and y_j.

For linear functions of random variables these equations are exact. Thus, we can, for example, calculate the variance of x when $x = AB$ and A and B are independent with variances σ_A^2 and σ_B^2. From [B3.1.14], therefore:

$$\sigma_x^2 = \left(\frac{\partial x}{\partial A} \sigma_A\right)^z + \left(\frac{\partial x}{\partial B} \sigma_B\right)^2$$

$$= (B\sigma_A)^2 + (A\sigma_B)^2$$

$$= B^2\sigma_A^2 + A^2\sigma_B^2$$ [B3.1.16]

The uncertainty arising from other *linear* functions can be derived in the same way.

The relative contributions of each uncertain input factor (and their interactions) can be evaluated by dividing through by σ_x^2 so that:

$$1 = \sum_{i=1}^{n} \left(\frac{\partial x}{\partial y_i} \frac{\sigma_{y_i}}{\sigma_x}\right)^2 + \sum_{i=1}^{n} \sum_{j=1, j \neq i}^{n} \left(z \frac{\partial x}{\partial y_i} \frac{\partial x}{\partial y_j} \frac{\sigma_{y_i y_j}}{\sigma_x^2}\right)$$ [B3.1.17]

A recent application of Gaussian error propagation to a simple ecological model may be found in Lo (2005). The Sobol' sensitivity analysis described in Section 3.5.2 is based on a similar decomposition of the total variance in the output variable, x. In applications to nonlinear models, however, it does not attempt to estimate σ_x^2, only to evaluate the relative sensitivities to each input factor.

There have been a number of techniques in which similar equations have been used to estimate the uncertainty associated with nonlinear models as an approximation, for example the First Order Reliability Method (FORM, e.g. Benjamin and Cornell, 1970; and applications such as Freissinet et al., 1999). The advantage of such a method is that it is simple to apply and computationally inexpensive in comparison, for example, to Monte Carlo sampling methods of Box 3.2. The disadvantage is that, even for mildly nonlinear models, the results may be rather inaccurate. FORM is based on a Taylor series expansion of the variation in a model prediction around a chosen value, retaining only the first-order gradient terms. Higher order approximations can also be used, but then lose some of the advantages of the simplicity of application of the first-order approximation.

Rosenblueth (1975) also provided a different form of easily implemented approximation for nonlinear models based on moment matching. This has also been used in applications with environmental models (e.g. Guymon et al., 1981; Binley et al., 1991), but again there is a danger that the approximation might be inaccurate. These types of approximate methods have largely been overtaken by Monte Carlo sampling of the model space as the computer power available to the modeller has increased.

B3.1.6 *Random variables and uncertainty in model outputs*

The probability operations described above (and more complex forms derived from them) are normally used in statistics for the propagation of uncertainty through linear functions of random variables. Here, we are much more interested in providing estimates of uncertainty in variables or combinations of variables that might be used as **inputs** to drive a nonlinear environmental model. It then follows that uncertainty in the outputs from such a model, either deterministic or **stochastic**, can be assessed by weighting the output values by the joint probability of the associated uncertain input variables.

In all these cases, the probability of the outputs of a model $M(\underline{I})$, conditional on the joint occurrence of a set of input variables or parameters $\underline{I}$, will still be given by the CDF for the output being less than some specific value X that results from the joint occurrence of the values of the vector of input variables $\underline{I}$. Thus:

$$F(M(\underline{I}) < X) = \int_{-\infty}^{X} p(\underline{I} \mid M(\underline{I}) < X)dX \qquad \text{([B3.1.18]}$$

In words, this equation states that to find the cumulative probability that a model prediction is less than some value X, we must integrate over all the associated input probabilities, $p(\underline{I})$, for which the model output is less than the chosen value X. The lower limit of integration can, of course, be zero if the model outputs are always greater than zero. Once the CDF has been calculated, any choice of prediction quantiles may be defined (e.g. the 5% and 95% prediction limits). Many studies have used this type of approach in forward uncertainty analysis. To give just one recent example, Smemoe et al. (2007) have used it to propagate uncertainty in upstream runoff predictions through a hydraulic model of flood inundation to provide probabilistic representations of the risk of inundation (but see also Pappenberger et al., 2006d, for a discussion of issues that arise when model uncertainties are conditioned on maps of actual inundation from past events).

This is a general form that can be used in the probabilistic evaluation of model prediction uncertainties. For nonlinear models, the actual evaluation might be carried out by Monte Carlo sampling which might require significant computation if the model is complex or if there are a large number of uncertain inputs. The way in which these uncertainties might be modified on the basis of a comparison of model outputs and observed responses is dealt with in Chapters 4 and 5 (see also Box 4.1).

B3.1.7 Interval and probability bounds representations of uncertainty

There are many applications where it can be difficult to specify probability (or fuzzy) distributions for variables but where we are pretty sure that values of the variables will lie within a certain interval or range. Uncertainty estimation is then a matter of propagating the intervals through a model, taking account of any interdependencies between variables. Similar issues can arise when the probability distributions are known only with a degree of uncertainty. Uncertainty estimation then needs to account for the propagation of the bounds on the uncertainties, taking account of any interdependencies between variables.

Interval analysis methods have been used since George Boole (1815–1864) in the 19th century (see Moore, 1979; Neumaier, 1990). More recently these methods have been integrated into imprecise probability and probability bounds analysis (Williamson and Downs, 1990; Walley, 1991; Ferson and Hajagos, 2004). Software, such as RAMAS Risk Calc (Ferson, 2002), is commercially available for such calculations (see Software Appendix).

In a probability bounds analysis, the uncertainty about the estimate of probability associated with the value of a variable is expressed in terms of intervals on the cumulative probability function (see Figure B3.1.3). A limiting form of this representation is when it is known only that the values of the variable might lie within a specified interval. The bounds form a p-box for each input variable. Propagation of uncertainties expressed in the form of a p-box is achieved by decomposing the intervals into a number of slices. Each slice is associated with an interval and a probability mass within that interval. The sum of all the probability masses should be equal to 1.

The intervals can then be used to convolve bounds on probability intervals for different variables assuming independence (Yager, 1986; Williamson and Downs, 1990). For any function, $f(A,B)$, of two variables A and B with n and m p-boxes with slice probability masses p_i and q_i respectively such that $\sum_{i=1}^{n} p_i = 1$ and $\sum_{i=1}^{n} q_i = 1$, then convolution results in a matrix of probability masses for each element $f(A_i, B_j)$ in the form of the product $p_i q_j$. This operation can be applied to addition, subtraction, multiplication, division, maximisation, minimisation, powers or other functions of A and B. Note that the slices for both A and B are in the form of intervals, so that the function is evaluated for the values of A and B at the end of each slice.

The algorithm is particularly simple if the probability mass of all the slices is kept constant. If, for example, each variable is divided into 100 slices, each with probability mass 0.01, then all 10,000 evaluations of the interval function $f(A_i, B_j)$ will have probability mass 0.0001. The final probability bounds can then be determined by sorting the left-hand bounds of each $f(A_i, B_j)$ and forming the CDF, then sorting the right-hand bounds and forming the equivalent CDF. This then forms the new p-box for $f(A_i, B_j)$ (e.g. Figure B3.1.4).

It turns out that the notion of independence for imprecise probabilities is more complex than for ordinary probability distributed variables. Couso et al. (2000) distinguish six types of independence for imprecise probabilities. Choosing different assumptions about independence could affect the calculation of the p-box for different functions $f(A_i, B_j)$ The product form $p_i q_i$ is, in their terminology, *random set* independence.

In environmental modelling, we will also often be interested in dependent variables. Correlation or interaction will also affect the resulting p-box for functions of two or more variables. The equivalent determination of the p-box for functions of dependent variables can be carried out using the copula sampling techniques described in Sections 3.3.5 and B3.2.2.

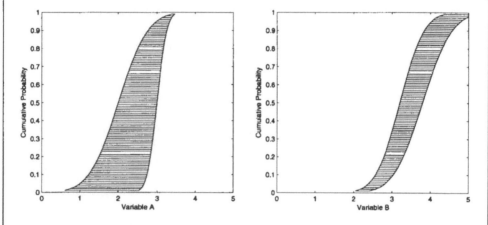

Figure B3.1.3 Representation of imprecise probabilities as p-boxes for two variables A and B

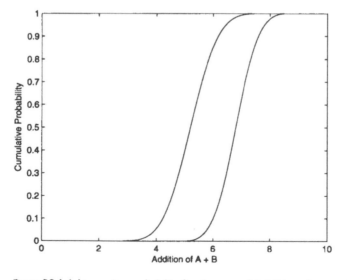

Figure B3.1.4 Imprecise probability for the sum of A + B by p-box method

Box 3.2 Monte Carlo sampling of a model space

The use of Monte Carlo techniques in environmental modelling has a long history. Press (1968), for example, carried out a Monte Carlo experiment involving five million simulations of a geophysical model of seismic wave propagation of the earth (even with the rather primitive computers at the time). He accepted only six of the models as providing adequate fits to the data available (although it is instructive that none of these is close to current models of the structure of the earth – science does progress to reduce prediction uncertainties but the lesson is that we should be wary of putting too much belief in the current generation of models!).

Essentially, in their simplest form, Monte Carlo model experiments involve choosing parameter sets (and perhaps representations of the model inputs) randomly and running multiple **realisations** of the model to determine the differing model responses (see Section B3.2.2 below). This is a way of exploring the response surface in a high-dimensional model space as an alternative to discrete interval sampling. It is particularly useful when the outputs of the model depend nonlinearly on the inputs and parameter values, so that analytical propagation of uncertainties is not possible. Nearly all environmental models are of this type.

Computational limitations apply in the same way as for discrete sampling (see Section 3.3.2 of main text): in a high-dimensional model space it may be difficult to make enough samples to represent adequately the shape of the response surface and, particularly, to identify the regions where good models are to be found. Thus, various forms of guided Monte Carlo search have been suggested, especially for the problem of evaluating likelihood surfaces in a model space, of which the most commonly used now is the class of Monte Carlo Markov Chain techniques (see Box 4.3).

Here we will consider the simpler problem of propagating input and parameter uncertainty through a nonlinear model using Monte Carlo sampling. In principle, this is a very simple process. It involves deciding on the distributions and co-variation of the uncertain inputs and parameter values, randomly sampling from those distributions in a way that is consistent with the specified co-variation to provide multiple realisations of the inputs and parameter values and running the model to get the resulting outputs. A cumulative density function (CDF) of the outputs can then be formed as the more general form of [B3.2.1]:

$$F(M(\underline{I}) \leq X) = \int_{-\infty}^{X} W(\underline{I} \mid M(\underline{I}) \leq X) dX \qquad [B3.2.1]$$

Here $M(\underline{I})$ is the output from a model driven by a vector of inputs $\underline{I}$ that is less than some particular value X. The output might be values of a particular variable, or it could be some performance or likelihood measure. $W(M(\underline{I}))$ is a weighting function for a particular run of the model that should reflect the relative occurrence of a particular vector $\underline{I}$. The cumulative sum of $W(M(\underline{I}))$ over all the Monte Carlo realisations should be scaled to sum to 1. In Box 3.1, equation [B3.1.2], the probabilistic form of [B3.2.1] was given, for which $W(M(\underline{I}))$ is expressed as a probability.

This is not the only way of specifying the inputs to the propagation of uncertainty using Monte Carlo realisations, however. The uncertainty in the inputs could equally be specified as a set of fuzzy measures or Info-Gap uncertainties (though the outputs might then be interpreted in terms of α-cuts over all realisations rather than the cumulative density function of [B3.2.1]).

B3.2.1 Choosing input distributions

The first stage in the process of setting up a Monte Carlo experiment is to decide on what variables will be randomly sampled and what probabilistic or fuzzy or other distribution will be used for each. This also needs to take account of any co-variation in the input variables. Box 3.1 discusses how to express probabilistic co-variation; Box 3.4 how to handle fuzzy co-variation.

There are many possibilities at this point, both in the form of distribution and form of any co-variation, especially if the number of inputs and parameters to be varied is large. The problem can be simplified if all the inputs and parameters can be considered to be independent. It is then necessary only to specify the marginal distribution for each parameter and generate samples from that. Even then there is a multitude of possible distributions that might be used, some of which have great flexibility in form depending on the choice of parameters (e.g. uniform, triangular, trapezoidal, exponential, gamma, normal, log normal, beta distributions; see Figure 3.2 in main text). The assumption of independence might give a misleading impression of the uncertainty in the model outputs, however, if co-variation is important for a particular application.

The difficulty is in knowing what parameter co-variation or interactions might be important *a priori* (we will deal with ways of assessing interactions *a posteriori* in Chapter 4). Indeed, it is very often the case that we have little or no idea of what form of distribution an input or parameter might take. It might be then possible only to specify some feasible range of values for that variable and choose to sample uniformly within that range, recognising that the wider the feasible range considered the less dense the sampling will be for a given number of computer runs. From a probabilistic point of view, uniform sampling within a range gives an equal prior probability to all values within the range, then, abruptly, that probability drops to zero at the edge of the range. Even where there is little information on what form of distribution should be used to represent a variable, most statisticians would choose to specify a range that drops off more gradually to the edge of the range.

B3.2.2 Generating samples

There is now a wide range of software packages for carrying out uncertainty propagation based on Monte Carlo realisations. These range from spreadsheet add-ons (such as @RISK and Crystal Ball for Excel) to stand-alone packages such as the Data Uncertainty Engine (DUE) of Brown and Heuvelink (2007) (see Software Appendix). They vary in what ranges of distributions and specification of co-variation are supported.

The method is also quite easy to program in any programming language. It requires simply an initialisation segment/subroutine to specify the ranges and distributions of the inputs to be varied, a loop to generate random values of the inputs, a call to the model for each realisation, and a summary subroutine to gather the results and present the uncertainty in the outputs. General programming and symbolic mathematics packages such as Matlab, Maple, Mathematica, Mathworks, and others now provide extensive routines for generating a number from a particular form of distribution, converting a number chosen uniformly from the range [0–1] to a sample from a particular form of distribution. Basic random number generators usually provide outputs in the form of the uniform distribution [0–1]. A discussion of random number generators will be found in Box 3.3.

In the case of complex interactions, an interesting recent technique for generating samples based on copulae seems promising (for example, see Section 3.3.5 in main text). A copula is a general transformation of a multivariate interaction on to a scaled space with each axis in the range [0–1]. A copula function then converts uniform samples along each axis into the required dependence structure. A variety of such functions are available. Figure B3.2.1 shows a variety of copula samples in the space of two beta-distributed variables with rank correlation of 0.8. Kurowicka and Cooke (2006) describe the UNICORN software that both supports the specification of interactions in high-dimensional problems and generates consistent samples using copulae. Similar facilities are provided in the RAMAS Risk Calc software of Ferson (2002) (See Software Appendix).

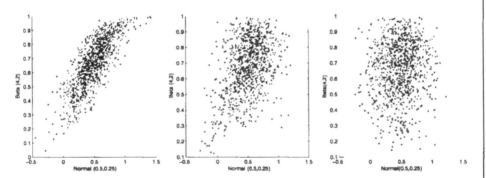

Figure B3.2.1 Copula samples for two beta-distributed variables with rank correlation 0.8 using the Gauss (left), Clayton (middle) and Frank (right) copulae

B3.2.3 Running Monte Carlo simulations on parallel computers

The possibility of carrying out Monte Carlo experiments involving large numbers of runs for a wider range of models has been made much easier in recent years by the availability of larger and cheaper parallel computers. Some parallel high-performance computers are still very expensive and dedicated to very large-scale modelling activities such as global circulation models, but there has also been the possibility of linking together cheap consumer PC machines into what is often called a Beowulf cluster, after the Beowulf project which was one of the first implementations of such a machine by Donald Becker and Thomas Sterling at the Goddard Space Flight Center in Greenbelt, Maryland in 1993. My own group starting using parallel computations on INMOS "transputer" boards (a processor that was specially designed for parallel computation in the 1980s, with its own parallel language called Occam) and have used Beowulf-type parallel machines for many years, running a Linux operating system. Such systems are easily set up for Monte Carlo experiments, at least when a model can be run on a single processor and its memory without the need for extensive memory paging that will slow the run down dramatically. Since the generation of the random inputs is often a relatively fast process compared with running the model itself, it is common to set up the parallel system in a master/slave configuration. The master keeps track of the runs being made and generates new sets of parameter values, the slaves run the model itself before sending the required outputs back to the master with a flag to say it is ready for a new run.

An interesting variation on the master/slave technique is when the slaves are actually networked PCs being used for other day-to-day purposes (such as word-processing, email, web surfing etc). There is now software available that allows a master to send model runs out to any machine on the web set up to accept them, where the job will run in the background until it is finished. The time it takes will depend, of course, on how fast the machine is, and on what other tasks the machine is being used for (how many spare clock cycles are available). The CONDOR software, for example, allows this on both Windows and Linux machines (see Software Appendix). The most famous projects to have used this type of distributed processing are the SETI (search for extra-terrestrial intelligence) that processes blocks of radio-astronomy data to search for signs of structure that might be due to intelligent emissions, and the *climateprediction .net* project (see Section 3.1) that has made possible tens of thousands of realisations of a global climate model. Both have used thousands of user machines worldwide, but systems such as CONDOR can equally be used on a network of local machines. More generally, in the future, the international GRID[1] computing initiative will allow the use of both distributed computing resources and databases of the GRID network, available to any single user as if they were attached to their own machine.

---

1 See the Open GRID Forum at www.ogf.org.

Box 3.3 Choosing a random number generator

B3.3.1 The background to pseudo-random number generators
For most environmental modelling problems the random number generators provided in a programming language will be adequate for Monte Carlo sampling. It is worth just noting, however, that no random number generator

implemented on a digital computer will be completely random – they are actually usually generated deterministically from a mathematical algorithm designed to give a long return period between sequences of the same numbers. Some generators will, however, be more random than others in the sense of giving a longer return period before repeats. The generation of random numbers has been the subject of extensive research in applied mathematics. Some guidance about methods to use is provided in Park and Miller (1988) and Press et al. (2007).

Many of the random number generators that are found as intrinsic functions in programming languages are based on the Linear Congruential Generator (LCG). These are very easily implemented and fast, having the form:

$$J_i = MOD(C + A^*J_{i-1}/M) \qquad\qquad [B3.3.1]$$

where C, A and M are constant integer parameters and i is an index. The randomness of this routine depends very much on the values of these constants.

A floating point uniform random variate in the range [0–1] can then be calculated as:

$$RAN = FLOAT(J_i)/FLOAT(M) \qquad\qquad [B3.3.2]$$

An integer random variate in the range $[J_{MIN} - J_{MAX}]$ can be simply calculated as:

$$J = J_{MIN} + (1 + J_{MAX} - J_{MIN})^*J_i/M \qquad\qquad [B3.3.3]$$

The outputs from some LCG routines, including some used as standard routines for some programming languages, have been queried as having poor randomness properties, in particular exhibiting serial correlation. This suggests that they may not be suitable for generating large Monte Carlo samples. Press et al. (2007) give a table of suitable values for the constants, for example, M = 217728, A = 84589 and C = 45989 giving a generator that will overflow at 2^{35} generated numbers. They also give a routine for additional shuffling of the outputs of a basic LCG or for using a combination of three different LCGs to improve randomness.

A modern random number generator that is widely used is the Mersenne Twister which, with suitable parameters, can give an enormous period before repeated numbers of $2^{19937} - 1$ (Matsumoto and Nishimura, 1998; see also Software Appendix).

It is sometimes useful to take advantage of the fact that random numbers are not totally random to check results for different runs in debugging programs. Many random number generators require the specification of a "seed" number at initialisation (for example, an initial value of J_i in the LCG outlined above). Using the same seed number should generate exactly the same sequence of random numbers because of the way in which they are generated deterministically. As noted in the main text, this can be useful in comparing results from different runs but not very useful if you forget to change the seed when making multiple runs in parallel to increase the number of samples (and I have to admit to mentioning this because I have wasted some very large sets of samples and very long computer runs in this way by forgetting to change the seed value for the random number generator when running the same program on parallel machines as a way of generating more realisations – moral, be very careful about initialisation of Monte Carlo realisations!). A common technique to ensure that each run has a different sequence of random numbers is to take a function of the modulus of the computer clock time at the start of a run as the seed.

More on the realisation effect of different sequences of pseudo-random numbers will be found in Section 3.3.3 in the main text.

Box 3.4 Fuzzy representations of uncertainty

Fuzzy approaches for uncertainty estimation are based on the theory of **fuzzy sets** pioneered by Lotfi Zadeh in the 1960s. Fuzzy sets can be used to describe variables that cannot easily be precisely defined. Instead of a definitive answer as to whether a value belongs to a certain set of values that would result from defining that set of values as a crisp set, the value is given a membership value (with a range from 0 to 1) to reflect the degree of membership of the set in a given context. It is easy to think of many environmental examples where a fuzzy definition might be useful (see, for example, the example of hydraulic conductivity in Figure 3.9 of the main text).

These differences and imprecisions can be described by fuzzy sets (see Figure B3.4.1) in a way that might be useful in such an analysis. It is worth noting straightaway that these definitions do not have to be represented by triangular membership functions; any shape appropriate to the problem can be used (see, for example, Figure B3.4.2). It is also worth noting that membership function values are not equivalent to probabilities. To make this distinction clear, the outcomes from a fuzzy analysis are often called relative **possibilities**.

A fuzzy set can therefore be used to represent a variable (or measurement or model parameter or model input data) subject to uncertainty in a rather flexible way. There are now a number of good texts that describe the application of fuzzy

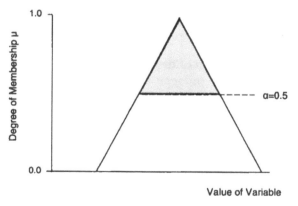

Figure B3.4.1 Triangular membership function for a fuzzy variable (e.g. temperature is "hot"), with α-cut at degree of membership 0.5

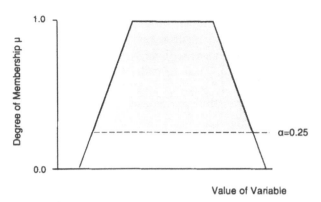

Figure B3.4.2 Trapezoidal membership function for a fuzzy variable, with α-cut at degree of membership 0.25

methods to scientific and engineering calculations (e.g. Bardossy and Duckstein, 1995; Ross, 2004; Siler and Buckley, 2005). Fuzzy methods can also be used in multicriteria model evaluation (see Section B3.4.4 below).

B3.4.1 *Fuzzy sets and degrees of membership*

We will consider a variable x that can take on a universal set of values X. Then, the fuzzy set of x is denoted by the **degree of membership** or membership function. $\mu(x):X = [0,1]$. Outside the range of current possible values of x the membership value (or possibility measure) will be zero. Inside that range it will have a positive real value up to a maximum of 1. Fuzzy sets that attain a membership value of 1 somewhere within the range of x are called normal sets. In most cases we expect the set of membership values of x to be convex (i.e. that a line drawn from any two points in the set is always within the set, as in all the cases of Figures B3.4.1 and B3.4.2), but it is quite possible for the set to be non-convex, as in Figure B.34.3. Fuzzy numbers are normal, convex, fuzzy sets representing a variable within some range to varying degrees.

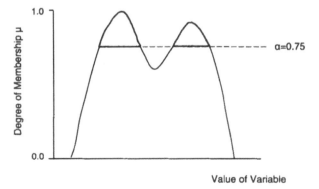

Figure B3.4.3 Non-convex membership function for a fuzzy variable, with α-cut at degree of membership 0.75

Evaluation of the different degrees of possibility for a set can be achieved by taking *a*-cuts across the range where the membership function has positive values (as in Figure B.3.4.3). This can also be done for the joint variation of fuzzy variables in two or more dimensions. Cuts with higher values of *a* are always subsets of cuts with lower values of *a*. Thus consider two sets *A* and *B* in the range of the variable *x*:

B will be a subset of *A* of higher possibility whenever

$$\mu_A(x) \le \mu_B(x) \tag{B3.4.1}$$

The notions of the complement, union and intersection of sets from the theory of crisp sets have their counterparts in fuzzy set theory. They are useful in combining different fuzzy sets or numbers in different ways. Thus, for a normal fuzzy set, the *complement* of the set *A* with membership function $\mu_A(x)$ is denoted as $\bar{A}$ (i.e. the set *"not A"*), defined by:

$$\mu_{\bar{A}}(x) = 1 - \mu_A(x) \tag{B3.4.2}$$

So that the complement of an element of *A* with membership $\mu_A(x) = 0.63$ will be a value of 0.37.

B3.4.2 Union and intersection of fuzzy sets

The union of two overlapping fuzzy sets *A* and *B*, denoted as $A \cup B$ (equivalent to a logical OR operator), is given as

$$\mu_{A \cup B}(x) = \max[\mu_A(x), \mu_B(x)] \tag{B3.4.3}$$

The intersection of two overlapping fuzzy sets *A* and *B*, denoted as $A \cap B$ (equivalent to a logical AND operator), is given as

$$\mu_{A \cap B}(x) = \min[\mu_A(x), \mu_B(x)] \tag{B3.4.4}$$

Figure B3.4.4 shows the result of the union and intersection operator for combining two trapezoidal degree of membership measures for a single fuzzy variable. Note that in this case the union operator produces a non-convex membership function.

Other intersection and union operators have been introduced, for example, by Dubois and Prade (1980). A general weighted mean operator scheme can also be implemented over *k* multiple measures with set of membership values $\{\mu_1, \mu_2, \mu_3, \dots \mu_k\}$ such that:

$$\bar{\mu}_\beta = \left(\frac{w_1 \mu_1^\beta + w_2 \mu_2^\beta + w_3 \mu_3^\beta + \dots + w \mu_k^\beta}{\sum_k w_i} \right)^{1/\beta} \tag{B3.4.5}$$

Figure B3.4.5 shows the results of combining fuzzy membership values for two fuzzy variables, each of which is described by a trapezoidal membership

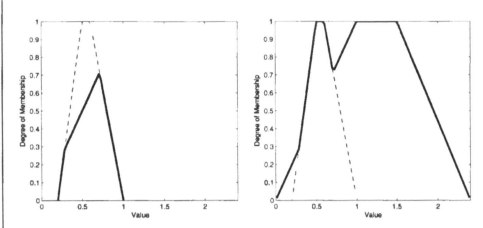

Figure B3.4.4 Union (left) and intersection (right) operators applied to combine two trapezoidal fuzzy measures (in short and long dashed lines) for the same variable. The
horizontal axis represents the value of the variable, the vertical axis the degree of
membership, the resulting degree of membership in each case is shown as the
solid black line

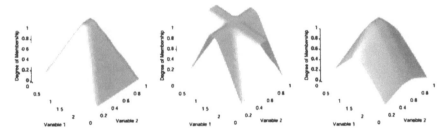

Figure B3.4.5 Union (left), intersection (centre) and weighted mean (left, with $w_1=1$, $w_2=0.5$ and
$\beta = 0.5$) operators applied to combine two trapezoidal fuzzy variables. The horizontal axis represents the value of the two variables, the vertical axis the resulting
degree of membership

function, using the union, intersection and weighted mean operators
respectively.

B3.4.3 Model outputs as functions of fuzzy input variables: the extension principle

Another concept that is useful in the working with fuzzy sets is the idea of
mapping of one set to another using the **extension principle**. The extension
principle states that if the fuzzy set A is defined by the membership values of a
discrete set of points in $\underline{X}$, $\{x_1, x_2, x_3, \ldots x_n\}$ with membership values $\{\mu(x_1),$
$\mu(x_2), \mu(x_3), \ldots \mu(x_n)\}$, then for any other fuzzy set $f(A)$ that is a function
of A, membership will be defined by the membership values
$\{\mu(f(x_1)), \mu(f(x_2)), \mu(f(x_3)), \ldots \mu(f(x_n))\}$. If more than one element of $\underline{X}$ is mapped
to the same element of $f(A)$, then the maximum of the membership grades is

taken to define $f(A)$. More details of the operators used in working with fuzzy sets can be found in an accessible form in Ross (2004).

For the purposes here, we will need to know how the possibility of an input variable or parameters to an environmental model might propagate through into the predictions of a model output, and how that output might be evaluated with respect to some observations that might also be represented by a fuzzy set. The model is here equivalent to $f(A)$ in that it can be treated as a function that maps the uncertain inputs and parameters, $A= \{A_1, A_2, A_3 \ldots A_N\}$, to some output set, $M(A)$, that is the response surface for the quantity of interest in the model space. By the extension principle, therefore, the membership of any element of $M(A)$ should be given the value of the joint membership of the elements of A. In general, of course, $M(A)$ will be a nonlinear mapping, so that there is an issue of how to carry out this mapping across the range of all the elements of A, in much the same way as there is an issue of how to carry out the integrations required for a statistical estimation of model uncertainty. The computational issues of exploring model spaces, using Monte Carlo methods, are dealt with in Box 3.2.

The result of this process will be a map of model outputs in the model space, each associated with a fuzzy possibility measure dependent on the degree of membership values for all the inputs used to drive the model. As noted in Section 3.3 of the main text, in principle, this model space can be filled in this way; in practice, we may only have a sample of values. We may then be interested in producing a summary of the model uncertainties based on the fuzzy measure associated with the outputs in much the same way as producing probabilistic prediction limits.

There are two ways of doing this. One is very similar to the probabilistic bounds in that we can form a cumulative possibility function for an output of interest by treating the possibilities as weighting functions in a way analogous to probabilities (compare Equations B3.1.2 and B3.2.1). Thus, for a discrete number of N samples:

$$F(M(A) < X) = \frac{\sum_{i=1}^{N} \mu(A \mid M(A) < X)}{\sum_{i=1}^{N} \mu(A)} \qquad [\text{B3.4.6}]$$

In words, this equation states that to find the cumulative possibility that a model prediction is less than some value X, we must take the sum of all the associated membership functions for which the model output is less than X, scaled by the total membership for all X. $F(M(A) < X)$, as with a cumulative probability function, will have the range [0–1] as X goes from the minimum to the maximum value of the variable of interest.

We can also look at summarising the fuzzy outputs in terms of α-cuts of the model space. Samples in the model space are here associated with both a value of a model output variable and its associated fuzzy membership value or possibility with range [0–1]. We can then look at the range of the output values in the crisp set defined by a threshold value of membership a. When a is large (close to 1) we would expect the range of outputs to be constrained, but as the a-cut is made with progressively lower values of a then more and more of the output values will be included. When $a = 0$ all the support of the fuzzy set, as expressed in terms of that output variable, will be included (see Figure B3.4.2 where $a = 0.25$ is shown).

B3.4.4 Sets of fuzzy rules

We can generalise the propagation of fuzzy inputs to fuzzy outputs as a representation of possibilistic uncertainties to sets of fuzzy rules (see, for example, Cox, 1994; Bardossy and Duckstein, 1995; Ross, 2004; Siler and Buckley, 2005). Fuzzy rule systems can be specified deductively or generated by induction from sets of observations. Such rule systems define a general formulation for fuzzy reasoning that is now used in control systems for a wide range of applications as a way of dealing with potential uncertainties in operation of the system. The basic fuzzy rule has the form:

> If <input> is <condition> then <output> is <result>

Here <input> can be defined quite generally as information about the system of interest. It can be a fuzzy variable defined by a membership function (with crisply defined values as a special case). It can be a linguistic or "string" variable ('hot", "very hot"). It can even be an estimate of a probability. For the case of any rule that has a non-zero truth value, the match to a defined <condition> will result in a new fuzzy variable, the <result> that defines the <output> condition. A variety of operators can be used in implementing systems of such rules, including union and intersection operators or a-cuts. It is also possible to add "hedging" functions to cases where additional uncertainty is expected, for example where an <input> is an estimated probability, but where the estimate of probability might itself have some uncertainty associated with it.

A fuzzy rule system may consist of a large number of such rules, all leading to a fuzzy estimate of a final decision or control variable. At this point it may be necessary to "defuzzify" the variable in order to take an action. There are also a number of defuzzification operators, but the most commonly used is to take an expected value (weighted average) across the range of the variable where the degree of membership is non-zero. For example, defuzzifying the result of the union operator shown in Figure 3.4.4 gives an expected value of the variable of 1.165. Defuzzification, of course, eliminates any uncertainty in the outcome which, in the context of this book, seems somewhat misguided so no examples will be given here! An example of using fuzzy rules to evaluate the uncertainty in catchment phosphorus export to Lakes Sempach and Baldegg in Switzerland is given by Schärer et al. (2006).

Fuzzy rules have also been used in assessing the performance of environmental models. Blazkova and Beven (2002), for example, used a rules-based system to assess the performance of a flood frequency model for assessing dam safety at a site in the Czech Republic. Three different fuzzy performance indicators were built into an 18-rule system with a final defuzzification to obtain a measure of possibility for a single model run. Many runs of the model with different parameter sets were made within the GLUE framework (see Section 4.5) and a possibility measure for each run was used to weight the prediction for each of the models for estimating the uncertainty in flood frequency characteristics over all models with non-zero possibility.

Chapter 4

Simulation with historical data available

The only relevant thing is uncertainty: the extent of our knowledge and ignorance. The actual fact of whether or not the events considered are in some sense determined, or known by other people, and so on is of no consequence.

Bruno De Finetti, 1974

A model structure for which parameters cannot be estimated is useless.

Jan Spriet, 1985

4.1 Model calibration and model conditioning

In the previous chapter, we addressed the problem of looking at the sensitivity and uncertainty of model predictions for cases where there were no historical data available. The results then are always conditional on the assumptions made prior to running the model and it was noted, particularly for complex model applications, how difficult it can be to be sure about some of those prior assumptions of model choice, ranges or distributions of parameter values and boundary conditions. Thus, while we may often be required to carry out purely a sensitivity analysis or forward uncertainty analysis, the problem of uncertainty estimation becomes much more interesting when there are some data available to be able to evaluate model performance and carry out an **inverse problem** of estimating model values (and in some cases, perhaps, of input uncertainties as well). Model calibration by **history matching** of an observed sequence of data has been the saving grace of most mechanistic environmental modelling. It allows a demonstration of success in modelling capability and it allows some degree of faith in model predictions. While, as we will see, there may still be no "right" answer to the inverse problem, at least we can use the data available to refine and hopefully constrain our estimates of the uncertainty associated with any model predictions.

Until relatively recently, many environmental modellers did not worry about the uncertainty associated with the inverse problem. Traditionally, the inverse problem was treated as a problem of find the "best" or "optimal" parameter set in a given model structure in terms of reproducing the available observations. At least for deterministic models, the optimal parameter set would then be used for prediction without allowing for any uncertainty. Models with stochastic components are sometimes used to allow for a random element in model prediction, but usually only evaluated around the single optimal model.

This is despite the widespread experience in using environmental models that the optimisation problem is not well posed. A huge amount of effort has been expended trying to develop more efficient global search algorithms, and multi-criteria optimisation methods, and response surface visualisation methods to allow interactive manual searches in finding an optimum model, but there simply may not be a clear optimum to be found in real applications of complex environmental models with limited data. In fact, if we think a bit more deeply about the inverse problem then the whole concept of an optimal model must be considered to be misguided. We know that, because of the multiple sources of uncertainty in the modelling process, any optimisation is going to be affected by errors in the data interacting with the model structure which, as we know only too well, can only be an approximate representation of the actual processes. Thus, any "optimum" model can only be *conditionally* optimal. It will be conditional on the particular set of calibration data (and their errors), the search algorithm used and the criteria used to evaluate the model performance. Change any one of these conditions and we would expect that there is a real possibility of finding a different optimal model. The idea that we can move towards "the" representation of an environmental system was shown in the discussion of Chapter 2 to be an aim underlying a strong realist philosophical stance rather than a real practical possibility.

Optimisation has, however, been central to nearly all approaches to model calibration, with the exception of some Monte Carlo set theoretic approaches (e.g. Spear and Hornberger, 1980; Hornberger and Spear, 1981; Beck and Halfon, 1990; Keesman and van Straten, 1990; van Straten and Keesman, 1991; Spear et al., 1994) where some set of models that are acceptable simulators of the system of interest is identified, an approach that may also be generalised to make use of fuzzy measures of acceptability (e.g. Franks et al., 1998; Aronica et al., 1998; Blazkova et al., 2002). Algorithms such as genetic evolution (including the shuffled complex evolution algorithm of Duan et al., 1992; Sorooshian et al., 1993 and Gan and Biftu, 1996); simulated annealing (Tarantola, 2005; Sen and Stoffa, 1995); Monte Carlo Markov Chain (see Box 4.3); or mixtures of these techniques (e.g. Vrugt et al., 2003) have taken advantage of the greater computing power available now to make more extensive searches of the parameter space and try to ensure that some global optimum parameter set is found rather than one of perhaps many local optima.

My own experience suggests that the concept of an optimum parameter set is severely compromised in most mechanistic models of environmental systems (Beven, 1989, 1993, 1996, 2002b, 2006a). Most such models are sufficiently complex that there may be many different sets of parameter values within a model structure that may be consistent with the data available for calibration (see, for example, Figure 1.3). There may, indeed, be many different model structures that may be compatible with the data (see also Draper, 1995, within a more traditional statistical framework). Certainly, one of those models/parameter sets will be the "optimum" according to some measure of goodness of fit or likelihood, but that optimum may not survive application to a different data set or different measure of goodness of fit. Parameter sets that give almost equally good fits may also be scattered throughout the parameter space. This is the *equifinality* concept discussed earlier (Beven, 2006a). The implication is that these modelling problems, while of great practical import, are not well posed (and may even be pathological) in terms of the available techniques for finding an optimum model.

We do have to remember that a primary reason why environmental modelling has been dominated by the concept of the optimal model in the past is that the number of runs that could be made with a model was limited by the available computing power. It clearly still is in many modelling domains, but it is also now often possible to carry out more complete searches of the model space to find out just how complex it is, how it changes with different performance measures and where there may be good models, almost as good as the apparent "optimal" model, in different parts of the space. In high-dimensional spaces the search may still be partial, but it is still possible to look much more widely than before to find models that might be useful in prediction.

This extended computing capability has led to the development of several different approaches to the inverse or calibration problem that allow the uncertainties in the modelling process to be accounted for more explicitly. Here we will look at the strengths and weaknesses of five different methodologies; an optimisation approach based on weighted regression (Section 4.2); formal Bayesian statistical methods (Section 4.3); the multi-objective optimisation approach of Pareto optimisation (Section 4.4); the rejectionist approach of the Generalised Likelihood Uncertainty Estimation method (Section 4.5); and fuzzy set methods (Section 4.6). A final section discusses the important question of the information content of data in conditioning prediction uncertainties (Section 4.7).

4.2 Weighted nonlinear regression approaches to model calibration

There is a class of model parameter calibration methods that are related to statistical inference of parameter values using regression techniques. The approach is similar to that used in trying to find empirical relationships between two or more sets of data using regression by fitting a parametric function such as the linear equation $y = ax + b$, where y is the dependent variable, x is the independent variable and a and b are parameters. The degree of fit provided by such an equation is normally expressed in terms of a correlation coefficient with tables for significance of the value of the correlation that depends on the number of (x,y) pairs of values used in the fitting. Linear regression of this type is taught in most introductory statistics classes. Higher-level courses extend to multiple regression with several independent variables and several a parameter values, and the possibility of fitting nonlinear relationships by making transformations of the original variables, such as the power law relationship $y = cx^a$ which can be fitted as $ln(y) = aln(x) + b$ with $c = \exp(b)$.

Most models used in environmental prediction are nonlinear, but where those models have been based on process equations the nonlinearities are not always easily transformed to a linear equation in this way. In this case the techniques of nonlinear regression can be used. The techniques for using nonlinear regression in environmental applications have been developed over many years and have resulted in extensive experience of what works and what does not that is beyond the scope of this book. However, a new book by Mary Hill and Clare Tiedeman (2007) gives an excellent introduction to the application of nonlinear regression to the calibration of groundwater models and to various software packages that can be used in other applications (see Software Appendix at the end of this book).

The essential basis of nonlinear regression is the same as in the simple linear case, to minimise an objective or cost function (usually based on the squared errors between model predictions and observations) subject to some constraints. Unlike the methods that follow in this chapter there is still an underlying idea in the use of regression techniques that there is an optimal model to be found. Thinking about the objective function as a response surface in the model space, minimising the squared errors is therefore equivalent to trying to find the point of lowest "elevation" on the whole surface. Uncertainties in the calibration of the model parameters and of the residual errors are then considered relative to the optimal model that minimises the objective function. Finding the very lowest point might depend on where you (or your search algorithm) start from and the smoothness of the surface. Some of the techniques used in the application of nonlinear regression are designed to help to ensure that the true optimum is found. In complex model problems this may not be easy, however, and it is generally good practice to repeat the process a number of times using different initial values.

The groundwater model calibration problem addressed by Hill and Tiedeman (2007) is a good problem with which to demonstrate the weighted nonlinear regression technique. This is because, in setting up models of groundwater systems, there are often conflicting requirements between a desire to represent the full complexity of the system and the limited number of observation sites that are usually available to inform the calibration process. Some of these types of problems can involve very large numbers of calculations in space and time and rather long computer run times even with modern-day resources. Where transport of pollutants is an issue, transport codes may be added to the predictions of the flow processes (some examples of conflicts between different models' representations of groundwater systems at the potential radionuclide disposal sites at Yucca Mountain in the US and Sellafield in the UK have already been mentioned in Section 1.7). In Denmark, twelve large-scale regional groundwater models have been set up, covering most of the country, for water resource management purposes (Henriksen et al., 2003). In this case, many hundreds of observation and water supply wells have been used to help the model calibration, but there are not always so many observation points available. In general, as is so often the case in environmental models, the perceived complexity of the system is much greater than can be identified from the data.

In a distributed groundwater model (and other distributed models in which the parameters might vary in space), this gives rise to the problem of how to represent the complexity in terms of the grid or element scale parameters of the model. In principle, every element could have a different parameter value (hydraulic conductivity; storage coefficients; dispersion and degradation coefficients if pollutant transport is being considered). There may be hundreds or thousands of elements, so there are potentially hundreds and thousands of parameters. Where there is information from observation wells, we might have some idea of the geological structure and the temporal changes in water table or piezometric head, but the number of observations will be very much smaller than the number of elements. We should expect, therefore, to be able to identify only a relatively small number of parameter values to reflect the complexity of the real system.

There are a number of ways of doing this. One way is to use a smaller number of zones, based on a conceptualisation of the structure, within which parameter values are considered constant (Hill and Teideman, 2007). A better method appears to be to use an interpolation function or regularisation function between observed values. This approach has been used widely in geophysical inversions (though often without any assessment of uncertainties). This is the approach taken in the PEST calibration and uncertainty estimation software (Doherty, 2005; Moore and Doherty, 2006; see also Box 4.1). Other types of interpolation could also be used, for example by assuming that a parameter is related to some other more easily mapped characteristic. The aim in each case is to reduce the dimensionality of the model space to be searched by still representing at least some of the local characteristics of the flow domain.

4.2.1 Choosing the cost (objective) function

Having decided on a parameterisation, the first requirement of the weighted regression approach is to define the objective function to be used. In fitting the simple linear regression mentioned above, $y = ax + b$, it is normal to assume that the errors in predicting the value of each observation are independent, additive and normally distributed (Gaussian) so that the objective function is calculated as the sum of the squared errors. Thus, the model that is actually fitted is

$$y = ax + b + \varepsilon \qquad [4.1]$$

where ε is the error term. The cost or objective function, J, is then simply calculated as:

$$J = \sum_{i=1}^{N} (y_i - \hat{y}_i)^2 \qquad [4.2]$$

where $\hat{y}_i$ is the predicted value at the ith observation and N is the number of observations. Other simple equations can be found in any basic statistics text to calculate the correlation coefficient, the variance of the estimates of the parameters a and b and the variance of any predicted value $\hat{y}$ given any value of the independent variable x.

The simple linear regression can be thought of as a special case of the more general problem of inferring parameter values given a nonlinear model and errors that are not necessarily independent (though the assumption of normally distributed errors is usually retained). In fitting an environmental model to data, we have to take account that not all the observations might be equally informative in the model calibration process. Some of the observations may be more uncertain in the measurement than others; some may show a high correlation to other measurements made nearby in space or closely in time. Weighted nonlinear regression allows this generalisation in which each residual contributing to the objective function is therefore weighted by an inverse function of the variance–covariance matrix. For the case where independence of the errors can be assumed, the equivalent of [4.2] will then be:

$$J = \sum_{i=1}^{N} W_i (y_i - \hat{y}_i)^2 \qquad [4.3]$$

where W_i is a weighting coefficient. For cases where the prediction errors cannot be assumed independent the weighting coefficients are best represented as a matrix. Equation [4.3] is then a special case in which only the diagonal elements of the matrix are non-zero (see Box 4.1 for details of the more general case).

Once an appropriate objective function has been chosen, it is minimised by searching for the lowest point on the equivalent response surface in the model space defined by the axes of all the parameters to be calibrated. In the case of simple linear regression, this optimum can be found analytically (the equations will be found in any statistics text that deals with linear regression). This is not the case for nonlinear environmental models. In this case, one of the available "hill-climbing" algorithms must be used to do this (though in the case of minimising a weighted sum of squared errors we wish to climb down the response surface to find the lowest valley rather than the highest peak). Hill and Tiedeman (2007) give details of using the Gauss–Newton gradient method but many other methods are available (see, for example, Tarantola, 2005). All such algorithms will work best where the response surface is smooth with a well-defined peak (not always the case in fitting environmental models, which is why we also consider formal Bayesian and GLUE approaches to model conditioning later in this chapter).

4.2.2 Evaluating parameter and prediction uncertainties

Once the optimum model has been found, then the theory of nonlinear regression allows uncertainties to be estimated for both parameters and model predictions. Details of how to do so will be found in Box 4.1. It is worth stressing two points here, however. The first is that the easiest way of estimating such uncertainties is by linearising the shape of the response surface of the objective function around the optimum. For a linear model, and assuming a Gaussian distribution of errors, such calculations provide analytically correct confidence limits for both parameter estimates and predicted variables. For a nonlinear model this is not necessarily the case but evaluating the true shape of the response surface in the vicinity of the optimum will take many more model runs and will therefore be computationally much more expensive, even using a guided search (e.g. Christiansen and Cooley, 1999). We will meet the same problem in discussing variational methods for data assimilation in the next chapter. Simple linear confidence limits on either parameter values or predicted variables should consequently be treated as only approximate and used with care in any further work.

The second point is that the user of such methods needs to recognise the difference between confidence intervals and prediction intervals for any predicted variable. In the theory of nonlinear regression, the estimation of confidence limits is based on assuming that the model is true and estimating the error of the model predicting the true response of the system. In considering what might be actually observed at a prediction point, it is therefore necessary to add an error component associated with the measurement error of an observation to obtain the prediction limits (again see Box 4.1 for more details).

4.2.3 Assessing the value of additional data

One important way of using prediction uncertainties is in assessing the value of new observations. Prediction uncertainties can be calculated for any model-predicted variable, not just the points for which there are observations available. We would expect therefore that, as we move away from an observation point in either space or time, the uncertainty associated with the predictions will get greater. Observation–prediction statistics can be used to assess the effect on the prediction uncertainties of adding or taking away certain observations (see Hill and Tiedeman, 2007, Section 8.3; Tiedeman et al., 2004).

This process can also be reversed to pose the question as to what new observation would have the greatest value in constraining the model parameter estimates. This is a technique called predictive calibration (or pre-posterior prior analysis, see Freeze et al., 1992) and essentially proceeds by assuming that an observation is known, with measurement error, at different potential measurement sites (but before any additional measurements are actually made) and then re-running the analysis to determine the potential effect of each new measurement on the model parameter and prediction uncertainties.

4.3 Formal Bayesian approaches to model conditioning

In the formal Bayesian approach to the inverse problem, prior estimates of parameter distributions are modified on the basis of a likelihood measure reflecting how well a model reproduces the available observations to calculate a posterior distribution of the parameters, including any co-variation between parameters in obtaining good fits (see Box 4.2 for full details). We can define Bayes equation in a form that, given a set of feasible models or hypotheses, and evidence or observations O, then the probability of any models M given O is given by

$$p(M|O) = p(M)\, p(O|M)\, /\, C \qquad\qquad [4.4]$$

where p(M) is some prior probability defined for all feasible models, p(O|M) is the likelihood of simulating the evidence given the models, and C is a scaling constant to ensure that the cumulative of the posterior probability density p(M|O) is unity. Bayes equation is effectively a formal learning strategy. Every time a new set of data is made available, the equation can be used to update the current distribution and determine a new posterior distribution. It can also be used to choose between different model structures using *Bayes factors*, and to combine the predictions of multiple model structures using *Bayesian model averaging* (see Box 4.2).

In this approach, the initial prior estimates of the parameter distributions will be chosen subjectively, but, as more informative data is added to the analysis and the posterior distribution is updated, the effect of the prior distribution will be reduced and the posterior distribution will approach the true joint distribution of the parameters – at least in ideal cases. The approach is therefore the most objective of those considered here to obtain valid estimates of joint probability distributions of parameters and residual model errors. Only in this way can we obtain the probability of predicting an observation conditional on the choice of a particular model and, as

Lindley (2006) and others suggest, probability is thought by many statisticians to be the *only* way to deal with uncertainty even if it might be difficult to find the correct form of the likelihood (see O'Hagan and Oakley, 2004).

The approach has become increasingly popular in a wide range of disciplines including ecology (Ellison, 1996; Omlin and Reichert, 1999; Wikle, 2003; Clark et al., 2005; Clark, 2006; van Oijen et al., 2006); environmental reconstruction (Toivonen et al., 2001); groundwater modelling (Marin et al., 1989; Sohn et al., 2000; Neuman, 2003; Feyen et al., 2003); water quality modelling (Dilks et al., 1992; Jackson et al., 2004), air quality modelling (Bergin and Milford, 2000); sea level rise projections (Patwardhan and Small, 1992); soil remediation (Dakins et al., 1996); flood frequency analysis (Cunnane and Nash, 1971); flood forecasting (Krzysztofowicz, 2002a,b); ensemble climate predictions (Tebaldi et al., 2005; Min et al., 2007); and rainfall-runoff modelling (Bates and Campbell, 2001; Vrugt et al., 2003; Engeland et al., 2005; Kuczera et al., 2006; Kaheil et al., 2006; Marshall et al., 2004; Yang et al., 2007). Bayes methods have a number of advantages: that information about parameters and other uncertainties can be used to convey prior information about the system; that the formal likelihood structure can be used to determine the information content of different data and to demonstrate the relative importance of different sources of uncertainty; that different model structures can be compared and combined; that once an appropriate likelihood function has been found then the process of applying Bayes equation is objective and coherent (coherent in this context means that, as more data is added, the solution should converge to the true solution in a well-defined away); and that it provides a predictive distribution of any variable of interest in terms of probability: the probability of predicting an observed value conditional on the model (e.g. Krzysztofowicz, 1999).

With all these advantages it might be puzzling to the reader as to why it is worth considering any other methods. Certainly, where the assumptions of the analysis are valid, and where the information content of the available data is such that the posterior distributions are not unduly affected by the subjectivity of choosing the priors, these advantages are likely to hold. Unfortunately, for many environmental models, this is not clearly the case and the definition of a formal likelihood measure can lead to misleading results if the assumptions on which it is based are not valid (see the discussions of Beven and Young, 2003; Beven, 2006a; Beven et al., 2008). At the heart of the issue is the question of whether the various sources of uncertainty can be represented adequately by a formal error structure. This structure, if valid, will define an appropriate likelihood function. For environmental models, subject to both input and model structural errors, it will generally only be possible to represent the complexity of the errors approximately. But this then means that the resulting likelihood function will also only be approximate, the resulting parameter estimates may be biased (see, for example, Beven et al., 2008) and some of the advantages might be lost.

From a statistical point of view, Kennedy and O'Hagan (2001) argue that it may be possible to represent model structural error by means of a model discrepancy function; while O'Hagan and Oakley (2004) suggest that the complexity of observed error series in some modelling problems does not mean that we should not use a formal likelihood approach, only that finding an appropriate likelihood function might be difficult. The problem is then that the more complex the model of the errors used, including in some cases for multiple sources of error, the more the number of statistical (non-physical, -chemical or -biological) parameters that must be estimated and, the greater the number of parameters, the greater the possibility of interaction between the actual model parameters and the statistical parameters.

It will be seen in what follows that, in some cases of complex error structures, the method can still be applied by using a transformation of the modelling errors such that the assumptions of a simple formal likelihood measure are more closely approximated, but it will also be argued that other less formal methods might still be useful in cases where the choice of a formal likelihood would be incoherent (in this context, would be *expected* to lead to biased or over-conditioned estimates of the model parameters).

4.3.1 Formal likelihood measures

The attraction of the Bayes method is in the objective definition of a likelihood function, and all the advantages of the method follow from this. The likelihood function follows from assumptions made about the sources of uncertainty, as evaluated either from strong prior information or from the differences between observed and predicted variables. The very simplest formulation is the additive error model (as seen earlier in formulating weighted nonlinear regression):

$$O(x,t) = M(\underline{\Theta}, \underline{I}, x, t) + \varepsilon(x,t) \qquad [4.5]$$

where $O(x,t)$ is some observation, $M(\underline{\Theta}, \underline{I}, x, t)$ is a model prediction dependent on the vector of model parameters, $\underline{\Theta}$, and vector of input variables, $\underline{I}$ (both generally assumed known perfectly), and $\varepsilon(x,t)$ is the residual error, often assumed to take a normal distribution and to be independent for each measurement. Box 4.2 shows how more complex forms of error structure can be developed into likelihood functions. The likelihood function can be used to calculate the *probability* of predicting $O(x,t)$ conditional on the model $M(\underline{\Theta}, \underline{I} x, t)$, *if* the assumptions about the error model are correct.

As already noted in the last section, the "if" is critical in that statement. There are two issues that arise in defining a formal likelihood measure in this way. The first is whether the effects of all the sources of error in the modelling process can be subsumed into a simple model of the errors (even if very often the lack of information about different sources of error forces us to make such an approximation). The second is what effect such an assumption will have on the estimation of parameter values and prediction uncertainties if it is not a good approximation. More complex assumptions can be included in this formulation, for example error in variables and model discrepancy functions (eg. Kennedy and O'Hagan, 2001).

The Bayesian formal likelihood approach has been shown to be useful in many situations. It can also be shown, for hypothetical "environmental" examples, that it can converge rapidly on the true parameter values and prediction uncertainties (since these are known in hypothetical cases – see, for example, Mantovan and Todini, 2006). The advantages are perhaps not quite so clear in applications to real systems where we do not know the answer, and where input and boundary condition errors and model structural errors might be an important source of uncertainty in the model predictions (Beven, 2006a). Or, more correctly, if we use a formal likelihood without specific knowledge of the nature of such errors, any effects of input and model structural error will have to be assumed to be implicitly included in the additive (or multiplicative) residual $\varepsilon(x,t)$.

As a simple thought experiment in this respect, consider a plant growth model that is intended to predict the observed responses of the plants to different concentrations of CO_2 in a greenhouse. Control of the gas concentrations is good, but imperfect mixing in the greenhouse means that the measured concentration (here an input to the model) is subject to error. Careful measurements show that this error has zero mean and a normal distribution (with some autocorrelation that dies away after a few minutes). In this situation, this information could be used to provide an uncertain input to the model (using the techniques of Chapter 3) but the growth model is non-linear in its responses to the CO_2 concentration, with complex interactions between carbon sequestration, stomatal controls, soil water content, ABA concentrations, temperature and other factors. Thus, even if there was no model structural error (unlikely with a plant growth model, see Landau et al., 1998, or the outcomes of the Project for the Intercomparison of Land Surface Schemes reported by Lohmann et al., 1998 and Nijssen et al., 2003), the output uncertainty arising from the input uncertainty would no longer be Gaussian, and the diffusive effects of the plant response to change would increase the autocorrelation length of the error series significantly. Since the nonlinear dynamics of the model would affect the processing of the input error in different ways under different conditions it might be difficult to find a suitable consistent transformation of the error into a form suitable for a formal likelihood (see Boxes 4.1 and 4.2 for more information about transforming errors).

Real modelling situations might be still more complex. There is then significant possibility for calibrated parameter values to compensate for different types of error, perhaps in complex ways. An obvious example is where it is attempted to adjust an input series in calibration, such as rainfall inputs to a rainfall-runoff model (e.g. Kavetski et al., 2005; Kuczera et al., 2006). At the end of a long dry period it is common for rainfall-runoff models to under-predict stream discharges during the wetting up period. An increase in the rainfalls for the storms during this period will result in smaller model errors (in a nonlinear way), but might also increase soil water storage too much, but this could be compensated by reducing rainfalls in later storms to reduce model errors. The estimated input errors may then be only partially related to real errors in the estimate of rainfall over the catchment area. To make the problem even more intractable, the compensatory effect may be dependent on the particular sequence of events or realisation of the different types of errors, such that asymptotic assumptions are not justified. Certainly, we generally find in optimisation studies that optimal parameter sets are dependent on the period of calibration data used. This has been demonstrated for the case of a hydrogeochemical model (MAGIC) to the long-term Birkenes catchment in Norway by Larssen et al. (2007). Different calibration periods resulted in reasonably well-defined parameter distributions using an MC^2 search see Box 4.3 but in some cases the posterior parameter distributions for different calibration periods were non-overlapping.

There does not appear to be a way around this problem without making some strong (and often difficult to justify) assumptions about the nature of the different sources of error. What it does imply, however, is that many different representations (model inputs, model structures, model parameter sets, model errors) might be consistent with the measurements with which the predictions are compared in calibration (allowing for the errors associated with those measurements). Equifinality is endemic to this type of environmental modelling. This would be the case even if we could be sure that we had a set of *equations* that were a good representation of the processes involved (the hypothetical "perfect model" of Beven, 2002a, noting that such perfection will never be achievable) but, as is normally the case, only limited information on which to estimate the *effective* parameter values of those equations in any particular application. The effect of all the complications that arise from different interacting sources of error will generally be that, if they are treated implicitly in an oversimplified likelihood function, the information content of the residuals will be overestimated. This is quite easily shown in hypothetical examples, where even slight modifications to the assumptions can be shown to lead to inaccurate parameter estimates (e.g. Beven et al., 2008).

In fact, there is an additional interesting issue that arises when we consider not only the suitability of an error model for a particular model run but whether that error model might be applicable generally in the model space. The structure of an error series is often only checked after the model with the highest likelihood has been found. It is definitely a good idea to make such a check but the fact that that model has the highest likelihood was dependent upon the choice of likelihood model in the first place. The assumptions of the likelihood model might prove to be a good description of the actual residuals but we can never be sure that a different likelihood function might not be more appropriate in another part of the model space. We might also find that the assumptions are not valid. A recent case can be found in Feyen et al. (2006), where a post-analysis check showed that the wrong likelihood function (ignoring residual autocorrelation, see Figure B4.2.2) had been used so that the parameter estimates were almost certainly biased. There are many other similar cases to be found in the literature. Lesson: always check that the actual residuals conform to the model assumed to define the likelihood function (for hydrological examples of good practice in this respect see Engeland et al., 2005, Case Study in Section 4.3.3 below, and Yang et al., 2007).

4.3.2 Markov Chain Monte Carlo search (MC²)

The Bayes approach, when properly formulated with a valid likelihood function, does have some nice features of a learning strategy as new data become available to refine the posterior distribution. It does still require that the posterior distribution can be integrated in the model space each time the likelihoods are updated. This is effectively a matter of searching the model space to characterise the likelihood response surface, and in particular those areas of high likelihood which dominate the posterior distribution. We have already seen how such searches can be difficult in complex high-dimensional model spaces. However, using formal Bayes likelihoods has the advantage that there are certain expectations about the shape of the surface (at least if the assumptions that underlie the error model are valid).

This has led to the development of strategies for the efficient identification of the likelihood surface. One class of widely used algorithms is called Markov Chain Monte Carlo search (MC²). There is now a wide range of MC² methods but all aim to choose samples with a density that varies through the model space, dependent on the likelihood. Regions of high likelihood are therefore sampled more frequently, areas of low likelihood much less frequently. In this way, the number of (sometimes expensive) runs of the model to characterise the surface is hopefully minimised (see the density-dependent sampling illustrated in Figure 3.5B). The aim of MC^2 methods is to set up a Markov chain of random samples whose marginal is that of the required distribution, which is here the likelihood response surface in the model space. Details of this type of sampling algorithm are given in Box 4.3.

4.3.3 Case study: Assessing uncertainties in a conceptual water balance model (Engeland et al., 2005)

In this study, Engeland et al. (2005) apply a simple water balance model, WASMOD, to the catchment area of Lake Mälaren in Sweden. The catchment area has 30 monitored sub-catchment areas from six to 4,000 km^2 of which 25 were simulated in the study. WASMOD has six parameters, two of which control snow accumulation and melt, one controls actual evapotranspiration, one the drainage of a fast-flow component and the last the drainage of a slow-flow component. The model is applied with a monthly time step with the aim of providing monthly discharge predictions.

The main reason for choosing to present this as a Case Study is the care with which the authors have assessed the nature of the simulation residuals. From past experience they know that the residual errors in hydrological models tend to be heteroscedastic (changing variance with changing magnitude of the prediction). They therefore start by transforming the residuals in order to stabilise the variance. The results of using a square root transform are shown in Figure 4.1 which shows the scaled residuals plotted for all 25 catchments and all simulated months. Some apparent tendency for variance to change with predicted flow remains (and possible bias at low flows) but there is no strong heteroscedasticity. The authors then develop a likelihood function based on assuming a Gaussian residual distribution, similar to Equation [B4.2.5] in Box 4.2 but truncated at zero, with zero autocorrelation. The last assumption was satisfactory for most of the catchments, but in about half the sample of basins the lag 1 autocorrelation was significant, even if not large (Figure 4.2). The autocorrelation

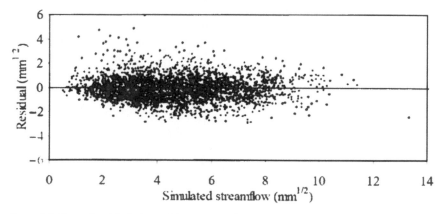

Figure 4.1 Plot of residuals (monthly time steps) against predicted streamflow in all time steps in all 25 basins after transformation using a square root transform

Source: Engeland et al., 2005, Copyright ©2005 IAHS Press

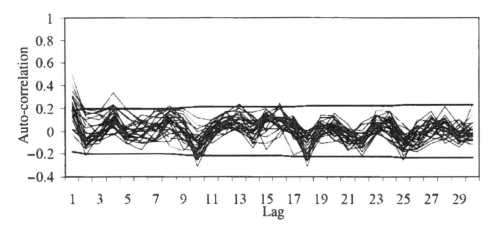

Figure 4.2 Autocorrelation for different lags (monthly time steps) with approximate significance bounds for all 25 basins

Source: Engeland et al., 2005, Copyright ©2005 IAHS Press, reproduced with permission

would normally be much stronger for hydrological models with shorter than monthly time steps in which case a likelihood function that accounts for autocorrelation should be used (see, for example, Yang et al., 2007). Under these assumptions, the parameter distributions are estimated using the Metropolis–Hastings MC^2 algorithm, tuned to have acceptance rates for new samples of about 40–50% for each parameter. The resulting prediction uncertainties can then be assessed for each basin (e.g. Figure 4.3).

The results show that, in common with very many hydrological modelling studies, there are some non-stationarities in the modelling process that may be due to either model structural input error or input error (e.g. around months 32–37 in Figure 4.3). From the results presented, the periods of model failure are not consistent across the basins. This results in a general widening of the residual variance to compensate, which means that the prediction uncertainties are generally high. Again, this is not unusual in this type of study.

In this study, the authors make no attempt to try to take separate account of input error or model structural error in their study. For hydrological examples of trying to do so in a Bayesian framework see Kavetski et al. (2005), Kuczera et al. (2006) and Marshall et al. (2004). The latter also provides an example of Bayesian model averaging over several model structures.

4.4 Pareto optimal sets

Compromise in model parameter calibration will be an issue in many environmental modelling applications. There are often different requirements, goodness of fit or likelihood measures for a model to satisfy and, almost certainly, different measures will suggest different optimal parameter values. Thus, trying to improve the fit on one measure might compromise the fit on another. This is the classical problem of multi-objective evaluation or decision making.

One interesting approach to multi-objective calibration, that allows that it may not

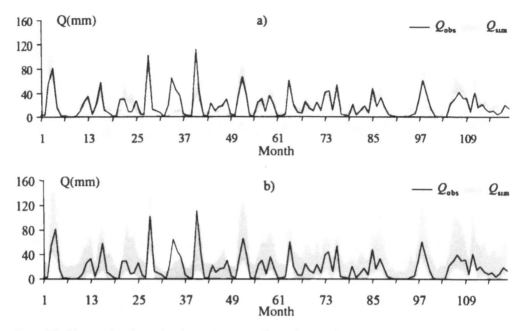

Figure 4.3 Observed and simulated monthly streamflows for the Stabbybäcken basin. Estimated parameter uncertainty a) and total uncertainty b) based on an additive (square root transformed) Gaussian error likelihood function

Source: Engeland et al., 2005, Copyright ©2005 IAHS Press, reproduced with permission

be necessary to decide on a single "optimal" model in the case of conflicting performance measures, is the Pareto optimal set approach. The Pareto optimal set is a set of models with different parameter sets that all have values of the various performance measures that are not inferior to any models outside the Pareto set. The models in the set are said to dominate those outside the set. This is most easily illustrated by a simple example involving two performance measures, both of which are required to be minimised in calibration. The models in the Pareto set will be found along a line, called the Pareto front, that reflects the trade-off between one measure and another (as shown earlier in Figure 1.2). In higher dimensions with more than two performance measures the Pareto front will be a complex surface but the principle will hold. In the case of only a single measure, then the Pareto set necessarily reduces to a single optimal parameter set. Thus, this is an approach that allows for some of the uncertainty in finding parameter values that are consistent with the available observations and performance measures used, but is still aimed at optimality.

The approach has been used in hydrological applications by, for example, Yapo et al. (1998); Gupta et al. (1998, 1999); Madsen (2003); Khu and Madsen (2005) and Madsen and Khu (2006). Khu and Madsen (2005) introduced a preference-ordering scheme that has a stronger concept of dominance than Pareto dominance. This gives preference to solutions that are Pareto optimal in sub-space combinations of the multiple objective functions. Madsen and Khu (2006) provide a demonstration of this approach in comparing two distributed hydrological models in an application to a

Danish catchment: one considering only the groundwater; the other a joint soil-groundwater model with more parameters. The models show quite different behaviour in the Pareto space (Figure 4.4). In particular, the joint model was able to provide better solutions (lower RMSE) for runoff predictions, while the groundwater model did better on predicting well levels despite the more complex recharge calculations considered in the joint model.

Finding the Pareto front in a high-dimensional model space, however, is a computationally challenging problem (though not as challenging as trying to completely characterise a multidimensional response surface throughout the model space). McIntyre et al. (2005) give a demonstration of the complexity of the trade-offs between objective functions in a semi-distributed water quality model (INCA-N). There have been a number of attempts to develop efficient search methods for identifying the Pareto front developed, but it seems that none has shown consistent advantages over all others over a wide range of test problems. A recent approach that combines a number of methods into a single search strategy seems to hold some promise for this type of optimisation (see Vrugt and Robinson, 2007a). But it is still a form of optimisation since it recognises only the models identified as being on the Pareto front as potential models of the system. In doing so, it is rejecting all the models that are close to the Pareto front, and which might have been at the Pareto front with a different period of calibration data or different realisation of the errors in the model inputs. In the next section we consider a methodology, based on the equifinality thesis, that takes a wider view of models that might be useful in prediction.

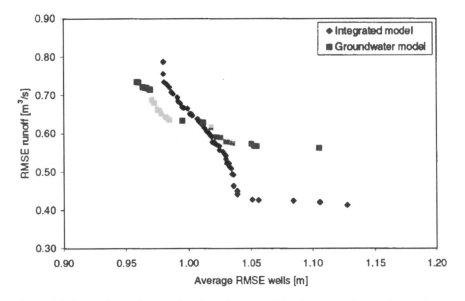

Figure 4.4 Pareto fronts for two distributed models of the Karup catchment, Denmark, using two criteria based on the root mean square errors in estimating observed stream runoff and well levels

Source: Madsen and Khu, 2006, Copyright ©2006 IAHS Press, reproduced with permission

4.5 Generalised Likelihood Uncertainty Estimation

With statistical methods of model calibration, and the Pareto optimal set approach, there is an underlying presumption that the data are adequate to identify an optimal model or Pareto optimal set of models (strictly this need not be the case for Bayesian methods that aim to identify the complete multi-parameter posterior distribution, but the oversimplification of likelihood functions often leads to this result (see, for example, discussions by Beven and Young, 2003; Beven, 2006a, Beven et al., 2008)). A fundamental problem with all optimisation techniques in applications to environmental models is that the optimal (or Pareto optimal) parameter sets are not always optimal when new performance measures or prediction periods are used. Indeed, there may be many parameterisations, differing in parameter values and/or structure that might produce simulations that are acceptably consistent with the available observations, especially if we make proper allowance for error in both model inputs and observations. In Section 1.5 this was referred to as the concept of *equifinality* to suggest that this is a generic problem, not just a problem of the difficulty of finding the true optimum. It is a result of accepting that, unlike traditional statistical inference, we cannot reliably assume that we have the correct model structure and therefore only need to find the true model parameters.

An acceptance of model equifinality is, in part, a recognition of possible model structural and input data errors. It allows for the fact that the formal model of equations that we are using to represent the system of interest may be, at times, a poor approximation to the perceptual model of the relevant processes (which itself may have limitations in its understanding and expression of how the system works). It also allows for the possibility that, even if we had a correct formal model, it may be difficult to specify accurately all the boundary conditions required to run the model. There is, however, not that much that can be done about model structural error since, if there were obvious improvements to be made, then there would be no reason why this should not be done (at least within the bounds of computational feasibility). There are many studies in the environmental literature that report on the difficulties of finding a single "true" model to represent a process. One recent ecological example that looked at multiple model structures as well as parameter sets is a study of simulating pollen dispersion in Kuparinen et al. (2007).

It seems that model structural error is something that will be endemic to most environmental models. We must live with it. This then suggests that an alternative approach to model calibration is required to allow for the effects of structural and data errors, even if these errors cannot necessarily be represented explicitly (as would be required in the formal Bayesian approach of Section 4.3). One alternative is to search for the set of models that are, in some sense, acceptable as simulators of the available data. This is the basis of various set theoretic approaches to model calibration. In fact model *conditioning* is a better phrase to describe a process that tries to find only those models that are acceptable or *behavioural* from the set of all possible models. The set of acceptable models will generally be much larger than the Pareto optimal set of the last section since we might expect that there will be many different models that are not on the Pareto front but which still give behavioural simulations. This larger set should hopefully be more robust to changes in calibration period or input data error, if, at least, the model is a reasonable representation of the system and not missing essential processes that might be important in prediction. Such approaches have generally been based on some form of Monte Carlo sampling from the population of feasible models, and, after making a run with each sample model, a qualitative or quantitative evaluation as to whether that model is accepted as behavioural or rejected as inconsistent with the data available in a way similar to the HSY generalised sensitivity analysis of Section 3.5.4.

This is the basis for the GLUE methodology. This was used for the first time by Beven and Binley (1992) in an application to a hydrological model, taking the possibility of multiple behavioural models into account. Predictions in this method are based on the set of behavioural models, weighted according to a likelihood measure that reflects how each model in the set has performed in calibration (see Box 4.4). Unlike the formal Bayesian likelihood methodology, the likelihood measure, which expresses a degree of belief in the predictions for each behavioural model, need not be based on a formal model of the errors, though the use of an informal or subjective likelihood measure will not then produce a probability of predicting an observation conditional on the model. It is, however, possible to use the formal likelihoods of Box 4.2 in GLUE in the special case that the formal assumptions can be shown to be valid (see Romanowicz et al., 1994, 1996, for example applications). As the name suggests, GLUE is generalised in that respect. The GLUE methodology, used with a formal error model and likelihood, should then give essentially identical results to the formal Bayesian likelihood approach (see Beven et al., 2008).

The GLUE approach has both advantages and disadvantages relative to the formal Bayesian likelihood method. An advantage is that the implicit handling of the modelling errors (which are effectively weighted along with the model predictions) does not force assumptions about the error structure that might be wrong and lead to over-conditioning. A further advantage is that model deficiencies are not allowed to be compensated by an error model. Thus, where a model *cannot* reproduce the behaviour of the real system as a result, for example, of a non-stationary bias, the failure will be evident in the prediction limits not being able to bracket the available observations in any consistent way. A disadvantage is that there is no theory to say what type of likelihood measure might be appropriate in different modelling applications or for different types of data, or how to decide between a model that is behavioural and one that will be considered non-behavioural (but see Section 4.5.3 for further discussion of this point). A second disadvantage is that, unless a formal representation of error structures is included as a model component (with its own parameters to be identified within GLUE), there is no real possibility of disaggregating the total error into different components or consequently of estimating the *probability* of predicting an observation. However, equally with formal methods, such a disaggregation might not be very secure when some sources of error are not included in the analysis or the strong formal assumptions are not valid. Formal does not always equate to correct.

4.5.1 The basis of the GLUE methodology

Conceptually, the basis of GLUE is very simple and can be summarised in terms of a number of decisions as follows:

1   Decide on an informal (or formal) likelihood measure or measures for use in

evaluating each model run, including the rejection criteria for a non-behavioural model which will given a likelihood of zero. Ideally this should be done before running the model, taking account of possible input and observation errors (Beven, 2006a).

2   Decide which model parameters and input variables are to be considered uncertain.

3   Decide on prior distributions from which those uncertain parameters and variables can be sampled.

4   Decide on a method of generating random realisations of models consistent with the assumptions in steps 1 and 2.

These decisions are, of course, similar to any of the other methods considered in this chapter. It is in decisions on generating realisations and the choice of likelihood measure and rejection criteria that the user has more flexibility in the GLUE methodology. In choosing a method of generating realisations, for example, since each model run will be associated with a likelihood weight, the response surface in the model space can be represented by structured or uniform random sampling (as shown earlier in Figure 3.5A) rather than attempting to generate density-dependent samples as is done in the MC^2 or other importance sampling techniques. For simple response surfaces, density-dependent samples can be much more efficient, but for complex surfaces that efficiency gain becomes less significant. If it is possible to make sufficient runs of a model, a uniform random sampling strategy plus likelihood weight approach might be better at identifying scattered regions of behavioural simulations on the response surface of a complex model space. The search might still be refined as the sampling progresses, such as using the regression tree method of subdividing the model space on the basis of the density of behavioural models suggested by Spear et al. (1994) (see Figure 3.10).

Even though the GLUE methodology will provide equivalent results to the Bayesian approach when used with a formal likelihood function, the approach remains controversial when used with informal measures and implicit error handling (see, for example, the critical views of Montanari, 2005, and Mantovan and Todini, 2006). This is because of the lack of a theory for deciding which models should be considered behavioural and which should be rejected (see Section 4.5.2 below) when an exploration of a model space will reveal those models that are "best" in some sense of performance in comparison with the available observations and those that provide obviously unacceptable predictions but no clear demarcation between these sets. If a formal error model can be defined, then the likelihood associated with a given parameter set will provide a direct estimate of the probability of predicting an observation using that parameter set (see Box 4.2). Thus, the likelihood provides a direct measure of the expected model performance and all models that are not useful in prediction should have a very low or near-zero likelihood (though never actually zero), and a threshold decision is not required.

If input error or model structural error is significant, however, a formal likelihood will tend to over-condition the difference in likelihood between the very best models and those that have a rather similar time series of errors or error variances (Beven et al., 2008). The crucial factor here is whether the real information content of the data for real applications rather than ideal problems can be properly reflected in a formal likelihood, or at least in the simple assumptions often made in applying Bayes methods. The question of information content is discussed in more detail in Section 4.5.5, where the reader will find a case for using common-sense model evaluations and informal likelihood methods within the GLUE methodology.

4.5.2 Deciding on whether a model is behavioural or not

Monte Carlo-based set-theoretic methods for model calibration and sensitivity analysis have been used in a variety of disciplines for some 50 years. The first use in geophysics was perhaps that of Press (1968) where a model of the structure of the earth was evaluated in the light of knowledge about 97 eigenperiods, travel times of compressional and shear waves, and the mass and moment of inertia of the earth. Parameters were selected randomly from within specified ranges for 23 different depths which were then interpolated to 88 layers within a spherical earth. Ranges of acceptabilty were set for the predictions to match these observational data. These were applied successively within a hierarchical sampling scheme for the compressional, stress and density parameters. Five million models were evaluated of which six passed all the tests (although of those three were then eliminated as implausible because of having a negligible density gradient in the deep mantle). The "standard model" of the time was also rejected on these tests. Subjective choices were made both of the sampling ranges for the parameters and for the multiple limits of acceptability. Those choices are made explicit, and are therefore open to discussion (indeed, Press discusses an additional constraint that might be evoked to refine the results to a single model but notes that "while reasonable, it is not founded in either theory or experiment", p. 5233).

It was only in the late 1970s that the Hornberger, Spear and Young generalised sensitivity analysis (see Section 3.5.3) started to recognise explicitly that there might not be a unique behavioural model. However, in some of the applications of this technique they found that it was difficult to have any clear objective ways of defining the threshold between behavioural and non-behavioural models. It will then often be the case that multiple runs of the model with different parameter sets reveal a range of performances from good to bad but with no clear differentiation of behavioural models from non-behavioural models.

This is easily illustrated by taking a large sample of all feasible models in the model space and evaluating the predictions of each model in terms of some likelihood measure that should reflect some relative degree of belief in each model. We can plot likelihood against individual parameter values to get an idea of how performance varies through the model space (see Figures 4.5 and 4.6). These plots have often been referred to as dotty plots, in which each dot represents one run of the model. The dotty plot represents a projection of points on the likelihood response surface, one point for each model run, projected onto each parameter axis. While such plots do not give a good impression of the complex interactions between parameters that control whether a model gives good or bad results, they do show quite clearly where there are values of each parameter associated with the highest likelihood runs. Very often, this is across the range sampled while, for any particular value of a parameter there will usually be a complete range of performance from the best models to those that would be considered as non-behavioural by any criterion. Empirically the best models must be

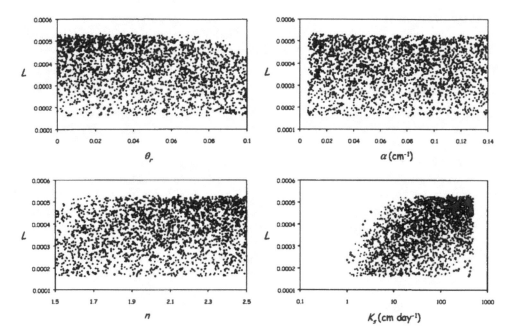

Figure 4.5 Dotty plots of informal likelihood measure against different parameter values for the van Genuchten soil moisture characteristic parameters in a study of predicting recharge to a chalk aquifer. The best models of the realisations simulated by uniform sampling in the model space are at the top

Source: Binley and Beven, 2003, reproduced with permission of Soil Science Society of America

considered behavioural (they are the best we have as yet and if we had other reasons to reject them we should have done so). The worst models will also usually be poor enough to reject as not useful in prediction. But where should the line be drawn so as to retain only those models that will be useful in prediction? How can an objective criterion for model rejection be defined? This is analogous to the problem of deciding on falsification criteria for models as hypotheses (a rather Popperian view; see Tarantola, 2006).

It is not, however, a simple question. It requires consideration of all the possible causes of model error. In the end, however, the answer must be related to the phrase "useful in model prediction". One interpretation of useful is that the range of model predictions should bracket the observations. Beven (2006a) has suggested an approach, not based directly on the choice of an informal likelihood but rather on the specification of limits of acceptability for all the observations of interest *prior* to making any model runs. The limits of acceptability should take account of measurement errors, incommensurability errors and the likely effects of input errors on the model runs, since, as noted earlier, even the perfect model might not be behavioural if driven by poor input data.

Thus, a model prediction $M(\Theta, X, t)$ will be classified as acceptable if:

$$Q_{min}(X,t) < M(\Theta, X, t) < Q_{max}(X,t) \text{ for all observed values } Q(X,t) \qquad [4.6]$$

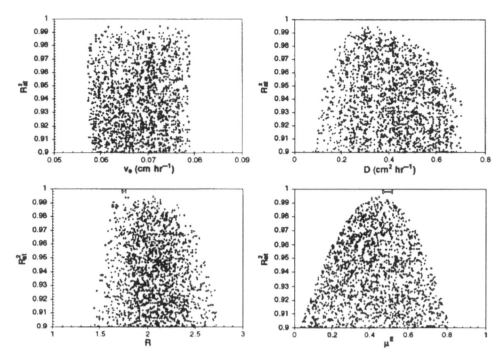

Figure 4.6 Dotty plots of a coefficient of determination in a pesticide transport model fitted to observed atrazine concentrations in a large undisturbed soil column. The four parameters are (top) an effective pore water velocity, a dispersion coefficient (the ranges for which were previously determined by fitting bromide concentration data assumed to be a near-conservative tracer on the same column), (bottom) a retardation coefficient and a degradation coefficient. The best models of the realisations simulated by uniform sampling in the model space are at the top of each plot. The error bars shown on the bottom plots are +/− 2 standard errors on the parameters estimated by nonlinear regression

Source: Zhang et al., 2006, reproduced with permission of Elsevier

Within the range, for all observations $Q(X,t)$, a positive weight could be assigned to the model predictions, $M(\Theta,X,t)$, according to its level of apparent performance. The simplest possible weighting scheme that need not be symmetric around the observed value, given an observation $Q(X,t)$ and the acceptable range $[Q_{min}(X,t), Q_{max}(X,t)]$, is the triangular relative weighting scheme (Figure 4.7A).

This is equivalent to a simple fuzzy membership function or relative likelihood measure for the set of all models providing predictions within the acceptable range. A core range of observational ambiguity could be added if required (Figure 4.7B). Other types of functions could also be used, including the beta function that is defined by Q_{min}, Q_{max} and two shape parameters (Figure 4.7C). These weights for individual data points can be combined in different ways to provide a single weight associated with a particular model. These weights can be used within the GLUE framework in forming prediction limits, reflecting the performance of each behavioural model resulting from this type of evaluation. Models that predict consistently close to the observational data will have a high weight in prediction; those that predict outside the acceptable

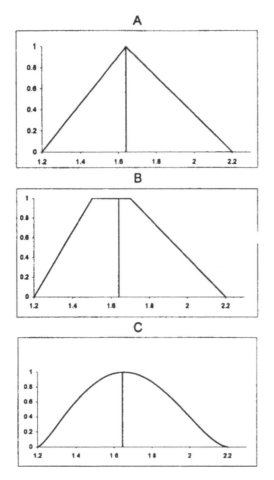

Figure 4.7 Use of a triangular (A), trapezoidal (B) and beta (C) weighting function within the limits of acceptability around a particular observation (here the observation is at 1.64 and the limits of acceptability have been set at 1.2 and 2.2)

Source: Beven, 2006a, reproduced with permission of Elsevier

effective observational error will be given zero weight. In forming prediction limits in this way, there is an implicit assumption (as in previous applications of GLUE) that the errors in prediction will be "similar" (in all their complexity) to those in the evaluation period.

Functions with infinite tails, such as the Gaussian distribution, would need to be truncated at the acceptable limits, otherwise the weighting function will also have infinite tails and a poor model would not be rejected, just given a very small likelihood or membership value. This might not be important in statistical inference when seeking an optimal model, but it is important in this context when trying to set limits for acceptable models. For those models that meet the criteria of [4.6] and are then retained as behavioural, all the methods for combining such measures available from fuzzy set theory are available (e.g. Klir and Folger, 1988; Ross, 1995; Box 3.4). Other possibilities of taking account of the local deviations between observed and predicted quantities for the behavioural models might also be used.

This methodology gives rise to some interesting possibilities. If a model does not provide predictions within the specified range, for any $Q(X,t)$, then it should be rejected as non-behavioural. Within this framework there is no possibility of a representation of model error being allowed to compensate for poor model performance, even for the "optimal" model. If there is no model that proves to be behavioural then it is an indication that there are formal, structural or data errors (though it may still be difficult to decide which is the most important). There is, perhaps, more possibility of learning from the modelling process on occasions when it proves necessary to reject *all* the models tried (see Section 4.5.9).

This then implies that consideration also has to be given to input and boundary condition errors, since, as noted before, even the "perfect" model might not provide behavioural predictions if it is driven with poor input data error. Thus, it should be the combination of input/boundary data realisation (within reasonable bounds) and model parameter set that should be evaluated against the observational error. The result will (hopefully) still be a set of behavioural models, each associated with some likelihood weight (Figure 4.8). Any compensation effect between an input realisation (and initial and boundary conditions) and model parameter set in achieving success in the calibration period will then be implicitly included in the set of behavioural models.

There is also the possibility that the behavioural models defined in this way do not provide predictions that span the range of the acceptable error around an observation (Figure 4.9). The behavioural models might, for example, provide simulations of an observed variable $Q(X,t)$ that *all* lie in the range $Q(X,t)$ to $Q_{max}(X,t)$, or even just a small part of it. They are all still acceptable but are apparently biased. This provides real information about the performance of the model (and/or other sources of error) that can be investigated and allowed for specifically at that site in prediction (the information on the quantile deviations of the behavioural models, as shown in Figure 4.9, can be preserved, for example). Time series of these quantile deviations might provide useful information on how the model is performing across a range of predictions.

This seems to provide a very natural approach to model conditioning and evaluation that avoids making difficult assumptions about the nature of the modelling errors other than specifying the acceptable effective observational error (and possible input realisations). It also focuses attention on the difference between a model predicted variable (as subject to input and boundary condition uncertainty) and what can actually be observed in the assessment of the effective observational error where this is appropriate; potential compensation between input and structural error; and the possibility of real model failure. It also allows model evaluations across multiple criteria to be handled naturally. It is also hoped, of course, that a set of models consistent with all the limits of acceptability across all the different types of observations would be found. Experience suggest that this will not always be the case.

There are certainly cases of the application of GLUE in the past where the prediction distributions fail to encompass the observational data (perhaps for good reasons) and model failure should have been considered (see, for example, the discussion of consistent model error following a wrong prediction of the onset of snowmelt in rainfall-runoff modelling of a small catchment in France in Freer et al., 1996, Figure 4.10).

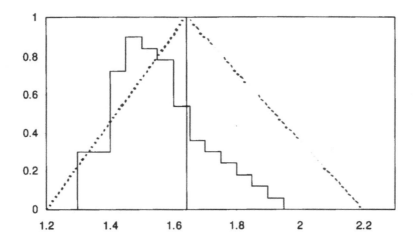

Figure 4.8 Histogram of predictions from a set of behavioural model runs in comparison with the original limits of acceptability. Dotted lines indicate the likelihood function used in weighting the model predictions for this observation. In this case the model predictions include the observation

Source: Beven, 2006a, reproduced with permission of Elsevier

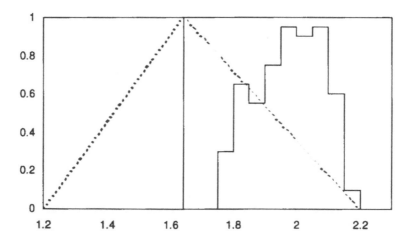

Figure 4.9 Histogram of predictions from a set of behavioural model runs in comparison with the original limits of acceptability. Dotted lines indicate the likelihood function used in weighting the model predictions for this observation. In this case the model predictions do not include the observation, although all the model predictions are consistent with the limits of predictability. In this case it appears that the model is always biased with respect to the observations

Source: Beven, 2006a, reproduced with permission of Elsevier

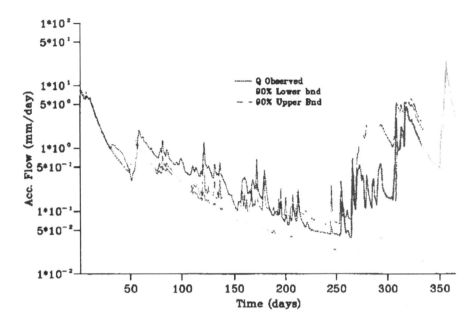

Figure 4.10 GLUE prediction limits for discharge prediction for 1991 in the small Ringelbach experimental catchment, Vosges, France. Likelihood weights based on global Nash–Sutcliffe efficiency values for each behavioural model (>0.5). The effects of a mismatch in predicting the start of snowmelt are obvious, together with wide uncertainty bounds during the wetting period at the end of summer

Source: Freer et al., 1996, Copyright ©1996 American Geophysical Union, reproduced with permission

There have also been cases where all the models considered have been rejected. This was the case for a distributed physically-based hydrological model in Parkin et al. (1996), where all parameter sets failed 10 out of 13 limits of acceptability, and for the applications of the TOPMODEL rainfall-runoff reported in Freer et al. (2003) and Choi and Beven (2007). It was also the case for a model of critical loads of atmospheric deposition in Zak and Beven (1999) and of a model of algal dynamics in Lake Veluwe reported in van Straten and Keesman (1991). They had to increase their limits of acceptability by 50% to obtain *any* behavioural realisations of the simplest model tried, "to accommodate the apparent structural error" (p.175) (their application may also have suffered from incommensurability and input realisation errors). An approach based on rejection rather than optimisation also tends to focus attention on particular parts of the record that are not well simulated or particular "outlier" errors. In this way we might learn more about model performance (and, hopefully, hypotheses about processes).

4.5.3 Equifinality, confidence limits, tolerance limits and prediction limits

In statistical inference, a number of different types of uncertainty limits are usually recognised. Hahn and Meeker (1991) for example suggest that confidence limits should contain a specified proportion of some unknown *characteristic* of a population or process (e.g. a parameter value); tolerance limits should contain some specified proportion of the *sampled* population or process (e.g. the population of an observed variable); prediction limits should contain a specified proportion of some *future* observations from a population or process. These simple definitions, underlain by probability theory, do not carry over easily to a situation that recognises multiple behavioural models and the possibility of model structural error.

Whenever predictions of future observations are required, the set of behavioural models can be used to give a prediction range of model variables as conditioned on the process of model evaluation. The fuzzy (possibilistic) or probabilistic weights associated with each model can be used to weight the predictions to reflect how well that particular model has performed in the past. The weights then control the form of a cumulative density (possibility) function for any predicted variable over the complete set of behavioural models, from which any desired prediction limits can be obtained (see [B3.2.1] and Box 4.4 for details of how to construct prediction limits in this way). The weights can be updated as new observations are used to refine the model evaluation. This is the essence of the GLUE methodology (e.g. Beven and Freer, 2001) and of other set theoretic approaches to model prediction. Figure 4.11 shows the results of applying this technique in comparison with a formal Bayesian calibration for a hypothetical rainfall-runoff example (Beven et al., 2007); Figure 4.12 for the real case of estimating the breakthrough curve of atrazine pesticide in a large undisturbed soil column in comparison with confidence limits based on nonlinear regression (Zhang et al., 2006). Beven et al. (2006) go on to show how these uncertain estimates at the column scale might be used to estimate pesticide transport at the field scale.

Note, however, that while it is necessary to assume that the behavioural models in calibration will also be behavioural in prediction, this procedure only (at best) gives the tolerance limits (in the calibration period) or the prediction limits over the weighted simulations of any variable. These prediction limits will be conditional on the choice of limits of acceptability; the choice of weighting function; the range of models considered; any prior weights used in sampling parameter sets; the treatment of input data error etc. All these components of estimating the uncertainty in the predictions must, at least, be made explicit. However, given the potential for input and model structural errors, they will *not* guarantee that a specified proportion of observations, either in calibration or future predictions, will lie within the tolerance or prediction limits (the aim, at least, of a statistical approach to uncertainty). Nor is this necessarily an aim in the proposed framework. In fact, it would be quite possible for the tolerance limits over all the behavioural models to contain not a single observed value in the calibration period (as in Figure 4.9), and yet for all of those models still to remain behavioural in the sense of being within some specified acceptable error limits for all observed quantities. The same could clearly be true in prediction of future observations, even if the assumption that the models remain behavioural in prediction is valid.

Similar considerations apply in respect of the confidence limits for a parameter of the model. Again, it is simple to calculate likelihood-weighted **marginal distributions** of any single parameter over all the behavioural models. The marginal distributions can have a useful role in assessing the sensitivity of model outputs to individual parameters (e.g. Hornberger and Spear, 1981; Young, 1983; Beven and Binley, 1992;

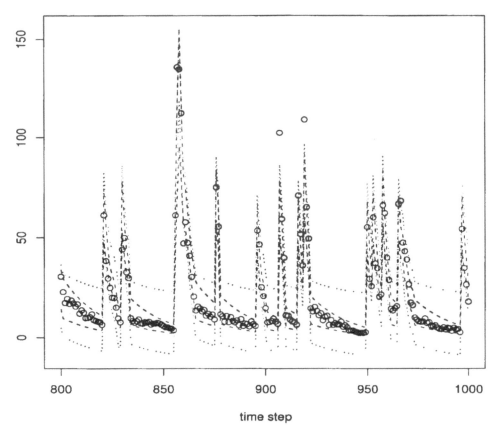

time step

Figure 4.11 GLUE prediction limits for the case of the Mantovan and Todini (2006) hypothetical
rainfall-runoff model where it is known that the model structure is correct. Observations
(o), dashed lines are 5% and 95% likelihood-weighted prediction limits over set of all
behavioural models, dotted lines are from a formal likelihood based on an additive Gaussian error model with Bayesian updating

Source: Beven et al., 2007, reproduced with permission of Elsevier

Beven and Freer, 2001). For each of those models, however, it is the parameter *set* that
results in acceptable behaviour. It is quite possible to envisage a situation in which a
parameter set based on the modal value of each of the parameter marginal distributions is not itself behavioural (even if this might be unlikely). Any confidence limits
for individual parameters derived from these marginal distributions therefore cannot
have the same meaning as in traditional inference (in the same way that the use of
likelihood has been generalised within this framework). Marginal parameter quantiles
can, however, be specified explicitly.

4.5.4 Equifinality and model validation

Model validation is a subject fraught with both practical and philosophical under-
tones (see Chapter 2). The GLUE limits of acceptability approach provides a natural

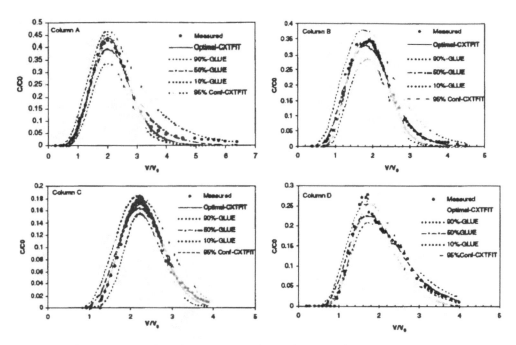

Figure 4.12 A comparison of prediction bounds from GLUE and CXTFIT (which uses a nonlinear regression method) after fitting parameters to atrazine pesticide breakthrough curves from four large undisturbed soil columns. The model is the same in both cases. Dots are observed data

Source: Zhang et al., 2006, reproduced with permission of Elsevier

approach to model validation or confirmation, even when faced with a large set of behavioural models. All the time that those models continue to provide predictions within the range of a consistent definition of the limits of acceptability (allowing for input data errors) they will continue to be validated in the sense of being behavioural. When they do not, they should be rejected as non-behavioural.

There are clearly, however, a number of degrees of freedom in this process. Stephenson and Freeze (1974) were perhaps the first in an environmental modelling context to point out that the dependence of model predictions on input and boundary conditions made strict model validation impossible for models used deterministically, since data for initial and boundary conditions could never be known precisely. The same holds within the methodology proposed here since whether a model is retained as behavioural depends on a realisation of input and boundary condition data.

There is also the question of defining the effective observational error. The more error that is considered allowable, the less likely it is that models will be rejected. Clearly, the error limits that are used in any particular study should be chosen on the basis of some reasoning about both the observed and predicted variables, rather than simply making the error limits wide enough to ensure that some models are retained (even if this might be considered an attractive possibility if otherwise all models are rejected). Strict falsification is not, however, so very useful when, in virtually all environmental modelling, there are good reasons to reject models when they are examined in detail (Beven, 2002a; see Freer et al., 2003; Choi and Beven, 2007, for examples). What we can say is that those models that survive successive evaluations suitable for the application are associated with increasing confirmation (even if not true validation).

4.5.5 Equifinality and model spaces: sampling efficiency issues

We have noted that acceptance of the equifinality thesis implies that there will be the possibility of different models from different parts of (a generally high-dimensional) model space that will provide acceptable simulations, but that the success of a model may depend on the input data sequence used. In one sense, therefore, the degrees of freedom in specifying input data sequences will give rise to additional dimensions in the model space.

There is, therefore, a real practical issue in working with the equifinality thesis of sampling the model space to find behavioural models (if they exist at all). Success in this endeavour will be dependent on the structure of where behavioural models are found in the space. There is an analogy here with the problem of finding an optimum model on a complex response surface in the model space. The problems of finding a global optimum, rather than local optima, have long been recognised and a variety of techniques have been developed to do so successfully. The equifinality thesis extends the problem: ideally we require a methodology that both robustly and efficiently identifies those (possibly arbitrarily distributed) regions of the parameter space containing behavioural models, but with the additional dimension that success on finding a behavioural model will depend on a particular realisation of the input variables required to drive the model.

As in any identification problem, including modern MC^2 methods (see Box 4.3) and importance sampling methods (see, for example, Cappé et al., 2004), the search can be made much more efficient by making strong assumptions about prior likelihoods for individual parameters and about the shape of the response surface. This seems a little problematic, however, in many environmental modelling problems when it may be very difficult to specify prior distributions for effective values of parameters and their co-variation. In the GLUE methodology, the usual (but not necessary) prior assumption has been to specify a feasible range for each parameter, to sample parameter values independently and uniformly within that range in forming parameter sets, and to allow the evaluation of the likelihood measure(s) to condition a posterior distribution of behavioural parameter sets that reflects any interaction between parameters in producing behavioural simulations. This is a simple, minimal assumption approach, but one that will be very inefficient if the distribution of behavioural models within the model space is highly structured or highly localised. It has the advantage that all the samples from the model space can be considered as independent, although this assumption is not invariant with respect to scale transforms of individual parameter dimensions (e.g. from an arithmetic to a log scale). It is also worth noting that, where a model is driven with different realisations of stochastically varying inputs or parameter values, then each point in the model space may be associated with a whole distribution of model outcomes.

There may be some possibilities of refining this type of search. The CART approach of Spear et al. (1994), for example, uses an initial set of sample model runs to eliminate regions of the model space where no behavioural models have been found from further sampling. This could, of course, be dangerous where the regions of behavioural models are small with respect to the initial sampling density, though by analogy with some simulating annealing, MC^2 and other forms of importance sampling methods, some safeguards against missing some behavioural regions could be ensured by reducing sampling density, rather than totally eliminating sampling in the apparently non-behavioural areas.

This problem will become less important as computer power increases, particularly since it is often easy to implement this type of model space sampling on cheap parallel processor machines. It certainly seems clear that, for the foreseeable future, computer power will increase much more quickly than any changes in modelling concepts in most domains of environmental science. Thus, we should expect that an increasing range of models will be able to be subjected to this type of analysis. Preliminary studies are already being carried out, for example, with distributed hydrological models such as SHE (Christiaens and Feyen, 2002) and distributed groundwater models (Feyen et al., 2001), albeit with reduced parameter dimensions.

4.5.6 Fuzzy measures in model evaluation

For the moment, let us assume that a mapping to the model outputs can be computed everywhere in the model space. This is always true in principle, though in practice we may be restricted by the number of runs that can be made to sample the space. On the basis of past experience, or prior assumption, we can associate each sample output in the model space with a measure of belief. This might be a formal likelihood measure, it might be a fuzzy membership value. We then effectively have, for any output variable from the model, a fuzzy set of possible model outputs. In model evaluation we will want to compare these values with a measurement, $O(x,t)$, which may also be represented as a fuzzy set. If these sets are non-overlapping, and all relevant sources of uncertainty have been taken into account, then the model could be considered to have failed as a hypothesis of how the system is working. If the sets are overlapping (as, for example, in Figure 4.13), then one of several fuzzy operators (such as the union or intersection operators of Box 3.4) can be used to determine the degree of membership for those models that provide predictions within the range of the observation. Where more than one observation is available for model evaluation, then there is a further choice as to how to combine the evaluations that result in membership values greater than zero. Intersection or union operators can be used to combine different rules (see Figure B3.4.5 in Box 3.4), or a generalised mean operator can be used. The uncertainty in the outputs can be assessed by taking nested a-cuts of the fuzzy set (see Box 3.4) or by using the membership value resulting from combining the different model evaluations as a possibilistic weighting function in GLUE. A recent application of fuzzy measures within the GLUE framework is provided by Jacquin and Shamseldin (2007) (albeit that they base their membership measures on a relative error variance of the time series of residuals and a relative absolute error in runoff volume). Past uses of fuzzy model evaluations within GLUE have included rainfall-runoff modelling (Franks et al., 1998; Freer et al., 2003, 2004); flood frequency modelling (Blazkova and Beven, 2002, Figure 4.13); soil nitrogen modelling (Schulz et al., 1999); water quality modelling (Page et al., 2003, 2007); air quality modelling (Page et al., 2004);

Flood freq. = 0.392   Flow dur. = 83.1   Snow w. eq. = 5.96   Fuzzy rel. likeli. = 350

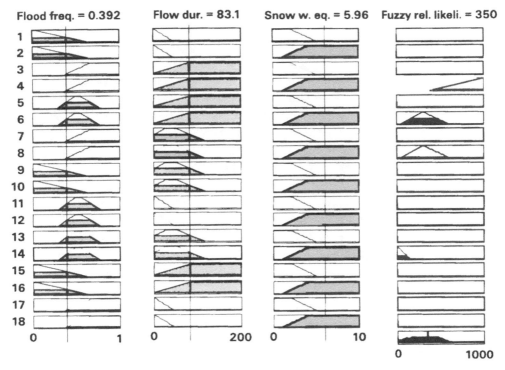

Figure 4.13 Fuzzy rule system for evaluating model performance across three different fuzzy evaluation measures for a single model run. A combination of all the relative likelihood values in
the last column is used as a weight for the model predictions

Source: Blazkova and Beven, 2002, Copyright ©2002 American Geophysical Union, reproduced with permission

flood inundation modelling (Romanowicz and Beven, 2003; Pappenberger et al.,
2006c); and modelling latent heat fluxes (Franks et al., 1999).

4.5.7 Case study: Hypothesis testing models of stream runoff generation using GLUE

Iorgulescu et al. (2007) provide an interesting example of the use of the GLUE methodology in testing different hypotheses about the nature of runoff generation in the
Haute Menthue catchment in Switzerland. Their study follows on from that of Iorgulescu et al. (2005) where a three-component mixing model had been used to estimate
different components of the stream hydrograph due to rainwater, soil water and
groundwater based on observations of silica and calcium concentrations. Piñol et al.
(1997) provide another example of hypothesis testing of this type in catchments of the
Prades mountains, Cataluña, but based only on trying to reproduce the stream discharges. In the Haute Menthue catchment, a three-component hydrograph separation
was possible because of the quite different silica and calcium concentration characteristics of the different water sources. The mixing model had more than the usual number of parameters because a characteristic time distribution for each component was

also estimated. In this study only 216 out of two billion simulations were considered behavioural on the basis of stringent limits of acceptability on the discharge and predicted silica and calcium concentrations in the stream. The discharge predictions from the model were very good (at least for a relatively short period of time for which the most detailed concentration observations were available; see Figure 4.14A).

The hydrograph separations allowed inferences to be drawn about the sources of runoff and in particular how the contribution of the soil water component increases in a highly nonlinear way as the soil wets up over a sequence of rainfall events. The results are, however, dependent on assumptions that the soil and groundwater

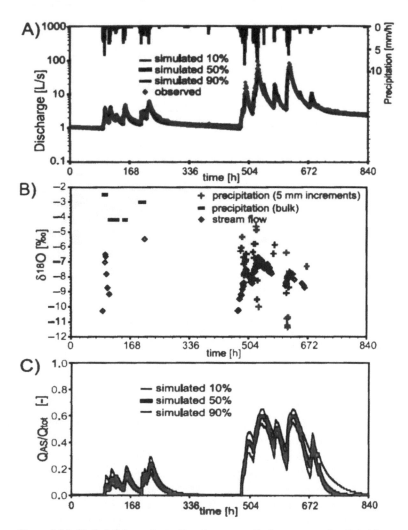

Figure 4.14 A) Rainfalls and predicted stream discharges on the Bois-Vuacoz sub-catchment, Haute-Menthue, Switzerland, using the 216 behavioural models from Iorgulescu et al. (2005). B) Measured oxygen isotope ratios in rainfall and streamflow. C) Predicted fractional contribution of the soil water component also based on the 216 behavioural models

concentrations are changing only slowly. This means that the model cannot reproduce the behaviour of a conservative tracer, here oxygen isotope ratios, without some additional assumptions. Oxygen forms part of the water molecule so that the isotope concentrations can be assumed to be conservative in the absence of strong fractionation processes. Different sets of assumptions about the behaviour of the soil water component and its interactions with the rainfall made up the different hypotheses to be tested. Again, stringent limits of acceptability were set at three times the standard laboratory analytical error for the isotope measurements. This was used as the effective observation error in applying the extended GLUE approach described above. Multiple realisations for the additional model parameters were run using each of the 216 behavioural discharge simulations. The most successful hypothesis was that which parameterises the additional mixing volumes for rainfall and soil water and the conversion of rainwater into soil water but without allowing for solute uptake to change the apparent characteristics of the rainfall in contributing to the stream discharge (Figure 4.15).

Interestingly, the outputs from the analysis regarding the effective mixing volumes associated with the direct precipitation and soil water components were rather uncertain (Figure 4.16), despite this strong constraint on the predicted isotope concentrations in the streamflow. Valuable inferences, however, could still be made, despite the uncertainty, since the effective mixing volumes for predicting the flow and the isotope concentrations were quite different. The much larger volumes involved in predicting the isotopes for both sources suggest that the prediction of the silica and calcium concentrations is dominated by displacement of soil water already stored in the system, while the rainfall component is subject to significant subsurface flow pathways.

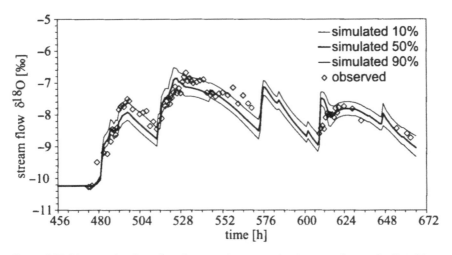

Figure 4.15 Measured and predicted oxygen isotope ratios in streamflow in the Bois-Vuacoz sub-catchment using the most acceptable hypothesis (H3)

Source: Iorgulescu et al., 2007, Copyright ©2007 American Geophysical Union, reproduced with permission

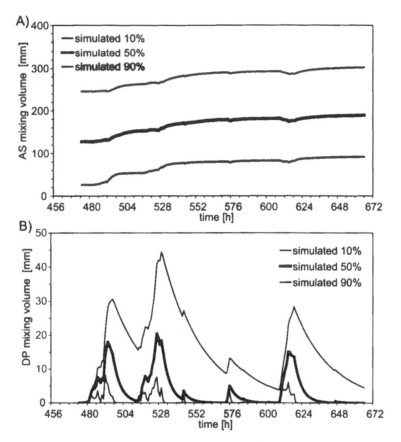

Figure 4.16 Estimated mixing volumes (in mm) for A) soil water and B) direct precipitation component for hypothesis H3

Source: Iorgulescu et al., 2007, Copyright ©2007 American Geophysical Union, reproduced with permission

4.5.8 Variants on the GLUE methodology

There have now been more than 100 applications of the GLUE methodology reported for a wide range of environmental modelling applications.[1] Examples range from studies of the parameters of hydrologic and hydraulic models (e.g. Lamb et al., 1998; Blazkova et al., 2002; Pappenberger et al., 2006c,d), modelling land surface to atmosphere fluxes (Schulz and Beven, 2003); modelling flood frequencies (Blazkova and Beven, 2002; Cameron et al., 2000; Cameron, 2006); to predicting the distributions of forest fires (e.g. Piñol et al., 2004, 2007; Figure 4.17).

The methodology has instigated a number of interesting variations. One of the most interesting is the dynamic identifiability analysis (DYNIA) of Wagener et al. (2003). DYNIA allows that different parameters in a model might be more or less identifiable

1 A list of published papers may be found at http://www.glue-uncertainty.org.

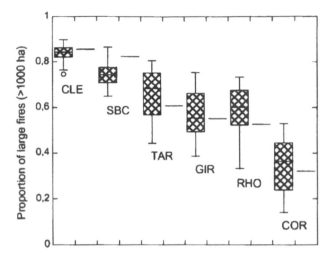

Figure 4.17 Box plots of the predicted proportion of large fires (>1000 ha) from 50 behavioural simulations for six different study areas in California, France and Spain. Observed mean values shown as horizontal lines to the right of each plot

Source: Piñol et al., 2007, reproduced with permission of Elsevier

under different conditions during a model run. It can be set up as a recursive analysis such that the marginal likelihood distribution for each parameter is updated after a certain number of time steps, taking account only of the model likelihood evaluations over a certain window of time. For those parameters that reflect short time-scale processes in the system this might be chosen to be a short window. For those parameters that reflect long time-scale processes this might be chosen to be a long window.

Figure 4.18 shows the results from such an analysis for three parameters of a simple rainfall-runoff model, taken from Wagener et al. (2003). The "rc" parameter is evaluated with a window of 101 daily time steps, the "bypass" parameter with a window of 41 days and the "rt(q)" parameter with a window of 11 days. The marginal distributions are updated after each time step in this case. The figures show quite clearly how information on the values and feasible ranges of the different parameters varies over time, and can be analysed to infer which combinations of parameters are important in producing good fits for different parts of the time series.

Other studies using GLUE that have looked at how well different sets of model parameters fit the observations under different conditions have been carried out by Freer et al. (2002) and Choi and Beven (2007). These were also applications to rainfall-runoff models. The first analysed different parts of the hydrograph separately (rising limb, recession periods), using a variety of likelihood measures. The second classified overlapping windows of the data set into wet, dry, wetting, and drying conditions (15 classes in all), and looked at the behavioural parameter sets for the different classes of periods. In both of these cases, no model parameter sets were found that were behavioural for all the periods and all of the performance measures. All the models could be rejected for one reason or another.

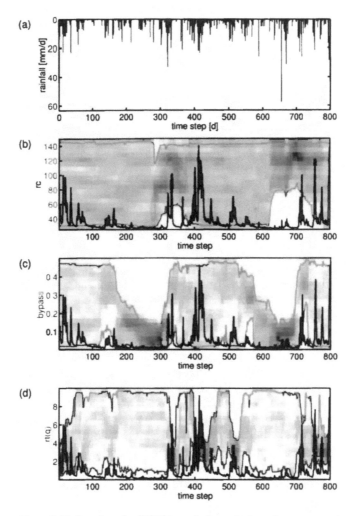

Figure 4.18 Results of a DYNIA analysis applied to three parameters of a rainfall-runoff model over a two-year period. (a) rainfall input over time; (b) changing likelihood distribution for "rc" parameter; (c) and (d) same for "bypass" and "rt(q)" parameters. Grey scales represent gradients of cumulative-likelihood distribution for increments of the parameter value. Darker values are steeper gradients. Grey lines are 90% quantiles taken from cumulative-likelihood distribution. Note how the effective behavioural range for each parameter can vary over time

Source: Wagener et al., 2003, Copyright ©2003 John Wiley and Sons Limited, reproduced with permission

4.5.9 What to do if you find that all your models can be rejected?

This is a very interesting question from a number of points of view. Our experience from a wide variety of GLUE applications is that it is often the case that, even with quite relaxed rejection criteria, all the models tried can be rejected as non-behavioural. More severe rejection criteria will, of course, increase the possibility of total rejection. Sometimes we find that a model can do well on one criterion but not on another (for an example where a model could not predict both runoff and sediment concentrations, see Brazier et al., 2000). The Pareto set multi-criteria optimisation approach of Section 4.4 is one response to this, but we really would prefer to have models that are not rejected on one or more criteria, even if we do not expect them to be consistently the best on each criterion tried, i.e. we should worry if our "best" models are not adequate in meeting reasonable expectations of acceptability in evaluation.

There are a number of responses to finding that all models are rejected:

1   Make sure that the model space has been searched adequately and that regions of potential behavioural models have not been missed. This is an obvious first response.

2   Make sure that the model is not failing because of deficiencies in the input data or the observations with which the model is being compared. We would wish to avoid making Type II (false negative) errors (rejecting a model that would be useful in prediction because of input or observation errors). This requires a more detailed analysis of the available data, and can be a rather ambiguous exercise, since very often the evidence that the data might be in error comes from error in the model predictions. The magnitude of the model residuals will also reflect the model structural error. However, what we can say is that the potential for errors in the data driving the model should be considered carefully.

3   Add a statistical error model or model discrepancy function to compensate for the model deficiencies. In the formal Bayesian approach that uses such statistical components, total model rejection will not normally occur since the statistical error variance will expand to reflect the larger residuals associated with a failing model. This will, of course, retain at least a semblance of predictability. It will also allow an assessment of the ability of the model to explain the variance in the data, relative to the remaining residual variance. This might still result in a conclusion that the model is inadequate.

4   Find a better model. This, of course, is easier said than done, but such a conclusion does suggest that there is some advantage in complete model rejection (rather than simply compensating for model deficiencies with a statistical error). All the time that a model is not rejected we do not really learn very much about whether it is getting the right results for the right (or wrong) reasons. We may be making a Type I (false positive) error (accepting an incorrect model as acceptable because of uncertainties in the input or output data). As soon as a model is rejected, however, we have to look more closely (at the data as well as the model structure) which will generally lead to some improvement in the representation of the system. This is a part of the learning process in modelling particular systems (Beven 2000; 2007). Additional data can be helpful in suggesting how model improvements might be found. The need for local calibration of a transmissivity parameter to match water table levels in the studies of Lamb et al. (1998) and Blazkova et al. (2002) suggested that the rainfall-runoff model used could, in fact, be simplified in their applications.

5   Use different models or parameter sets to predict different variables for which they do give behavioural simulations. This is the fall-back strategy when all the parameter sets tried with a given model structure do not satisfy the acceptability criteria across all the observables with which they are compared. Ideally of course

(as realists) we would like our models to simulate all the variables satisfactorily in space and time but (as pragmatic realists) we realise that this might not always be possible. Choi and Beven (2007) for example found that different parameter sets were necessary to get behavioural models in wet and dry periods in predicting stream discharge. They therefore provided a method for classifying a time step into one or more classes of hydrological conditions and weighting the consequent predictions using different parameter sets for each class. Pappenberger et al. (2006d) found a similar problem in applying a flood inundation model to the Alzette river in Luxembourg. It proved to be impossible to find parameter sets that fit the historical flood inundation maps everywhere. They therefore suggested using different parameter sets according to which type of prediction was required, where the type of prediction was expressed in terms of the vulnerability of different communities, or the road system. The result is a form of purpose-specific uncertainty estimation using models that are behavioural for that purpose.

4.6 Fuzzy systems: conditioning fuzzy rules using data

Fuzzy rules are a non-statistical way of representing uncertainty. In Box 3.4, the concept of using systems of fuzzy rules to represent the propagation of uncertainty in complex systems was introduced. In the previous section, the concept of using fuzzy measures in model evaluation with the extended GLUE framework of Beven (2006a) was demonstrated. Fuzzy rules can also be used as an inductive model of data that can be used for making predictions directly. They are used in much the same way as artificial neural networks, or other inductive modelling methods, without any requirement to make any process interpretation of the resulting rules. In the latter case, the fuzzy rules must be calibrated (or trained) in much the same way as any other model, by comparing the model predictions with historical data. The advantage over artificial neural networks is that the uncertainty inherent in the representation of data can be included explicitly within the fuzzy rules (although some attempts are being made to combine neural networks with uncertainty estimation in environmental applications: see Guan et al., 1997; Salas et al., 2000; Markus et al., 2003). This is not, however, always the case. There are many fuzzy prediction systems that are used with a final defuzzification step to provide a crisp value of a predicted variable without uncertainty. This seems to throw away some of the advantages of using fuzzy methods.

More interesting are the methods based on fuzzy sets and rules that predict an associated possibilistic uncertainty. Environmental examples that include conditioning on observations are the fuzzy regression methods of Bardossy et al. (1993) and Hojati et al. (2005). Once the degree of membership for a predicted variable has been established throughout the model space, the uncertainty in the outputs can be assessed by taking nested α-cuts of the fuzzy set. A recent study by Shrestha et al. (2007) is of this type, using fuzzy nonlinear regression to evaluate uncertainty in the stage-discharge relationship for a reach of the River Neckar in Germany. These uncertainties are then used to determine uncertainty in flood risk along the river by running α-cuts for the predicted discharges through a hydraulic model.

4.7 Comparing methods for model conditioning: coherence and the information content of data

In all real applications, whether using a nonlinear regression, formal Bayes, Pareto optimisation, GLUE or fuzzy approach, given some historical observations with which to compare model predictions, the modeller has to choose *what* likelihood functions or performance measures he/she will implement *without* strong prior information about the correctness of the model structure and the nature of the errors. We suspect that all models will have some limitations as a description of the true response; we have to estimate multiple parameters; we expect errors in the input and outputs; and we know that any errors in the input data will be processed through the model in a nonlinear way. Therefore, we would *expect* that it would be very difficult to find an adequate formal error model for use, for example, in the Bayes likelihood methodology. Sometimes, of course, we might be lucky and find an error model or models that give a good approximation to the error series for all models in the model space and for all the different types of observations that we wish to use in evaluation, and in such cases the Bayes methodology would probably be the method of choice.

It would be even better if we could also find a way of characterising the input data error and the model structural error in a formal way (e.g. Krzysztofowicz, 1999; Kennedy and O'Hagan, 2001; Oakley and O'Hagan, 2004). If this were the case then again the Bayesian methodology would probably be the method of choice. But in most real cases, it does not seem that there is real hope of being able to justify all the assumptions required for this succession of formal error models, nor estimate the parameters required (since the different error sources will certainly interact). Where such real difficulties exist, therefore, there may be scope for using one of the other approaches in model conditioning, especially a GLUE or fuzzy approach.

These, however, appear to many people to lack the objectivity of formal Bayesian methods. In fact, Mantovan and Todini (2006) accuse the GLUE methodology of being incoherent in a formal statistical sense, since it does not take account of all the information content of the residuals as new observations become available. However, they argue the case on the basis of a hypothetical example in which the model structure in known to be correct. Beven et al. (2008) respond by showing that even small deviations away from this ideal case, introducing input data error, model structural error and "unknown" residual error structure, results in the formal Bayes approach producing biased parameter estimates (this is already a well-known result in statistics for linear models). This is a result of over-conditioning resulting from making too strong (or wrong) assumptions about the information content of the observations. In more realistic applications, Beven et al. (2008) argue that it may be incoherent to choose a formal likelihood measure when we expect that there is error in the input data and model structure. In such cases it might be more robust to *under-condition* in the model calibration process.

Coherence in this context has a technical meaning. A very general definition of coherence states that the modeller should not knowingly choose a measure of the information of an experiment that is less informative than some other measure. In the Mantovan and Todini (2006) study, cited above, a much stronger form is used. They require that for some measure of discrepancy between some prior, g_o, and a posterior given n+m observations, $d[g_{n+m}, g_o]$

$$d[g_{n+m}, g_o] > \max\{d[g_n, g_o], d[g_m, g_o]\}$$

for any n and m (including the case of m = 1, i.e. a single observation is being added). Implicit in this principle is that new observations should *always* be informative in the conditioning process (that $d_n = d[g_n, g_o]$ should be monotone for any *n*), and that the results of any analysis should not depend on the order in which the observations are used.

Any real application, however, cannot possibly meet this criterion when faced with the possibility of input error and model structural error. Disinformation (in terms of identifying acceptable models or likelihood distributions) might be produced when significant input errors are processed nonlinearly through the model and, because of the time (and space) scales of the system response, this might affect the effective information associated with many observations. What is then needed (that is in real applications of environmental models) is a likelihood measure that properly reflects the information content of the observations in constraining the uncertainty in the predictions of an imperfect model run with imperfect input data and with imperfect knowledge of the nature of the different sources of error, which will often be complex.

Whatever that likelihood measure might be! Stating the principle is much more difficult than deciding on an appropriate measure that properly reflects the information content in the data in conditioning a model of the system. But when there are multiple sources of error, disaggregation of the effects of different sources within a nonlinear modelling problem is known to be a very poorly posed problem. There is, at least as yet, no general theory of the information content of data for these realistic cases, particularly for cases where model structure error is likely to be significant. There is therefore no right or unambiguous answer, although, if the residual errors do have a simple structure, probabilistic estimates of predicting an observation conditional on the model as assumed correct might still be possible and useful. Even then, the use of measures based on squared errors is not the only, or necessarily best, choice (Tarantola, 2005, 2006). This seems to be the next major advance required in our understanding of the uncertainty issue in environmental modelling.

The argument that simple formal measures might be incoherent in real cases, however, does provide some suggestions as to what informal measures might need to look like in order to reflect the information content of additional observations more realistically. It seems quite possible that *single* observations might introduce disinformation (in terms of constraining the uncertainty in model prediction) in the face of input and model structural errors. In that case we will clearly need to average or filter the performance of the model over some blocks of information, where the length of a block should be great enough to integrate over the response time of the system, and any longer time sampling over potential distributions of input error. Clearly, for environmental models, the potential time scales involved could vary widely depending on the residence times in the system and the nature of the input errors. However, the type of block evaluations that have been used previously in GLUE are clearly of this type, even if any formal assessment of the real information content of additional observations of different types is, as yet, lacking.

Ultimately, we all want to do good environmental science and make good predictions in real applications. What is then needed is either a deeper understanding of the real information content of data sets (with all their shortcomings), or new measurement techniques that will increase the real information about inputs and improvements to model structure available to the modeller (Beven, 2006c). Until such an understanding is forthcoming there will be little basis upon which to decide what is a coherent and what is an incoherent choice of uncertainty estimation methodology. The problem is evidently deeper and more complex than the search for the "right" formal likelihood measure, and needs further research.

4.8 Summary of Chapter 4

This chapter has looked at the use of observations about the response of environmental systems in calibrating and conditioning environmental models. The conditioning process was presented as a form of mapping of the real system into a hyperdimensional model space of model structures and parameter sets. A number of different approaches to the conditioning problem and searching the parameter space were considered, including weighted nonlinear regression, formal Bayes methods, multi-criteria Pareto set methods, the GLUE methodology and fuzzy set methods.

The most important points of the chapter may be summarised as follows:

- Optimisation of environmental models cannot be considered a good strategy when the optimum model found may depend on input and model structural errors. An acceptance of the possibility of multiple behavioural models (the equifinality thesis) is recommended.
- Statistical methods (weighted regression and formal Bayes methods) require the specification of a formal model of the errors. Very often, especially if all sources of error are lumped into an additive error term representing the total model residuals, it may be difficult to verify the assumptions made. Where an incorrect model of the errors is used, or where the uncertainty in the observations with which the model is being compared is neglected, it will generally lead to over-conditioning of the parameter values.
- In such cases, more flexible, but less formal, model conditioning approaches might still be useful, such as GLUE or fuzzy set methods. While these require more subjective (if common sense) choices about model evaluation, they allow the possibility of learning from model rejection.
- There is a lack of a theory of information content of new observations for real applications with multiple sources of error. Until the problem of assessing the real information content in such cases is addressed it will be difficult to draw conclusions about the coherence of the different methods for model conditioning and uncertainty estimation.

Box 4.1 Weighted nonlinear regression

Weighted nonlinear regression is a methodology for calibrating the parameters of a nonlinear model. It is a methodology that lies firmly within the optimisation paradigm, but can be used to provide uncertainties in both parameter estimates and model predicted variables. It is a methodology that has been used widely in environmental modelling and has been incorporated into both parameter estimation routines for particular model codes (e.g. MODFLOW groundwater code of Hill et al., 2000; the CXTFIT steady state transport code of Toride et al., 1995) as well as into stand-alone parameter estimation codes such as UCODE_2005 (Poeter et al., 2005) and PEST (Doherty, 2005) (See Software Appendix at the end of this book). The recent book by Hill and Teideman (2007) gives a full exposition of the techniques of nonlinear weighted regression as applied to groundwater modelling problems based on many years of experience.

In the weighted nonlinear regression approach, the optimisation problem is always set up in the form

$$\underline{O}(x,t) = \underline{M}(\Theta, \underline{I}, x,t) + \varepsilon(x,t) \qquad [B4.1.1]$$

with the $\underline{O}(x,t)$ as a vector of observations in space, x, and time, t; $\underline{M}(\Theta, \underline{I}, x,t)$ as the equivalent vector of model predictions that depend on a set of model parameters $\underline{\Theta}$ and inputs $\underline{I}$; and the aim of minimising the vector of residual errors $\underline{\varepsilon}(x,t)$, that are expected to vary in space and time. The optimisation is carried out conditional on the assumption that the model is a true representation of the system response, and that the residual errors can be made to conform (perhaps after a mathematical transformation) to certain statistical assumptions.

B4.1.1 Choosing the cost (objective) function

In the main text, the cost or objective function for the case of weighted independent prediction errors was presented as a summation (Equation [4.3]). Here we generalise the cost function to the case where co-variation of the error terms must be considered. It is then most compact to represent the errors and objective function in matrix notation such that:

$$J = \{y - \hat{y}\}^T \, \mathbf{W} \{y - \hat{y}\} = \underline{\varepsilon}^T \, \mathbf{W} \underline{\varepsilon} \qquad [B4.1.2]$$

where y is an observation, $\hat{y}$ is the model prediction, { } indicates a vector, $\mathbf{W}$ is a square matrix of weighting coefficients and ε is a prediction error. For cases where the prediction errors can be assumed independent the weighting matrix will have non-zero values only along the diagonal. Other forms of objective function are possible, including maximising a likelihood function rather than minimising the sum of squares in [B4.1.2] (see Draper and Smith, 1998; Hill and Tiedeman, 2007). The log likelihood objective function then has the form:

$$\ln(J') = N_o \ln(2\pi) - \ln|\mathbf{W}| + \underline{\varepsilon}^T \, \mathbf{W} \, \underline{\varepsilon} \qquad [B4.1.3]$$

where N_o is the number of observations, and $\ln|\mathbf{W}|$ is the log of the determinant of the weight matrix. The use of a weighted least squares objective function has a number of implicit assumptions that need to be met if the results of the parameter inference are to be accurate. These are that the true errors are random and unbiased (have zero mean), and that the weighted true errors are independent so that the weighting matrix can be taken as inversely proportional to the estimated constant variance–covariance matrix of the true errors. As noted above, there is a problem here in that we do not know what the true errors are; we can only assess the actual total error between the model predictions and observations, treating the model as if it were a true representation of the system. Thus, it is probable that the ideal assumptions will not be met, and that the user should take care to evaluate the validity of those assumptions in using this approach.

Additional information can be added to the objective function. For example, there may be some prior information about the value of some parameter at a particular location (in a groundwater modelling example, perhaps as the result of a pumping test to determine local hydraulic conductivity or transmissivity values at a particular observation well). In this case we may wish to impose some constraints on such values by using a weighted addition to the objective function for the difference between an observed and predicted transmissivity or some other constraint. This is easily done.

Simple functions of the errors can also be incorporated. In many environmental modelling problems, for example, it is found that the errors are **heteroscedastic**, which means that their variance changes with the magnitude of the prediction. This is in conflict with the assumption of having a constant variance–covariance matrix. Thus, it is necessary to transform the error in some way. A number of general transforms have been suggested as a way of transforming a set of residuals to a simpler Gaussian structure with constant variance so that a standard form of objective function can be used. For the simplest possible case it is assumed that the observed value y is equal to the true value of the variable + some error that is a simple function of the magnitude of the variable so that:

$$y = y^{true} + y^{true}\varepsilon = y^{true}(1 + \varepsilon) \tag{B4.1.4}$$

This is equivalent to using a logarithmic transform of the observed values to return to the form of a simple additive error, i.e.

$$\ln(y) = 1 + \ln(y^{true}) + \varepsilon \tag{B4.1.5}$$

Other transformations are based on square root transforms and Box–Cox transform for controlling for heteroscedasticity and the meta-Gaussian transform for controlling for non-Gaussian distributions of residuals.

The simplest square root transform, for a model residual ε, is:

$$y = y^{true} + z \text{ with } z = \varepsilon^{0.5} \tag{B4.1.6}$$

where z is a variable that (it is hoped) will be Gaussian. Freeman and Tukey (1950) proposed a variant on this such that

$$y = y^{true} + z \text{ with } z = \{1000\,\varepsilon\,/n\}^{0.5} + \{1000(\varepsilon + 1)/n\}^{0.5} \tag{B4.1.7}$$

Box and Cox (1964) introduced a transform that has the log transform as a special case. It has the form:

$$z\,(\lambda,\alpha) = \{(\varepsilon - \alpha)^{\lambda} - 1\}/\lambda \qquad ; \lambda \neq 0$$
$$= \ln(\varepsilon - \alpha) \qquad ; \lambda = 0 \qquad\qquad [B4.1.8]$$

with parameters α and λ.

B4.1.2 Iterative optimisation

In the weighted nonlinear regression methodology, once an objective function has been decided upon (complete with any transformations and constraints), calibration of the model parameters becomes an optimisation problem. In essence, we wish to search the response surface in the model space to find the set of parameter values that minimises the objective function. In linear regression this can be done analytically (as in the simple case of $y = ax + b$ given a set of (x,y) paired values). In the nonlinear case the optimisation must be done numerically using a so-called "hill-climbing" method. In the weighted nonlinear case the weight matrix in [B4.1.2] will vary depending on the parameter values since the parameter values will determine the error variance–covariance each time the model is run. An iterative approach is then required in which local linearization is used to decide on which direction to search. Hill and Tiedeman (2007) explain both the methods available and the issues involved in their use. They recommend an iterative Gauss–Newton gradient search algorithm and this has been incorporated into the parameter identification routines of the groundwater modelling code MODFLOW and the general parameter identification code UCODE_2005. The PEST software (e.g. Moore and Doherty, 2006; Gallagher and Doherty, 2007) uses an efficient variant of the Gauss–Marquardt–Levenberg hill-climbing method.

B4.1.3 Checking the assumptions

The weighted nonlinear regression approach involves a significant number of assumptions and these should be checked for validity before the results of a parameter identification exercise are accepted. In particular, diagnostic tests should be used to check for the randomness (lack of bias) of the residuals and the form of the distribution of the (transformed) residuals. Both graphical (comparing plots of the actual residual errors for normality) and statistical checks (checking distribution statistics for bias, skew and correlations) can be made (see Hill and Tiedeman (2007, pp100–112)). Checks can also be made for over-parameterisation of the fitted model using statistics developed in maximum likelihood parameter estimation in the time series analysis literature but applied successfully to other situations (e.g. Carrera and Neuman, 1986, in groundwater modelling). Burnham and Anderson (2004) suggest use of the corrected Akaike information criterion (AIC) which has the form:

$$\text{AIC} = \ln(J') + 2N_P + (2N_P(N_P + 1))/(N_O + N_P - 1) \qquad [B4.1.9]$$

where $\ln(J')$ is the log likelihood function from [B4.1.3], N_O is the number of observations, and N_P is the number of parameters. This gets rapidly larger (less negative in log) as the number of parameters increases relative to the information content in the observations.

B4.1.4 Parameter uncertainty

Once an optimum set of parameter values has been found, together with the associated weighting matrix $\mathbf{W}$, then the uncertainty in the identified parameter values and model predictions can be assessed. In the case of nonlinear regression, estimates of parameter uncertainty are made based purely on the local shape of the response surface at the optimum as reflected in the covariance matrix (if there are other, almost as good, models elsewhere in the model space they might have quite different covariance, but this is neglected in this methodology – but see Section 4.5).

The form of the parameter variance–covariance matrix (analogous to [B3.1.14] in Gaussian error propagation) is as follows:

$$[\text{Cov}<\theta_i\theta_j>] = \sigma^2\,(\mathbf{L}^T\mathbf{W}\mathbf{L})^{-1} \qquad [B4.1.10]$$

where σ^2 is the common error variance, the θ are the parameter values, $\mathbf{W}$ is the weight matrix, and $\mathbf{L}$ is a square (N_P by N_P) sensitivity or *Jacobian* matrix that here defines the gradient of the objective function with joint changes in parameter values i and j (i.e. $\partial J/\partial\theta_i\partial\theta_j$). The diagonal elements of [$\text{Cov}<\theta_i\theta_j>$] represent the variance of each individual parameter so that:

$$\sigma^2_i = \text{Cov}<\theta_i\theta_i> \qquad [B4.1.11]$$

The off-diagonal elements represent the co-variation between parameters from which a correlation coefficient, ρ_{ij}, for parameter interaction in affecting the objective function J, can be calculated as:

$$\rho_{ij} = \text{Cov}<\theta_i\theta_j> / (\sigma_i\,\sigma_j) \qquad [B4.1.12]$$

For some simple models the Jacobian gradient matrix, $\partial J/\partial\theta_i\partial\theta_j$, or its transpose $\mathbf{L}^T$, which is known as the *adjoint* model, can be calculated analytically. For larger codes, the numerical implementation of the equations can also be differentiated to provide local gradients analytically but this can double the size of the model code (as in recent versions of the MODFLOW groundwater code, for example). It can also be calculated numerically after the optimum has been identified by linearising around the optimum parameter set (this form of linearisation is also repeatedly used in the Gauss–Newton algorithm mentioned earlier to determine the local gradient of the cost function response surface as the optimisation proceeds).

The standard deviations of [B4.1.11] can be used to estimate confidence limits on the parameters using standard t statistics. Thus, for any chosen level of confidence, α, e.g. $\alpha=0.05$ for 90% confidence limits, the confidence interval can be calculated as:

$$\theta_i - t(n, 1-\alpha)\,\sigma_i < \theta_i < \theta_i + t(n, 1-\alpha)\,\sigma_i \qquad [B4.1.13]$$

where n is the number of degrees of freedom and the value of t is tabulated in many statistics texts. For n > 30, the normal distribution confidence limits can be used such that for 90% and 95% confidence limits t(n, 1-α) can be replaced by values of 1.68 and 1.96 respectively.

However, [B4.1.13] and the large sample normal distribution multipliers are both linear estimates of the confidence in the parameter values and the actual shape of the response surface may not conform well to the assumption of a constant linear gradient (i.e. $\partial J/\partial \theta_i \partial \theta_i$ may change rapidly as we move away from the optimum point on the response surface). More accurate nonlinear methods for finding confidence in the parameter values are available (e.g. Hahn and Meeker, 1991; Christensen and Cooley, 1999) but are computationally much more expensive. Linearisation is therefore a common assumption, but it should be remembered that [B4.1.10] to [B4.1.13] are only an approximation and might be misleading in certain nonlinear cases.

B4.1.5 Prediction uncertainty

The uncertainty in the model predictions after parameter calibration is of as much interest as the parameter uncertainty in real applications. The equivalent of [B4.1.10] and [B4.1.11] for the case of a single prediction, $\hat{y}$, is given by

$$\sigma_{\hat{y}} = \left[\sigma^2 \left(\{x\}(L^T WL)^{-1} \{x\}^T \right) \right]^{1/2} \qquad [B4.1.14]$$

where the vector $\{x\}$ is a further gradient term representing the rate of change of the prediction with respect to each of the θ_i parameters, $i = 1, \ldots, N_p$. This also therefore requires a linearisation around the point of interest (here N the optimum parameter set) meaning that again the resulting estimates of $\sigma_{\hat{y}}$ should only be used with care in calculating confidence limits for the predictions.

We also need to differentiate between the confidence limits for a model prediction and the prediction limits for an observation. The confidence limits will be based on the weighted sum of squared errors between observation and prediction but in the theory of nonlinear regression this is used to get an estimate of the error of the model predicting the true response of the system. In considering what might be observed at a prediction point, it is therefore necessary to add an error component associated with the measurement error of an observation. This difference is also often overlooked. In some cases this measurement error component might be expected to be small but in other cases it might be significant and should not be forgotten. Linear confidence limit estimates will then be given by:

$$y - t(n, 1-\alpha) (\sigma_{\hat{y}}^2 + \sigma_y^2)^{1/2} < y < y + t(n, 1-\alpha) (\sigma_{\hat{y}}^2 + \sigma_y^2)^{1/2} \qquad [B4.1.15]$$

where σ_y is an estimate of the measurement error variance associated with any predicted value $\hat{y}$.

Since confidence limits of the form of [B4.1.13] and [B4.1.15] are so quick and easy to calculate they are very often provided by the standard nonlinear regression codes and the potential inaccuracy of making such a linear approximation is often overlooked by users.

B4.1.6 Regularisation of distributed parameter problems

As already noted, the weighted least squares approach has been widely used in the identification of parameters for distributed models, particularly groundwater models. These types of models often have many more distributed elements than there are observations to constrain the parameter values. Parameter identification in such a situation is a poorly posed, underdetermined problem and obtaining a solution will only be possible by imposing additional constraints. One rather general approach is to impose regularisation conditions, an approach widely used in the inversion of geophysical data. Moore and Doherty (2006) give a clear outline of the use of Tikhonov regularisation in the PEST software. Tikhonov regularisation uses a constrained cost function minimisation with [B4.1.2] reformulated in the form

$$J_z = (Z\underline{\Theta} - \underline{z})^T \, W_z \, (Z\underline{\Theta} - \underline{z}) \qquad \qquad \text{[B4.1.16]}$$

where $\underline{\Theta}$ is the vector of N_p parameter values, Z is a matrix of regularisation conditions, W_z is a weighting matrix and $\underline{z}$ is a vector of regularisation observations indicating preferred parameter states at particular points in the system. [B4.1.16] is minimised subject to the constraint that

$$J_y = (L\underline{\Theta} - \underline{y})^T \, W_y \, (L\underline{\Theta} - \underline{y}) \leq J_y^L \qquad \qquad \text{[B4.1.17]}$$

where W_y is the observation weight matrix as before, $\underline{y}$ is the vector of observations, and L is the Jacobian gradient matrix, and J_y^L is some limiting value of the observation objective function, defined as a parameter of the optimisation.

This formulation is general but does not say anything yet about the regularisation conditions, the regularisation observations or the regularisation weighting matrix. It is also clear that, while regularisation might allow a unique solution for the field of parameters for this type of otherwise underdetermined problem, the solution will depend on the choices made for these conditions. In the type of groundwater application described by Moore and Doherty (2006), there will normally be quite a lot of information available at certain points of the flow domain where observation wells have been drilled. In this case local values of the parameters will normally be available, either estimated on the basis of the aquifer materials or directly measured by pumping tests. These can then be used as "pilot points" in the regularisation, known only subject to measurement errors. The regularisation conditions then define how parameters for the whole domain should be determined from these regularisation observations. A common strategy is to determine or assume some interpolation algorithm between the regularisation observations. A **Kriging** estimator, given a known **variogram**, for example, can be used to determine both the regularisation conditions and the weight matrix (which will be the inverse of the Kriging covariance estimates at every element in the domain). Effectively the regularisation conditions are providing a low-dimensional means of interpolating from the pilot points, where good parameter estimates can be assumed *a priori* to the complete field required in the distributed model. Kriging is a linear interpolation methodology. In principle, the regularisation conditions could also be nonlinear, but then it might be necessary to re-estimate the Z and W_z matrices at every iteration in the calibration.

Moore and Doherty (2006) show how this methodology can be used to provide information on how well the resulting parameter estimates provide estimates of the true parameters.

Box 4.2 Formal Bayes methods

As noted in Chapter 2, the origins of Bayesian statistics lie in a paper found amongst the papers of the Rev. Thomas Bayes (?1701–1761) after his death. Presented at the Royal Society of London by his friend Richard Price in 1763, the *Essay towards solving a problem in the doctrine of chances* contained the first expression of what is now called Bayes theorem. A more general discrete form was developed, apparently independently, by Pierre-Simon Laplace (1749–1827) and published in France in 1774.

We can define Bayes theorem in a form that, given a set of feasible models as hypotheses H and evidence E, then the probability of any H given E is given by

$$P(H|E) = P(H)\, P(E|H) \,/\, C \qquad\qquad [B4.2.1]$$

where $P(H)$ is some prior probability of H in the range of feasible hypotheses (or models), $P(E|H)$ is the likelihood of simulating the evidence given the hypotheses, and C is a scaling constant to ensure that the cumulative of the posterior probability density $P(H|E)$ is unity.

Bayes theorem represents a form of statistical learning process. When applied to models it provides a rigorous basis for the expression of degrees of confirmation of different model predictions, expressed as probabilities, as long as the different components of [B4.2.1] can be defined adequately.

This learning process starts with the definition of prior distributions for the factors that will be considered uncertain. There is a common perception of Bayes statistics that the choice of prior distributions introduces subjectivity into the analysis, but both Bayes and Laplace originally applied their methods using priors that were *non-informative*, given equal chances to all possible outcomes until some evidence became available. Berger (2006) points out that Bayes equation underlay the practice of statistics for some 200 years, without it being considered necessary to be too specific about the prior distributions, and that the development of frequentist statistical theory in the 20th century from R. A. Fisher onwards was, in part, a response to dissatisfaction with the constant prior assumption.

The choice of subjective prior distributions is a relatively recent innovation, but is now argued strongly for by many Bayesian statisticians (see Goldstein, 2006, and the additional comments on the Berger and Goldstein papers by Draper, 2006, and O'Hagan, 2006b). In principle, such priors can be defined by the modeller, independently of the evidence E, on the basis of expert elicitation, past experience or subjective judgement. This subjectivity remains difficult to accept for some people and the so-called "objective Bayes" methods have been used to try and reduce the subjectivity of the prior by selecting an appropriate distribution or by modifying it on the basis of the model error (the evidence in [B4.2.1] in the environmental modelling case) once some observations are available. Berger (2006) cites a number of different "objective" Bayes methods and there are examples of the use of such techniques in environmental modelling (see ver Hoef, 1996; Fortin et al., 1997). O'Hagan (2006b), in particular, feels strongly that the choice of an "objective" method for the determination of priors is itself a subjective choice. This argument is likely to continue amongst academic statisticians for the foreseeable future!

The prior should, however, only be really important if there is only a very limited amount of data or, at least, of good data. This is because, as more observations become available, and more evidence about the likelihood of a model is gained by repeated application of [B4.2.1], then the prior will come to be dominated by P(E|M) and the relative degree of belief in different models should become more secure (Box and Tiao, 1973; Bernado and Smith, 1994; Howson and Urbach, 1994). For this reason, when there is no strong evidence to express greater belief in the outputs of one model over another, a non-informative prior is often used. How to define a non-informative prior is the subject of extensive discussion in the statistics literature; in environmental modelling it often means sampling a parameter uniformly across some chosen feasible range. This is not totally non-informative of course. The modeller is still imposing a range for the *effective values* of the parameters required by the model, which may not be commensurate with, for example, measured values, and the sampling will depend on the chosen scale (e.g. arithmetic or log scale) though this can be avoided by using, for example, the Jeffreys' prior (Bernardo and Smith, 1994).

A more critical issue in the application of Bayes equation is, however, the definition of the likelihood P(E|H). Bayesian statistics is predicated on the assumption that a likelihood function can be defined that formally represents the probability of predicting the evidence E given a model as hypothesis H. Here we concentrate on formal methods for defining a likelihood function (but see Section 4.5 and Box 4.3 for arguments for using informal likelihood measures in Bayes equation within the GLUE methodology).

B4.2.1 Simple additive error models

The formal definition of the likelihood function follows from assumptions made about the structure of the errors. For example, consider the very simplest model of the errors that is useful, that is that the model errors are normally (Gauss) distributed, with zero mean, and independent one from the other. Thus, the prediction of an observation can be presented in the linear additive form as:

$$O = M(\underline{\Theta}, \underline{I}) + \varepsilon \; ; \; \varepsilon = N[0, \sigma_\varepsilon] \qquad \text{[B4.2.2]}$$

where O is a measurement, $M(\underline{\Theta}, \underline{I})$ is the model prediction of that measurement, given parameters, $\underline{\Theta}$, and set of input data, $\underline{I}$, and ε is the model residual. Measurement errors over a number of independent samples are often assumed to be of this form. This error model has only one parameter which is the variance of the residuals, σ_ε.

As soon as a new observation becomes available then a new residual error can be assessed by putting [B4.2.2] into the form

$$\varepsilon = O - M(\underline{\Theta}, \underline{I})$$

Assuming (for the moment) the model to be unbiased and that an estimate of σ_ε^2 is available, the contribution to likelihood function from a single residual is assumed to be given by

$$L(\varepsilon \mid M(\underline{\Theta}, \underline{I})) \propto \exp\left[-\frac{\varepsilon^2}{\sigma_\varepsilon^2}\right] \qquad\qquad [B4.2.3]$$

Over n such residuals, assuming independence, the contributions can be multiplied so that:

$$L(\varepsilon \mid M(\underline{\Theta}, \underline{I})) \propto \prod^{n} \exp\left[-\frac{\varepsilon^2}{\sigma_\varepsilon^2}\right] \qquad\qquad [B4.2.4]$$

For the assumption of the Gaussian-distributed errors, the final form of the likelihood function is given by

$$L(\varepsilon \mid M(\underline{\Theta}, \underline{I})) = (2\pi\sigma_\varepsilon^2)^{-n/2} \exp\left[-\frac{1}{2\sigma_\varepsilon^2}\left\{\sum_{i=1}^{n}[\varepsilon_i]^2\right\}\right] \qquad\qquad [B4.2.5]$$

Note that as n gets very large (as it often will if we are dealing with models of time series of observations or predictions in many different spatial locations), then the inverse of the error variance (usually the dominant term in [B4.2.5]) is being raised to a very large power ($n/2$). This means that models that have very similar error variance might have very different likelihoods because, under this model, each additional residual is presumed to carry significant weight in differentiating between models. It also provides computational difficulties because of the potential for rounding errors in raising what might be very small numbers to very large powers. Fortunately, this problem can be avoided in practice by taking the log of [B4.2.5], so that the power becomes a simple multiplication of $\ln(\sigma_\varepsilon)$ and will result in the correct relative likelihoods after back-transformation.

It is perhaps worth noting here that the tradition of weighting residuals according to their squared value is a tradition that has developed partly because of the mathematical advantages of the normal distribution. There is another longer tradition, derived from Laplace, to use the absolute error (Tarantola, 2006). This is the same as assuming that the errors have a probability that are distributed as $p(\varepsilon) = \exp(-|\varepsilon|)$, rather than as $\exp(-\varepsilon^2)$ for the Gauss distribution. While this was much less mathematically convenient in the past since analytical solutions were less tractable, such an approach could perhaps be advantageous in environmental modelling problems because the resulting distribution is not so greatly affected by (is more robust to) extreme residuals or "outliers" (see also Box 4.3). Tarantola (2005) presents a general exposition of different L-norms in model identification, where the L_1 norm is the Laplace absolute error measure, the L_2 norm is the Gauss squared error measure, and the L_∞ norm is a max-min criterion (other choices are clearly possible).

B4.2.2 Multiplicative error models

Sometimes we observe that the residuals get larger as the magnitude of the prediction gets larger. This means that the variance of the residuals is changing with the prediction. As previously noted in Box 4.1 this is called *heteroscedasticity* of the residuals. If it can be assumed that the heteroscedasticity can be treated as a simple increasing function of the model outputs, then it is simply allowed for by using a multiplicative error that is a function of the model outputs, $f(M(\underline{\Theta}, \underline{I}))$, rather than an additive error. This is easily done by using the logarithms of the predictions since:

$$O = M(\underline{\Theta}, \underline{I})f(M(\underline{\Theta}, \underline{I}))$$

can be represented as

$$\ln(O) = \ln(M(\underline{\Theta}, \underline{I})) + \varepsilon \; ; \; \varepsilon = \ln(f(M(\underline{\Theta}, \underline{I}))) \qquad [B4.2.6]$$

The simplest case is again that when $\varepsilon = N[0, \sigma_\varepsilon]$. When the model is applied, Equation [B4.2.6] can be inverted after the calculation of prediction bounds based on the magnitude of σ_ε to give prediction bounds in the original arithmetic scales (which will then not be symmetrical around the best estimate). There is also another advantage of using the multiplicative form of [B4.2.6]. In cases of large σ_ε the log transformation means that the resulting prediction bounds cannot be negative (whereas this is quite possible for the additive error in [B4.2.6], and there are many papers in hydrology, for example, that show error bounds on predicted discharges that are less than or cut off at zero, even though it is not quite clear what such negative flow predictions might mean physically).

Some other simple forms of error transform have been discussed in Box 4.1.

B4.2.3 Meta-Gaussian distribution transformation

The meta-Gaussian transform is a more recent, rather neat, idea based on using a quantile–quantile plot of the cumulative distribution of observed residuals to transform what might be a highly non-Gaussian distribution to the equivalent cumulative density value of a Gaussian distribution (e.g. Figure B4.2.1). Then simple Gaussian-based likelihood functions can be applied. Again, once the Gaussian prediction bounds have been determined they can be transformed back to the original distribution before plotting the results. The meta-Gaussian distribution, first used in a hydrological context by Kelly and Krzysztofowicz (1997), has been used with formal likelihood functions in flood forecasting by Krzysztofowicz (2002a,b) and in rainfall-runoff modelling by Montanari and Brath (2004). The disadvantage of this is that the data on which the transformation is based may not cover the full range required in prediction so that care should be taken in predicting and transforming extreme values to the original distribution.

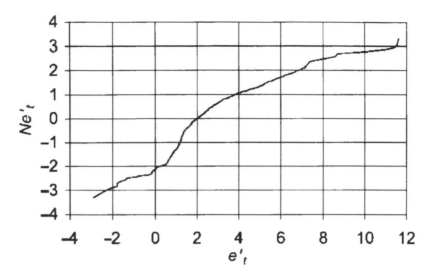

Figure B4.2.1 Meta-Gaussian transform from the CDF of a set of model residuals, e'_t, to a normalised Gaussian variate, Ne'_t (z-score scale in standard deviation units). In this case, the model residuals had already been subjected to a nonlinear transformation to try and stabilise the error variance

Source: Montanari and Brath, 2004, copyright © 2004, American Geophysical Union, reproduced by permission

Where more than one variable is involved, and they cannot be assumed independent, then the type of copula transformations into a multi-Gaussian space described in Section 3.3.5 might be useful.

B4.2.4 Autocorrelation in the errors

None of the transforms above deals with correlation in the residuals, either in time or space (or for models making distributed time series predictions in space and time). In environmental models, particularly models predicting time series of outputs, correlation in residuals is often found (e.g Figure B4.2.2 from Feyen et al., 2007). Ignoring the correlation by using an error model similar to [B4.2.3] or [B4.2.6] will overestimate the information content of each new observation included in the analysis and lead to over-conditioning of the parameter estimates.

Effectively if there is a strong correlation between successive residuals, some of the information content at one time step (or spatial location) has already been taken into account in evaluating the model at the previous time step (or location). We might also want to introduce the possibility that a model has a constant mean bias in its residuals, μ. That means that we should use a more complicated error model of the form (here assuming a first-order autocorrelation correlation in time with coefficient, ρ):

$$O = M(\underline{\Theta}, \underline{I}) + \mu + \varepsilon_t + \rho(\varepsilon_{t-1} - \mu); \ \varepsilon = N[\mu, \sigma_\varepsilon] \qquad [B4.2.7]$$

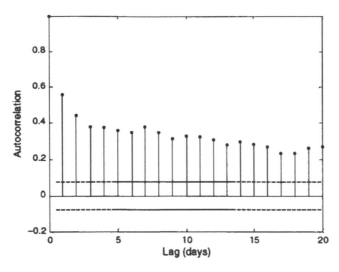

Figure B4.2.2 Autocorrelation in the residuals of a rainfall-runoff model run with a daily time step for the 21,000 km^2 Meuse catchment in Belgium, France and the Netherlands using the LISFLOOD model with a 5-km spatial grid. Dotted lines represent approximate significance limits

Source: Feyen et al., 2007. Reproduced by permission of Elsevier

This model yields a likelihood function of the form:

$$L(\varepsilon \mid M (\underline{\Theta}, I)) = (2\pi\sigma^2)^{-T/2} (1 - \rho^2)^{1/2}$$

$$\exp\left[-\frac{1}{2\sigma^2} \left\{ (1 - \rho^2)(\varepsilon_1 - \mu)^2 + \sum_{t=2}^{T} [\varepsilon_t - \mu - \rho(\varepsilon_{t-1} - \mu)]^2 \right\} \right] \quad [B4.2.8]$$

Now there are three statistical parameters to be estimated, μ, ρ and σ_ε. If there is evidence of longer time-scale correlations (as shown in Figure B4.2.2) then the model can be extended further, but there will then be additional ρ coefficients to be estimated.

It is possible to add additional terms to allow for structured model error (e.g. the model inadequacy function of Kennedy and O'Hagan, 2001). As noted in the main text, such error can arise from both model deficiencies and errors in the input data and it may be questionable as to whether such errors can be easily be represented by a simple additive function. It is also debatable in environmental problems as to whether it is better to compensate for model and input deficiencies in this way, or better to improve one or the other. The most complex form of model deficiency function used by Kennedy and O'Hagan (2001) was a constant mean bias in the residuals (which is already included in [B4.2.8]). More complex forms could clearly be used but will add additional statistical parameters to be estimated.

B4.2.5 Model selection using Bayes factors

In some modelling problems there may be a choice of model structures. This is true for even simple modelling problems such as the choice of a particular regression function (e.g. Draper, 1995; Smith et al., 2006) or of a frequency distribution to represent the extreme magnitudes of a particular environmental process with only limited data (e.g. flood frequency, avalanche frequency, or earthquake frequency). Within the Bayes methodology, each model structure can be given a prior probability, and differentiating between model structures can be achieved using Bayes factors.

Consider the set of competing model structures $\{M_1(\Theta_1), M_2(\Theta_2), \ldots M_N(\Theta_N)\}$ with parameters for which we can assign a set of prior probabilities $\{p_o(\Theta_1), p_o(\Theta_2), \ldots p_o(\Theta_N)\}$. The model structures may have different numbers of parameters or incommensurate effective parameters with the same name. Given a set of data, $\underline{I}$, that is used to evaluate the predictions for each model, we can determine the likelihood of predicting the observations given each model and parameter set as discussed above. The marginal distribution for the predictions for each model, averaged over the prior distributions of all the parameters for that model, is then given by

$$P(O \mid M_i(\underline{\theta}_i, \underline{I})) = \int L(O \mid M_i (\underline{\theta}_i, \underline{I}))\, p_o\, (\underline{\theta}_i) d\theta_i \qquad \text{[B4.2.9]}$$

Models can then be compared in terms of these marginal distributions. The Bayes factor for comparing two models i and j is given by

$$BF_{ij} = \frac{P(O \mid M_i(\underline{\theta}_i, \underline{I}))}{P(O \mid M_j(\underline{\theta}_j, \underline{I}))} \qquad \text{[B4.2.10]}$$

If we can also define some prior probability or belief in each model as the true model, $p_o(M_i)$, we can also use Bayes equation to define a posterior probability of each model given the observations as:

$$P(M_i \mid O) = \frac{P(O \mid M_i) p_o\, (M_i)}{\displaystyle\sum_j P(O \mid M_j) p_o\, (M_j)} \qquad \text{[B4.2.11]}$$

The denominator ensures that the cumulative posterior probability over all models is unity. The Bayes factor can then be shown to be equal to the ratios of the posterior to the prior odds for each model so that:

$$BF_{ij} = \frac{P(M_i \mid O) / P(M_j \mid O)}{p_o\, (M_i) / p_o\, (M_j)} \qquad \text{[B4.2.12]}$$

A clear presentation of the use of Bayes factors in model choice in the ecological context of simple models of brown trout returns after tagging in South Island, New Zealand, is given by Link and Barker (2006). An example from the use of ensemble climate predictions is provided by Min et al. (2007).

B4.2.6 Combining models using Bayesian model averaging
There is another possibility when there are several competing model structures under consideration. That is to combine the predictions of all the models (or at least all that are thought to be useful in making predictions) to try and achieve better predictions using the combined model. The idea is to combine the best aspects of all the models in the useful set. This is called Bayesian model averaging (BMA) or Bayesian melding (Draper, 1995; Poole and Raftery, 2000). It has been quite widely used in econometric modelling, and is increasingly being used in environmental modelling.

In the Bayesian model averaging approach, each of the competing model structures is given a weight directly related to the Bayes factors defined above. Thus, to produce a prediction, y, averaged over all model structures:

$$P(y \mid O) = \sum_i P(y \mid O, M_i) P(M_i \mid O) \qquad \text{[B4.2.13]}$$

An example is provided by Raftery et al. (2005) who applied BMA to ensembles of meteorological forecasts and, more recently, the application by Min et al. (2007) to climate change ensembles. Ye et al. (2004) also applied the method in the spatial problem of predicting the variability of parameters in a groundwater model (where the alternative weighted regression techniques of Box 4.1 are often used in model calibration). In such problems, information might be available at only a limited number of points in space since boreholes are expensive to drill. It is then necessary to try and infer the nature of the spatial variability in a way consistent with the available data while allowing for the uncertainty in the regions between data points. Geostatistics provides a useful set of mathematical tools for this type of problem but requires the specification of a variogram to describe the spatial correlation structure of the parameters. There are a number of different models for the variogram, and it may be difficult to distinguish between them with only limited data. Thus, what Ye et al. (2004) suggested was to combine the variogram models in a Bayesian optimal way.

The method has also been applied to rainfall-runoff modelling by Ajami et al. (2007) and Duan et al. (2007) who show how a nine-member ensemble of three models, each evaluated using three different performance measures, when combined using Bayesian model averaging, produced more skilful and reliable model predictions than any single model.

Box 4.3 Markov Chain and Population Monte Carlo methods

One of the most important issues in any form of uncertainty analysis is how to sample the model space to find those models and predictions that are of most interest. In problems of estimating the uncertainty in model predictions, of most interest generally means of greatest likelihood, since we are not really interested in sampling locations in the model space that contribute little to the posterior distribution of any variable we are interested in. Determining the posterior distribution of a variable is equivalent to a multidimensional integration of the likelihood associated with different values of that variable. In terms we have used before in describing the model space, we are effectively wishing to integrate under the likelihood response surface in the model space.

As noted in the main text, this is not a simple problem, especially in high-dimensional spaces, since we know that environmental models often show quite complex response surfaces in the model space. That is why strategies of simple random Monte Carlo sampling or discrete structured sampling are still used. It would clearly be advantageous in minimising the number of model runs required for an uncertainty analysis to sample only the regions of the model space that are of interest. In particular, rather than representing the response surface by means of weights (the "height" of the response surface for each point in the model space) associated with each sample, a much more efficient strategy would be to have samples that were distributed such that the density of sampling was proportional to the height of the response surface. Each sample could then be given equal weight. The simpler the nature of the response surface, the more efficient such a strategy will be.

The most commonly used method of density-dependent sampling strategy is a class of methods that go under the general name of Markov Chain Monte Carlo (MC^2) methods (e.g. Gamerman, 1997; Robert and Casella, 2004). These are methods for learning about the nature of the response surface, especially likelihood surfaces, and gradually refining the sampling until convergence to a density-dependent sampling. They do so within a random sampling framework that continues to allow some sampling in areas of apparently low likelihood to avoid missing local areas of interest by refining the search too quickly. The development of MC^2 methods has been closely linked with the development of Bayesian likelihood methods (see Box 4.1) as an effective way of integrating under the likelihood surface in forming the posterior distribution of model predictions. The best known MC^2 algorithms are the Metropolis–Hastings sampler (Section B4.3.1) and the Gibbs sampler (Section B4.3.2)

The concept behind MC^2 methods is quite simple. The method starts with a proposal distribution or transitional kernel for the Markov Chain. Random samples of model parameters are generated in the model space consistent with the proposal distribution. At each sample point, the model is run to determine the value of likelihood at that point. The initial sample of points is then used as the basis for selection of new samples, using the proposal distribution to choose new points around each current sample. After a further sampling sequence, the method is checked to see if it is converging on a consistent posterior distribution. If not, another iteration is carried out. The approach is analogous to a collection of random walks over the likelihood surface, with shorter steps where the likelihood on the surface is higher, so that the sampling density is then greater. The final result should be an accurate representation of the likelihood response surface and, consequently, the posterior distribution of the predictions associated with those likelihoods.

An alternative to MC^2 methods is provided by the Population Monte Carlo (PMC) approach. This does not require any specification of prior distributional forms for the parameters or proposal distributions for sampling, but is a form of iterated importance sampling. An initial set of samples is used to suggest the shape of the response surface. A new set of samples is then generated based on the form suggested by the previous iteration. In this way, more samples will be generated in areas of high likelihood and the form of the surface should become better defined at each successive iteration. PMC is described in more detail in Section B4.3.4 below. It is also closely related to the Particle Filtering methods of data assimilation described in Chapter 5.

B4.3.1 The Metropolis–Hastings algorithm

The earliest form of MC^2 search was suggested by Metropolis et al. (1953) and generalised by Hastings (1970). The method generates a Markov Chain in which the set of samples at step t is dependent on the samples at step $t - 1$. At each step the samples are randomly generated from a proposal distribution that depends on the results found at step t.

The critical step is then the choice of whether a new point should be accepted or rejected, i.e. whether the model should be run to evaluate the likelihood at the new point or not. Essentially this decision is made on the basis of the likelihood evaluated at the new point, while allowing for some random chance that a new point should be selected even if its likelihood is not so high, just in case it provides a pathway to a new high likelihood region in the random walk. The new point is selected if a random number, u, chosen from a uniform distribution in the range [0–1] satisfies the condition:

$$u < \frac{L(x')Q(x^t,x')}{L(x^t)Q(x',x^t)} \qquad [B4.3.1]$$

where $L(x')$ is the likelihood determined by running the model at the new point, $L(x^t)$ is the likelihood at the previous step in the chain, $Q(x^t,x')$ is the probability in the proposal distribution of moving from x^t to x' and $Q(x',x^t)$ is the probability of moving from x' to x^t. If this condition is not met then the current value is retained such that $x^{t+1} = x^t$. [B4.3.1] is the product of two ratios: the likelihood ratio of the proposed new point x' relative to the current point x^t, and the proposal density ratio in moving from x^t to x' to that in moving from x' to x^t. For symmetric proposal distributions this ratio $Q(x^t,x')/Q(x',x^t)$ is equal to 1, but if the proposal distribution is non-symmetric then the ratio might be greater than or less than 1.

The efficiency of the method will depend on the complexity of the response surface. If the form of the expected posterior distribution is known (for example from the assumptions made in setting up a formal likelihood function such as those in Box 4.1) then the proposal distribution can be given the same form. In applying the method, the first batch of samples is generated using the proposal distribution. The chain is then run until the effects of this starting sample are negligible (known as the burn-in period) before the chain starts to be checked for convergence.

Most proposal distributions have some free parameters that need to be tuned during the process because, in general, the exact form of the posterior distribution will not be known *a priori*. For example, if a multivariate normal distribution is used as the proposal distribution, the variance can be tuned to try to optimise the convergence rate. If this variance is too low, the step sizes will be small and the chain will sample the space only slowly. If the variance is too high, the step size will be too large, and there will be too many samples generated that produce low likelihoods and which therefore result in a very low acceptance rate. As the chain progresses, we would expect step sizes to decrease as the samples converge on the high likelihood region in the model space. Roberts and Rosenthal (2001), for example, suggest a scheme for modifying the variance of the proposal distribution as the chain progresses. Vrugt et al. (2003) have combined Metropolis–Hastings sampling with a shuffled complex evolution algorithm in a way that allows the proposal distributions for multiple complexes of models to evolve over successive iterations. They demonstrate the method on a hydrological modelling example. Testing for convergence of the chain is considered, for example, by Roberts and Smith (1994) and Brooks and Gelman (1998).

The Metropolis–Hastings algorithm has also been used in rainfall-runoff modelling by Kuczera and Parent (1998), Bates and Campbell (2001), Marshall et al. (2004), Engeland et al. (2005) and Gallagher and Doherty (2007). Other environmental modelling applications include forest growth models (van Oijen et al., 2005) and a catchment nitrogen model (Jackson et al., 2004).

B4.3.2 Other MC^2 algorithms

A variant on the Metropolis–Hastings algorithm is called the Gibbs sampler, first introduced by Geman and Geman (1984) and named after the physicist J. W. Gibbs because the process is analogous to random walks in statistical physics. In the Gibbs sampler it is assumed that the form of the distribution of each parameter can be assumed known *a priori*. The algorithm then chooses a random parameter (or cycles through the parameters in turn), picks a new value for that parameter from the current estimate of the marginal distribution and continues to the next parameter. The process can also be shown to be a form of Markov Chain that converges to a posterior distribution consistent with the assumptions about the distributions of the individual parameters (Casella, 1992). It is useful if, for example, all the distributions can be assumed to be normally distributed, but the marginal distributions and covariance structure are unknown. The Gibbs sampler has been used in water quality modelling by Qian et al. (2003) and tracer test analysis by Fienen et al. (2006).

B4.3.3 Reversible jump methods

Green (1995) introduced an MC^2 algorithm that could cope with more than one model with arbitrary numbers of parameters. This is equivalent to finding the posterior likelihood response surface in a model space that has model structure as well as parameter dimensions. In particular, it can be used to assess whether the data provide sufficient information to allow preference for one model over another.

Consider the case where $m = 1, 2, 3, \ldots .M$ models out of the set of all possible models are to be evaluated. The variable m is then used as an index variable in the chain. The models have parameter dimensions d_m where the number of dimensions in each model need not be the same.

The algorithm proceeds in a similar way to the MC^2 search strategies described above. A step is proposed based on some proposal distribution and evaluated for acceptance or rejection. In this case, however, the step can also include a jump from one model index to another.

B4.3.4 Population Monte Carlo methods

Population Monte Carlo (PMC) are an alternative to MC^2 sampling strategies that can avoid assumptions about proposal or parameter distributional forms. Effectively, it is an iterative importance sampling strategy that uses the empirical information gained from a set of samples at one iteration, to guide the sampling at the next iteration. At each iteration, importance weights associated with each sample are modified to reflect the additional information gained about the nature of the target distribution (which is here the shape of the likelihood response surface in the model space). In this way, the estimate of any summary integral statistics required from the distribution can be determined by a discrete weighted summation over all the samples such that

$$\hat{Z}^k = \sum_{i=1}^{n} w_i^k \, G(x_i^k); \sum_{i=1}^{n} w_i^k = 1 \tag{B4.3.2}$$

where $\hat{Z}^k$ is the integrand of interest at iteration k, n is the current number of samples, the w_i^k are the weights associated with each sample in the model space x_i^k and $G(\ )$ is some function of interest. In the case of a model-predicted variable the simplest case would be that $G(\ )$ is the predicted variable, and $\hat{Z}^k$ is the current estimate of the mean of that variable.

At each iteration, the weights associated with each sample are corrected for not using the correct target distribution by

$$w_i^k = \frac{L(x_i^k)}{Q(x_i^{k-1})C} \tag{B4.3.3}$$

where $Q(x_i^{k-1})$ is the distribution used to generate the samples x_i at the kth iteration, on the basis of the information available at the $k-1^{th}$ iteration, $L(x_i^k)$ is the value of the likelihood evaluated at x_i at the kth iteration, and C is a scaling constant chosen such that the sum of the weights is unity. As the iterations progress, the generating distribution should become closer to the target distribution and the weights should become closer to being equal, meaning that the samples are being generated with a density proportional to the target distribution.

The estimates of $\hat{Z}^k$ can be shown to be asymptotically unbiased while the evolving sampling distributions $Q(x_i^t)$ are naturally centred on the parameter distributions (although other choices are possible, for example, if there is particular interest in getting good estimates of the extreme values of the function $G(\ )$ of interest).

B4.3.5 More information and applications of MC^2 and PMC methods
There are a number of texts on MC^2 methods. Gamerman (1997) and Congdon (2006) provide a good introduction. Robert and Casella (2004) also explain importance sampling and MC^2 methods in detail. A good introduction to PMC is given in Cappé et al. (2004).

Some sources of software for MC^2 methods are suggested in the Software Appendix at the back of this book.

Box 4.4 Generalised Likelihood Uncertainty Estimation (GLUE)

GLUE is a form of model conditioning methodology that can be used without the need to define a formal structure for the errors. It was developed by Beven and Binley (1992) to deal with multiple acceptable models (the equifinality thesis of Beven, 1993, 2006a) and the complex and non-stationary error series that are commonly found in applications of environmental models to real data sets. In GLUE, the errors associated with a particular model run are handled implicitly by assuming the characteristics of the errors associated with that model in calibration will be "similar" in prediction. Both model predictions, and implicitly the error series, are weighted by a likelihood measure that expresses a degree of belief in that model (and its implied residuals) as a useful simulator of the system being studied. GLUE can use formal likelihood methods in which case the degree of belief can be expressed as a probability and the assumed error model included in the analysis (see Romanowicz, 1994, 1996).

In the main text, the decisions required in applying the GLUE methodology are outlined, and their subjectivity discussed. Once those decisions about the parameters to be varied, the prior distributions of those parameters, the method of sampling the model space and the likelihood measure to be used have been made, a large sample of model realisations is generated by Monte Carlo simulation. This is normally a sample of randomly chosen independent parameter sets, consistent with any prior information about parameter distributions, within a single model structure, but the concept is easily extended to the evaluation of multiple model structures.

Each simulation is run and evaluated with respect to any observed variables or other information that may be available using the chosen likelihood measure or measures to reflect the performance of individual models in reproducing the behaviour of the system under study. The only formal requirements on the likelihood measure are that it be zero for models considered unacceptable or non-behavioural, should increase with increasing levels of performance, and should scale to sum to unity over all the models retained as behavioural (Beven et al., 2000; Beven and Freer, 2001). Application of Bayes equation (or some other way of combining different likelihoods, see below) then allows a posterior likelihood to be calculated for each model realisation.

Combining likelihood weights in this way provides a formal way of learning from more observations. Within GLUE it may be applied to multiple model structures as well as multiple parameter sets provided that the likelihood measures used in the evaluation of each model can be directly compared. In all cases, the model (structure and parameter set) is treated as an entity. While it may be desired to extract sensitivity information on individual models or parameters by calculating appropriate marginal distributions of the likelihood measure (see below), it is not necessary to do so. It is the model as a whole that gives a good or bad performance in simulating the observations, and marginal distributions on individual components may not always be very informative in applications where there is a variety of models that give a good fit to the available data.

Given the posterior likelihood values, the predictions from each realisation can then be weighted by the associated likelihood value to calculate prediction quantiles over N_B behavioural models as:

$$P(\hat{Z}_t < z_t) = \sum_{i=1}^{i=N_s} \{L[M(\underline{\Theta}_i, \underline{I})]| \hat{Z}_t < z_t\}$$ [B4.4.1]

where $\hat{Z}_t$ is the value of variable Z at time t simulated by model i. Within this framework, accuracy in estimating such prediction quantiles will depend on having an adequate sample of models to represent the behavioural part of the model space.

B4.4.1 Choosing a likelihood measure

It is worth re-emphasising that GLUE can use the type of formal likelihood measures described in Box 4.2 in formulating the posterior likelihoods required for the application of [4.4.1]. In this case, by treating the error model as an extension of the model space, GLUE will be formally equivalent to the formal Bayesian methods of Section 4.2. However, other informal likelihood measures can also be used within this framework, including fuzzy possibilistic measures, or binary (yes/no, behavioural/non-behavioural) evaluations based on "soft" or qualitative data.

This remains controversial, since the choice of such measures appears to be so subjective. Formal statistical likelihood measures, on the other hand, are objective and allow an objective probability of predicting an observation to be assessed, but *only* if the assumptions on which they are based are valid (see discussion in Section 4.7). In applying Bayes equation it is quite acceptable that the prior distribution of parameter distributions is subjective. Thus, the posterior must also be subjective until the likelihood arising from model evaluation against the observations dominates the prior likelihood (Bayesian statistics has itself been criticised for such subjectivity). There is absolutely nothing in Bayes, however, that suggests that the likelihood arising from model evaluation should not also be subjective; it is just that the posterior will then be a belief measure rather than a probability, conditional on all the assumptions (the decisions of

Section 4.5.1) that underlie it. Clearly it is possible to choose common-sense measures of belief for use as likelihood weights within this framework. Since we are still lacking a theory of the information content of observations in conditioning models subject to different sources of error in real applications (see Section 4.7), whether such measures of belief are judged to provide an adequate approach must be a matter of personal preference.

In what follows, we describe some of the error series-based measures that have been used in past applications in GLUE. In the main text a different approach based on limits of acceptability is described. The origins of GLUE lie in the observation that optimisation and formal likelihood methods have the effect of separating models that have very similar error variances. If performance was to be considered purely as a matter of the variance of the residuals, then it seemed to be a common-sense idea that we should not have greatly differing degrees of belief or likelihood in models with similar residual variance (pure optimisation is the most extreme case, here; the best or lowest error variance model is effectively given a likelihood of one and all others a likelihood of zero).

The first set of measures therefore are based on functions of the residual variance, σ_i^2. One possibility that has been used in a number of studies is to use a power function of the error variance with shaping parameter N, where N is ideally chosen to reflect the effective information content of the observations (Box and Tiao, 1973). Thus, the likelihood measure for the vector of predictions for the ith model defined over a vector of observations, $\underline{O}$, is defined as:

$$L\left(\underline{O}| \underline{M}(\underline{\Theta}_i,\underline{I})\right) = (1 / \sigma_i^2)^N / C \qquad \text{[B4.4.2]}$$

where C is a scaling constant. In practice the value of N may be chosen empirically to control the peakiness of the response surface in the model space and how far the model prediction uncertainty brackets the observations. Note, however, that use of [B4.4.2] implies an expectation that no perfect simulation will be found since in that case this measure would go to infinity (it is also possible that a near perfect model would dominate the sample of behavioural models in the cumulative likelihood). Studies using this type of likelihood measure in rainfall-runoff modelling applications have shown that, even with time series with large number of time steps, the appropriate value of N is not high, reflecting the lack of constraint provided by measurements in the calibration of many environmental models.

Another form of performance measure based on the error variance, commonly used in environmental modelling studies, is the proportion of the observed variance explained by the model (sometimes called the Nash–Sutcliffe *efficiency* measure after Nash and Sutcliffe, 1970)

$$L\left(\underline{O}| \underline{M}(\underline{\Theta}_i,\underline{I})\right) = (1 - \sigma_i^2/\sigma_o^2)^N / C \qquad \text{[B4.4.3]}$$

where σ_o^2 is the variance of the observed data and C is again a scaling constant. Again a shaping parameter has been added in the form of a power N. For N equal to 1, [B4.4.3] is analogous to a coefficient of determination. Clearly equation [B4.4.3] gives a different likelihood scaling for any given σ_i^2 than [B4.4.2].

A further function of the error variance with some advantages in combining different likelihood measures for different periods or different types of observation is:

$$L\left(\underline{O}|\ \underline{M}(\Theta_i,\underline{I})\right) = \exp\left(-\ N\sigma_i^2/\sigma_o^2\right)/\ C \qquad\qquad [\text{B4.4.4}]$$

where N is again a shaping parameter and C a scaling constant.

Variance based measures of this type tend to give greater weight to errors associated with the highest magnitude predictions, where the errors are potentially greater, due to the way in which each residual is squared. This effect can be reduced in a number of ways. One is to take the sum of the absolute values of the errors such that:

$$L\left(\underline{O}|\ \underline{M}(\Theta_i,\underline{I})\right) = \Sigma|\underline{M}(\Theta_i,\underline{I},x,t) - \underline{O}(x,t)|/\ C \qquad\qquad [\text{B4.4.5}]$$

Clearly many other functional forms are possible, depending on what seems a sensible measure of degree of belief in a particular application. A particularly interesting type of evaluation is the binary measure (yes/no) evaluation based on "soft" data. This can be applied, for example, on the basis of information about the processes in the model and in the real system, including the type of perceptual understanding outlined in Chapter 1. Thus, if a process that is thought to be important in the real system is not predicted by the model, then that model might be rejected even if it might be given a high likelihood on the basis of a residual measure like those above. Other residual-based measures, and a summary of applications of the GLUE methodology may be found in Beven and Freer (2001).

B4.4.2 Deciding which models should be considered behavioural

One of the decisions that must be made in applying the GLUE methodology is how to decide whether a model is behavioural or non-behavioural. As noted in Section 4.5.2, this might not be an easy decision because of the smooth range of model performances from best to bad that is found in most applications. However, again, a common-sense approach can be taken to this question, allowing models that can be considered fit for purpose to be retained in prediction, and the rest rejected as non-behavioural and given a likelihood of zero. The definition of fit for purpose will clearly be problem-specific.

It is also possible to evaluate the prediction limits for a range of different decisions about the performance threshold or limits of acceptability for a model to be considered behavioural, as discussed in the main text. In this way, the limits or threshold can be set to bracket a chosen proportion of the observations, or to explore whether even the best models can be considered to be acceptable simulators (in terms of thinking of them as hypotheses about how the system is functioning). In GLUE, it is quite possible to reject all the models considered. This can be a useful tool in a learning process (See section 4.5.9) in that it can lead to a re-evaluation of the data used to drive the model, the model structure or the way of evaluating the predictions. The formal likelihood measures of Box 4.2 can compensate for such failings by treating the residual deviations as a purely statistical distribution (if an appropriate structural model of the errors can be found), but this might not necessarily be the best strategy in considering whether a model is an adequate hypothesis.

B4.4.3 Combining multiple model evaluations

There will often be more than one possibility of model evaluation, in terms of both quantitative and qualitative performance. These may range from the calculation of different likelihood measures based on different simulated variables to rejection criteria based on whether the way in which the model is simulating a certain process properly reflects some qualitative understanding of the system responses. Each criterion or additional information should allow some refinement of the likelihood distribution associated with each model, and in particular the rejection of some models that had previously been considered as behavioural.

The addition of different types of information in this context may be handled in several ways. One is by repeated use of Bayes equation in the form:

$$L(\underline{M}(\Theta_i,\underline{I})|\underline{O}) = L(\underline{M}(\Theta_i,\underline{I})) \, L \, (\underline{O}| \, \underline{M}(\Theta_i,\underline{I}))/ \, C \qquad [B4.4.6]$$

where $L(\underline{M}(\Theta_i,\underline{I}))$ is some prior probability defined for the range of feasible models with parameter set Θ_i and set of inputs $\underline{I}$; $L(\underline{O}| \, M(\Theta_i,\underline{I}))$ is the likelihood of simulating the vector of observations, $\underline{O}$, conditional on the set of behavioural models; and C is a scaling constant to ensure that the cumulative of the posterior probability density $L(M(\Theta_i,\underline{I})|\underline{O})$ over all the $i = 1, 2, \ldots .K$ behavioural models is unity.

The application of Bayes equation in this form requires a certain orthogonality of the samples which is why the parameter sets should be chosen to be independent samples from the parameter space. To get a good definition in a parameter space of high dimensions will therefore require a large number of samples (and consequently a large amount of computer time). Simple random sampling may not be the most efficient way of defining the response surface where that surface has a simple form (see Section 4.3.2 and Box 4.3). However, where it has a complex form, the simplicity inherent in this methodology may be considered an advantage.

At each stage, a posterior likelihood is calculated conditioned on the additional data. The multiplicative nature of [B4.4.6] will mean that if at any stage a model is given a likelihood of zero, it will be rejected from further consideration as non-behavioural. In cases where a model cannot predict adequately all available observables, this will lead to the rejection of *all* the model realisations (see the discussion of Section 4.5.9).

The multiplicative nature of [B4.4.6] also means that the order of application of different likelihood measures might be important. This can be seen when the additional information is derived from conditioning based on observations of the same variable but for different time periods. If a likelihood measure is calculated as a function of the inverse error variance for each period then the repeated application of [B4.4.6] will produce different posterior likelihood values than using the inverse error variance of the whole period treated as a single series (this is used as a criticism of GLUE in Mantovan and Todini, 2006, but in fact simply reflects different choices about how to evaluate the models and, by implication, different assumptions about the real information content of the available observations; see the discussion of Section 4.7). This difference may, of course, be a desirable property in cases where there is an expectation that the system (and therefore the parameter values) may be changing over time. It may be avoided by using an alternative likelihood measure, such as the negative exponential function of the error variance [B4.4.4], such that repeated applications of [B4.4.6] will be equivalent for multiple (equally shaped) subdivisions of the whole period, i.e.

$$L\ (\underline{O}|\ M(\underline{\Theta}_i,\underline{I}) = [\exp\ (-N\sigma_1^2/\sigma_o^2)\ \exp\ (-N\sigma_2^2/\sigma_o^2)\ \exp\ (-N\sigma_3^2/\sigma_o^2)\ \ldots]\ /\ C$$

$$= [\exp\ (-N\sigma_1^2/\sigma_o^2 - N\sigma_2^2/\sigma_o^2 - N\sigma_3^2/\sigma_o^2 - \ldots .)]/C$$

$$= [\exp\ (-N(\sigma_1^2 + \sigma_2^2 + \sigma_3^2 + \ldots .)/\ \sigma_o^2)]/C \qquad\qquad [B4.4.7]$$

In the most general case, defining the error itself may be difficult where the observations are imprecise or fuzzy in nature. In this case, it may be more appropriate to use a limits of acceptability approach or fuzzy measure in the model evaluation or conditioning. The values of the fuzzy measure can still be used to weight the predictions of a sample models to create prediction quantiles, as in Equation [B4.4.1]. They essentially therefore serve in the same way as the likelihood measures within the GLUE methodology. The use of fuzzy measures, however, also expands the ways in which measures might be combined by utilising operators from fuzzy set theory. Given m different fuzzy measures of performance, L_a, L_b, ... L_m, the two simplest ways of combining two fuzzy measures are by fuzzy union and fuzzy intersection, and general fuzzy weighted mean [see Box 3.4].

B4.4.4 The meaning of GLUE prediction limits
There are thus a wide variety of ways in which to approach the evaluation and conditioning of models within the GLUE framework. The choice of a measure will be generally a subjective choice, but argued and reasonable for the model purpose. The resulting prediction quantiles will therefore also be dependent on this choice. Thus, any prediction bounds produced using the GLUE methodology are conditional on the choices made about the range and distribution of parameter values considered, the model structure or structures considered, the likelihood measure or measures used in defining belief in the behavioural models and the way of combining the different measures. These are all subjective choices but must be made explicit (and can be debated or justified if necessary).

The prediction bounds are then taken from the quantiles of the cumulative likelihood weight distribution of predictions over all the behavioural models as in [B4.4.1]. They may be considered as empirical probabilities (or possibilities) of the set of behavioural model predictions. They have the disadvantage that unless a formal error model is used (e.g. using [B4.2.5] or [B4.2.8] where the assumptions are justified) they will not provide formal estimates of the probability of estimating any particular observation conditional on the set of models; they have the advantage that the equifinality of models as hypotheses, non-stationarities in the residual errors, and model failures are more clearly revealed.

They reflect what the model can say about the response of the system, after conditioning on past data. It is then up to the user to decide whether the representation is adequate or not.

Experience suggests that for ideal cases, where we can be sure that the model is a good representation of the system (unfortunately in the case of environmental systems this is normally only the case in hypothetical computer experiments), the GLUE methodology with its implicit treatment of residuals can provide good bracketing of observations (e.g. Figure 4.11 in the main text). The effects of strong input error and model structural error, however, may mean that it is simply not possible for the range of responses in the model space to span the observations consistently. This is, however, valuable information. Since the residual errors in such cases are not necessarily random or stationary, it may not be appropriate to represent them as a random error model.

Some sources of software for applying the GLUE methodology are suggested in the Software Appendix at the back of this book.

Chapter 5

Forecasting the near future

The only thing that makes life possible is permanent, intolerable uncertainty; not knowing what comes next.

Ursula LeGuin

Life is uncertain. Eat dessert first.

Ernestine Ulmer

5.1 Real-time data assimilation

It was noted in Chapter 1 that the problem of forecasting the near future is different from the calibration of a simulation model of the system. In the forecasting case it is not necessary to simulate all aspects of the system. The requirement is to provide predictions at the lead time of interest with minimum uncertainty. The forecasts need to be available with sufficient lead time to be useful in making decisions, and data assimilation can be used to improve the predictions and reduce the forecast uncertainty as a particular situation or event evolves.

There are a number of issues that arise in trying to make forecasts with sufficient lead time. The first is whether there are data available to be able to make a prediction at all. A forecast is always dependent both on having an adequate model of the system and on the availability of input data to drive that model. The initiation of the tsunami of 26th December 2004, for example, was caused by an earthquake of 9.1 on the Richter Scale just north of Simeulue Island off Sumatra in the Indian Ocean. The earthquake was the second largest ever recorded on a seismograph and caused waves up to 30m high. There were, however, no tsunami sensor systems and no forecast models implemented for the Indian Ocean. No forecasts were made and no warnings were issued. A total of over 220,000 people were lost across the countries affected, including Sri Lanka, Thailand, India and Indonesia, but damage and deaths were recorded as far away as South Africa, 8,000 km away. It is probable that if the equivalent event had happened in the Pacific Ocean, where there is a tsunami forecasting system in place, the death toll would have been very much less because warnings would have reached many places in time. A forecasting system is now being implemented by the Indian Ocean states.

There are a number of issues that arise in trying to make forecasts with sufficient lead time. The first is whether the natural time delay in the system is sufficient. The Boscastle flood, on the north coast of Devon, England, on 16th August 2004, occurred after some 200 mm of rain on a small steep catchment and caused 58 properties to be flooded and over £2 m of damage. The flood was caused by a succession of small intense convective cells along a squall line moving from the Irish Sea over the catchment upstream of the village. The time delay between the heavy rainfall being measured at rainfall gauging sites and the peak discharge was too short for an adequate warning to be issued. During the event a flood wave 3 m high moved down the valley, probably as a result of a sudden release of water following failure of a dam of debris. The speed of rise of the river took many people by surprise. Fortunately, the flood took place during daytime and seven helicopters were mobilised that picked up over 100 people stranded on rooftops or cars. Nobody was killed. Fifty years before, on the night of the 15th August 1952, in a similar flood in the same area in the village of Lynmouth, a flood wave caused by up to 300 mm of rain in the catchments of the East and West Lyn rivers, swept through the village, killed 34 people and deposited 200,000 tons of debris. These events are not unprecedented; previous damaging summer floods of this type have been recorded at Boscastle in 1824, 1847, 1950, 1957, 1958 and 1963.

In systems with short response times, therefore, the only way of increasing the lead time of a forecast to be able to give a warning would be to forecast the input to the system. There are systems for forecasting rainfall inputs, using numerical weather prediction (NWP) models or extrapolations of rainfall radar data, but both, as yet, give only very uncertain predictions for any particular location (see Collier, 2007, for a recent review of rainfall forecasting). There is a danger, therefore, that given the uncertainties either a warning will be ordered, but without the expected danger actually occurring, or a warning will not be ordered and a flood or other extreme event will occur. This was the case in the Vaison-la-Romaine flood, in the south of France in 1992. In this case, on the basis of the atmospheric model forecasts, two warnings that flood-producing rainfalls might occur in the Department of Vaucluse, which contains Vaison-la-Romaine were issued by Météo France ahead of the event. However, nobody was sure which of the catchment areas within the Department might be affected, so no local warnings were given (there are some flood-producing rainfalls in this area in late August or September nearly every year, but they are often localised). Some 35 people lost their lives, many of whom had been staying in riverside campsites. The river overtopped the Roman bridge in Vaison-la-Romaine, and reached some 18 m above its normal level. Estimates of peak discharge varied between 600 and 1,200 m^3s^{-1}.

A third issue is that the speed with which model predictions can actually be made is important. This is currently an issue with numerical weather prediction models. The forecasters would like to do two things to improve their predictions. They would like to refine the grid scale of the models, and they would like to increase the number of ensemble runs of the model for each forecast. An ensemble is made up of a number of different runs of the model (see Section 5.2 below), each with different patterns of initial conditions, reflecting the uncertainty in the knowledge of the atmosphere at the start of the run. At the European Centre for Medium-Range Weather Forecasts (ECMWF) 50 forecast ensembles are run, twice every day, for comparison with the control run, but these have coarser resolution in both the horizontal and vertical than the deterministic forecast (Figure 5.1). Both decreasing the grid scale and increasing the number of runs would increase the computer run time. Undoubtedly both will be

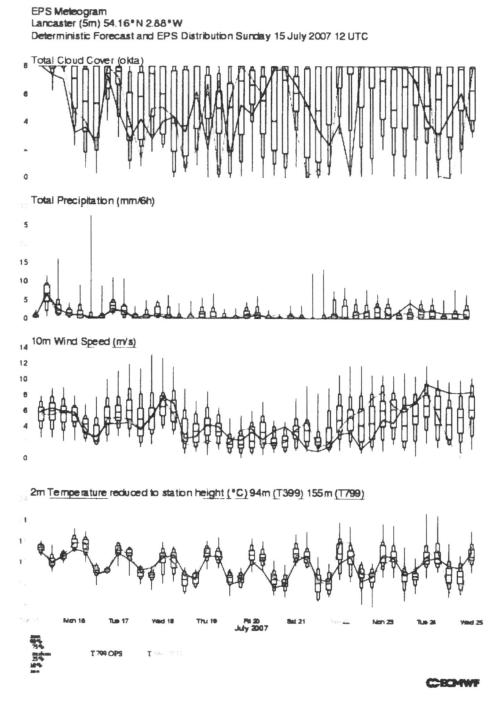

Figure 5.1 ECMWF Control and Ensemble Predictions as Box Plots for cloud cover, precipitation, 10 m wind speed and 2 m temperature for the period starting 15th July 2007 for Lancaster, UK (reproduced by authorisation of ECMWF)

improved in the future as the models are implemented on the next generation of computers (and, as a hydrologist, it is to be hoped that the representation of rainfall formation and runoff generation processes will also improve), but limitations will remain so as to not have total computational times that reduce the necessary lead times for decision making.

Even where future inputs are extremely uncertain, however, model forecasts of the near future can generally be improved by the use of data assimilation or model updating. All data assimilation techniques try to correct, in some way, for departures between the forecast and what actually happens, in making a new forecast. Thus, at each forecasting step, a comparison is made between the previous forecast and the available data on the current state of the system (for example, temperatures, pressures and wind fields in weather forecasting and river levels in flood forecasting). This requires that such data be available to the forecaster in "real time" so that they can be used in this way. With modern telemetry and communications technology this is becoming both easier and cheaper and the range of data available to the forecaster is increasing all the time.

As such data arrive at a forecasting centre it can be seen whether the past model forecast was under-predicting or over-predicting the observed changes. The magnitude of the under-prediction or over-prediction can then be used to adjust the next forecast, hopefully to make it more accurate, although there is no guarantee that, given uncertain inputs, an under-prediction at this forecasting step will continue at the next forecasting step. Thus, data assimilation is inherently uncertain and should ideally be carried out in a way that allows the forecast uncertainty to be assessed. Indeed, the aim of any data assimilation methodology should be to minimize the forecast bias and prediction variance at the required "N step ahead" lead time for which a decision might be required. This is an important consideration in assessing forecasting and updating techniques, since there are many, many papers in the literature that present only one step ahead forecasts. Where the length of such a time step is short relative to the decision time frame or required lead time, then one step ahead forecasts might not actually be very useful, however accurate they are.

A simple real-time forecasting problem, in this case the prediction of flood water levels in a river, is illustrated in Figure 5.2. Given information about what has happened up to time now, t_o, how can we best predict what will happen in the future? Figure 5.2 shows the predictions (with uncertainty estimates) originating from different initiation points on the flood wave over the natural lead time of the system. It can be seen how the estimate of the forecast variance changes over time and, in particular, for every individual forecast the forecast variance gets larger as the forecast time gets longer. At each initiation time, the state of the system (here water level at some downstream site) is known and has been used to update the forecast model. Operationally these types of calculations would be updated every time a new observation is received (typically every 15 to 60 minutes for flood forecasting).

Such predictions involve two important sources of uncertainty; the first is how to correct for the recent deviations between our model prediction and the available observations, the second is how to take account of the potential future inputs into the forecasting system (for example, future rainfall predictions in the case of flood forecasting). Many forecasting problems in different domains are analogous to this. Even where a model has been calibrated on the past behaviour of a system, we know that it

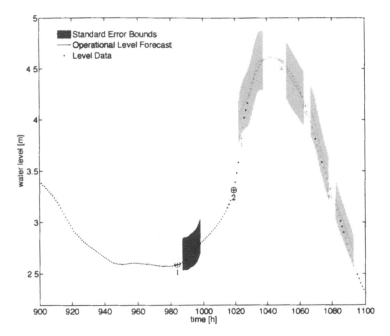

Figure 5.2 Flood forecasting on the River Severn, UK. 14 hour ahead forecasts of water level at Bridgnorth, with estimates of uncertainty (90% prediction limits) illustrated for different initial forecast times

Source: Peter Young, reproduced with permission

will not predict perfectly (for all the reasons discussed in Chapters 1 and 2), and we are often quite ignorant about the probability of future inputs. Indeed, it may not be possible to specify future inputs in terms of probabilities at all, in which case forecasting lead times may be limited to the natural time scale of response of the system. Flood forecasting is usually like this. There are techniques for forecasting rainfalls, but none of them are very accurate, and in Norway, for example, an operational forecasting system has been based on analogues from past events as a way of trying to assess possible (but uncertain) future rainfalls (see Skaugen et al., 2005). This might improve in future as the rainfall forecasts of atmospheric forecasting models improve but, as yet, they have some way to go.

The problem of future inputs is not such a problem where the forecast lead time required for a useful decision to be made is no greater than the natural time scale of the dynamics of the system. In flood forecasting, for example, as the scale of an upstream catchment area increases, so does the time delay between an input of rainfall and the peak of a flood at a point at risk of inundation. Thus, for catchment areas of the order of 500–1000 km^2, depending on the flashiness of the system, it may be possible to get forecasts with a lead time of six hours or more without needing to predict future rainfalls. For small catchments, this will not be possible, which is why it is still very difficult to provide flood warnings for flash floods on small catchments even though, in some parts of the world, they are one of the most important causes of death by natural hazards. Similar considerations of forecast lead time relative to the system response time will apply for other forms of natural hazard.

Here, we will consider situations where the natural time delays are sufficient to allow adequate lead times in making decisions, and where, as in Figure 5.2, inputs to the system and observations on the state of the system are available to the forecaster in real time, if only up to time "now" (the t_o of the forecast). Thus, past inputs to the system can be used to drive the forecast model and a comparison can be made between the predicted state of the system (here water level at a certain location) and the observation of that state. For this simple one variable example, it is immediately obvious as to whether the forecast model has been under-predicting or over-predicting and we would have an expectation that our future predictions at the required lead time might be improved if we can learn from this past behaviour. At the next forecast time step, we will be able to reassess any adjustments we have made in again forecasting to the required lead time. Predictions further into the future will be increasingly uncertain, particularly once predictions start to be made for longer lead times than the natural delay in the system, but might not be required for making decisions. Thus, any data assimilation or updating system should concentrate on trying to minimise the bias and the error variance of the forecasts at the required lead time. Given the sequence of residuals between the forecasts and the observations up to time now, there are two simple strategies that might be used in updating. The first is to let the forecast model run deterministically and try to model the residual error directly. The second is to update the parameters of the forecast model itself, in particular those that affect the gain of the model.

The problem of data assimilation becomes more difficult as more complex and nonlinear models start to be used and where it might be necessary to consider updating the internal states as well as the parameters of the model. In such cases, a real question arises as to whether the information content of the data being assimilated is sufficient to support the modification of both states and parameters, since there may be complex interactions between states and parameters in nonlinear and spatially distributed models. For many environmental applications this might be a very poorly posed problem mathematically and data assimilation algorithms have to be implemented with great care. After an introduction to the principles of least squares error correction, two types of techniques in operational forecasting use will be considered here, the updating techniques based on the Kalman filter (EnKF), and the three- and four-dimensional variational techniques that have been used widely in weather forecasting. For simplicity, the techniques are developed in the main text for a single scalar variable, with extension to the multivariate case in Box 5.1 and Box 5.2 respectively.

5.2 Least squares error correction models

In any data assimilation problem, as new observations are made available, we can compare the value of each observation with the corresponding prediction of a model. For the moment, we will assume that the observed and predicted values are directly comparable (commensurate), though, as we have noted before, this is not always the case. Consider first a single scalar comparison between an observed state and its model prediction (Figure 5.3).

Observed/Forecast Variable

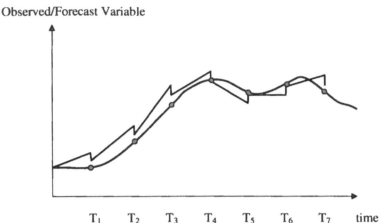

T_1 T_2 T_3 T_4 T_5 T_6 T_7 time

Figure 5.3 Updating of a forecast (with adaptive step at each time increment) as new observed data about the predicted variable (smooth curve) become available at times T_1 to T_7

They will almost certainly be different, suggesting that for real-time forecasting we might get better future predictions if we "correct" the model in some way depending on whether it is over-predicting or under-predicting. We will refer to this corrected set of states as the target for the correction procedure, S^*. A new estimate of a target state in the model can be obtained from a weighted correction based on the difference between the predicted or background state, S_b, and the observation, O, such that for the simple scalar case:

$$S^* = S_b + W(O - S_b) \qquad [5.1]$$

where S^* is the updated target value of state S, S_b is the background value, O is the observed value and W is a weighting coefficient. Equation [5.1] is a form of *innovation* equation and the difference $(O - S_b)$ is the *observation innovation*. Clearly the important question that then arises is how to determine the appropriate weight. Very often, the solution to this problem is based on minimising a squared error cost function. If the least squares function is minimised (or likelihood is maximised) the optimal weights will be given by

$$W = \sigma_b^2 / (\sigma_b^2 + \sigma_O^2) \qquad [5.2]$$

It can also be shown that the variance of S^* given by

$$\sigma^{*2} = (1 - W)\sigma_b^2 \qquad [5.3]$$

Since W and σ^{*2} depend on both the background variance and the observation variance, the ratio of these two variances acts as an important control on the damping of how much account is taken of the observations relative to the background values. The greater the observation variance relative to the background variance, the less effect the observation innovation will have on the new estimate of the state S^*. These equations provide a simple form of recursive estimation that are easily extended to the multivariate case (see Box 5.1). It is also worth noting that there is a direct analogy between this approach to data assimilation and the weighted least squares approach to model calibration outlined in Section 4.1 of the previous chapter.

These least squares equations can be applied directly to obtain optimal weights in cases where σ_b^2 and σ_o^2 can be assumed known. Unfortunately, this is rarely the case. We might have some idea of what the observation error might be on the basis of past studies in cases where the observed and predicted states really are commensurate, but the background variance of the predicted states is not so easily assessed and will certainly change as the forecast progresses. The Kalman filter and variational data assimilation techniques of the sections that follow provide ways of estimating the appropriate variances, including allowing for observations that might not be directly comparable with the predicted states of the model.

5.3 The Kalman filter

The Kalman filter (KF) derives from work by Kalman (1960) on the optimal control of linear systems. It is a way of *recursively* (step by step) updating a model as new data about the behaviour of the system are received. In the original control problem formulation it was the states of the system that were of most interest, under the assumption that the input data and the model were correct. The Kalman filter has also been used in updating the states of environmental models as a way of improving forecasts. It does so in a way that aims to maximise the correlation between observed and predicted residuals and minimise the forecast error variance. A generalised formulation of the Kalman filter method of data assimilation is based on writing the model in a discrete *state space* form:

$$\underline{S}_{t+1} = M[\underline{S}_t, \underline{U}_t, \underline{\Theta}] + \underline{\varepsilon}_t \qquad [5.4]$$

where the states at the next time step $\underline{S}_{t+1}$ are predicted by the model $M[\ ]$ as a function of the states at time t, a vector of inputs $\underline{U}_t$ at time t and past time steps, and a parameter vector $\underline{\Theta}$, assumed constant. To apply the Kalman filter the model $M[\ ]$ must be assumed to be *linear* (i.e. doubling the inputs to the model will produce exactly double the outputs). The term $\underline{\varepsilon}_t$ represents a dynamic model error term. In this case, the outputs of the model will be a linear function of the model states and the input forcing, conditional on the assumed values of the model parameters. It is then further assumed that the true value of an observable state, S_t^*, can be represented as a function of the model output and a random error such that, for any single state:

$$S_t^* = M[S_t] + \eta_t \qquad [5.5]$$

where S_t is the system state at time t and η_t is a random error, normally assumed to have zero mean and to be normally distributed (although it is worth noting that Kalman's original formulation was developed in terms of orthogonal projections and did not require an assumption of Gaussian errors, only an expectation of a symmetrical distribution of forecasting errors, see Young, 1984).

Then, as a new actual observation of a state, $\tilde{S}_{t_o}$ becomes available at time t_o, the forecast error from the true state can be represented in two ways, noting that the prediction at time t_o is a function only of the past states.

$$\tilde{S}_{t_o} - S'_{t_o} = M[S_{t_o}] - S'_{t_o} + \eta_{t_o} \qquad [5.6a]$$

or

$$\tilde{S}_{t_o} - M[S_{t_o}] = \eta_{t_o} \qquad [5.6b]$$

The Kalman filter then directly updates the forecasts to correct for this error using a two-step process, a *predictor step* followed by a *correction step*. In the predictor step, an updated estimate of the states is obtained using the innovation equation:

$$\hat{S}_{t_o + 1|t} = S_{t_o} + K_{t_o} (\tilde{S}_{t_o} - M[S_{t_o}]) \qquad [5.7]$$

where $\hat{S}_{t_o + 1|t}$ are new conditional estimates of the states and K_t, called the *Kalman gain*, is the equivalent to the simple weight of [5.2]. In the Kalman filter, however, it is no longer necessary to assume that the background variance is known since this can be estimated and updated as the forecasts proceed (see Box 5.1). The size of the Kalman gain will depend on the magnitude of the prediction error. It is also updated each time a new observation is available. The prediction is then updated in the correction step as

$$\hat{S}_{t_o + 1} = M[\hat{S}_{t_o + 1|t}] \qquad [5.8]$$

If a forecast is required more than one time step ahead then the prediction equations can be applied at successive time steps until the required lead time is attained (noting that this requires making some assumptions about the inputs during the lead time, since these have not yet been measured). The correction step can only be applied, however, as a new observation becomes available to improve the estimation of the states (and thereby correct for over- or under-prediction as in Figure 5.3). Extension of the Kalman filter equations to the case of multiple states will be found in Box 5.1.

5.3.1 Updating a model of the residual errors

When a complex model is being used for forecasting involving many parameters and many states, a way of applying the Kalman filter is to allow a forecasting model to run deterministically but then to use a simple stochastic model of the residual errors to update the forecast in real time. The Kalman filter can then be used to update the model of the errors as each new observation becomes available. This is the simplest case, because very often there will only be a single state variable at each forecast site (effectively the deviation from the prediction of the deterministic forecast model). There have been a number of operational hydrological forecasting systems that have used this approach in the past.

The approach demands, however, an appropriate model of the errors. This is often taken as the form of an autoregressive model (AR) described by Box and Jenkins (1976). AR models are linear models. Mathematically, an AR model can be written in terms of a weighted sum of the past residual errors. Thus, in this case, the model forecast at the lead time t_L will be the sum of an output from the forecast model and an estimate of the error at lead time $t_o + L$ which will be a function of the known past residuals up to time t_o:

$$\hat{\varepsilon}_{t_o + L} = a_o + a_1\varepsilon_{t_o} + a_2\varepsilon_{t_o - 1} + \dots a_N\varepsilon_{t_o - N + 1} + \eta_{t_o} \qquad [5.9]$$

where η_{t_o} represents a purely random component. This has the form of [5.5] above and can easily be implemented in the Kalman filter. Note that the prediction of the residual error at t_L can only be a function of the residuals at t_o and earlier, since for time steps between t_o and t_L no information about the magnitude of the residual is available since no observations have yet been received. This equation has a set of parameters that must be estimated prior to making a forecast. Normally a study of the forecast behaviour in past events allows such a calibration. Such a study could also show that the residual errors might be a function of some other input variables and, in this case, the model of the residuals might be extended to an ARMA (autoregressive moving average) model or transfer function (see Young, 1984). This is also a linear model if the model coefficients are assumed constant. The states that are modified in the Kalman filter in these cases are the past residuals, ε, between the observations and values predicted by the deterministic model.

5.3.2 Updating the gain on a forecasting model

In most forecasting problems it will not always be a good assumption that the model is correct and that the forecast residuals are only a result of not knowing the true values of the system state. Very often we suspect that both the input data and the parameters of the model might also be in error and there is no reason why the parameters of the model should be constant from event to event (unless we could demonstrate that this is a good assumption by analysis of a large number of past events). Thus, in many forecasting problems, simply updating the model of the deviations between a deterministic forecast and the observations may not be sufficient if the model itself is not that accurate a predictor (for all the different reasons considered in previous chapters).

Fortunately, given that the number of observations available for data assimilation is often limited (though see the weather forecasting application in Section 5.4 below), it does not appear necessary to correct all sources of model error (inputs, parameters and states) in the updating. We can take advantage of the fact that many of these sources of error interact in producing a particular forecast error. For example, an observed input quantity that underestimates the true input to the system could be compensated by an increase in a model parameter that controls the gain in the system such that, if S_t is the forecast *output* state from a model, it is assumed that this can be corrected to the true output by a gain multiplier, G_t, together with some additive error. Thus:

$$S_t^* = G_t(S_t) + \eta_t \qquad [5.10]$$

This again has the form of [5.5] and can be easily implemented in a simple scalar Kalman filter. The approach was used for some years in the Dumfries flood warning system on the River Nith catchment in Scotland (Lees et al. 1994; see also Beven, 2001a) with transfer function models in predicting rainfall-flow, upstream to downstream flow routing, and tidal influences (with both upstream and seaward inputs to predict water levels downstream of the tidal limit). They applied [5.10] to the outputs from the transfer function model. They also implemented another forecasting "trick" that might be useful in other circumstances. The authority that commissioned the flood warning system required a six hour lead time, but the data suggested that the natural lead time of the system was only five hours. By fitting transfer function models with an "artificial" time lag of six hours, *simulation* of the system was not so good, but the six hour ahead *forecasts*, with the KF updating, could be improved.

In [5.10] the variable that is updated in the KF is the gain coefficient, G_t, which is treated as a single state in the filter algorithm This will have an initial value of unity, but after that is allowed to evolve as a random variable, with values that will reflect whether the model is over- or under-predicting the observed variable. Thus, given an estimated value of G at time t_o, [5.7] can be written to update G as

$$\hat{G}_{t_o+1} = G_{t_o} + K_{t_o} (\bar{S}_{t_o} - G_{t_o} (S_{t_o}))$$
[5.11]

In updating the Kalman gain (equations [5.9] to [5.11]), no estimate of the variance σ_η^2 is readily available, but this can be treated as a tuning parameter of the updating algorithm. It is effectively a *noise-variance ratio* (NVR) and the larger the value, the slower the algorithm will change the gain in response to the observation innovation. Small values on the other hand will result in a very rapid response. The NVR can therefore be tuned for the system of interest to give a suitably damped response to prediction errors given the time step and response time of the system. A good summary of the application of the gain updating approach to the flood forecasting problem is given by Young (2002).

5.3.3 Case study: Flood forecasting on the River Severn

The approach has been extended by Romanowicz et al. (2006) in an application to a flood forecasting system for the River Severn, UK (Figure 5.4), to allow for the fact that the relationship between rainfalls and flood water levels might be highly nonlinear. The approach is an extension of the Dumfries forecasting system (Lees et al., 1994) and uses a cascade of linear stochastic transfer function models for modelling rainfall to river level in the upstream parts of the catchment and for routing the flood wave in the downstream river reaches. The stochastic transfer functions take the form (Young, 1984, 2002; Taylor et al., 2007):

$$y_t = \frac{B(z^{-1})}{A(z^{-1})} u_{t-\delta} + \varepsilon_t$$

where y_t is the forecast water level at time t, $u_{t-\delta}$ is an input at the time $t-\delta$, δ is a time delay, ε_t is a stochastic error term, and $A(z^{-1})$ and $B(z^{-1})$ are polynomials in the backward shift operator, z^{-1}, where $y_t z^{-1} = y_{t-1}$, of the following form:

$$A(z^{-1}) = 1 + a_1 z^{-1} + a_2 z^{-2} + \ldots + a_n z^{-n}$$

$$B(z^{-1}) = b_o + b_1 z^{-1} + b_2 z^{-2} + \ldots + b_m z^{-m}$$

Past experience suggests that normally only second-order models are required for flood forecasting (with 2 *a* parameters and 1 to 3 *b* parameters) for these types of forecasting problems. A model with 2 *a* parameters can often be decomposed into two flow pathways, one representing a fast component, the other a slow component (Young, 1984, 2002; Beven, 2001a). The linear transfer function models are implemented using the Kalman filter with updating of the fast and slow model states as each new water level observation becomes available in real time.

There are a number of features of the River Severn flood forecasting system that make it interesting. The first is that it forecasts water levels directly, not river discharges (there is generally a nonlinear relationship between water level and discharge because of the shape of the river cross-section and the way in which river velocities change with water level, particularly when the river starts to flood). In hydrological terms this means that there is no requirement to maintain any mass balance, either in the rainfall–river level components or the flood wave routing components. The transfer function models are not therefore attempting to simulate the flow processes in any way, only to mimic the response of the system. In physical terms, this ought to be a disadvantage; there is no reason why a rainfall to water level model should work well when it ignores the basic hydrological principle of mass balance.

In practice, however, this approach can actually produce good forecasts for a number of reasons. One is that the mass balance principle is difficult to apply since there are known to be errors in estimating the rainfall inputs over a catchment area from either rain gauge or radar rainfall data. There are also known to be errors in estimating river discharges from measurements of water levels, particularly at flood flows.

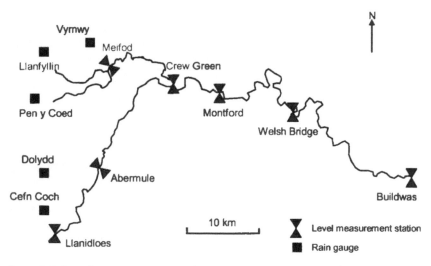

Figure 5.4 River-flow gauging sites and raingauge sites used in the flood forecasting model for part of the River Severn catchment, UK. The Welsh Bridge site is in the middle of the town of Shrewsbury which has been subject to a number of floods in recent years

Water level is, however, a direct observation with relatively small error and is also the variable of interest in flood forecasting since it is water level that determines whether a flood embankment will be overtopped or whether houses will be flooded.

A second feature of interest is that the system recognises that the relationship between rainfall and water level and also in routing the flood wave in the downstream reaches is nonlinear. It handles this by defining an optimal transformation of the *inputs* to the stochastic transfer functions during calibration and by adjusting a model gain at each time step. The gain adjustment is implemented using a form of the recursive least squares algorithm discussed in Section 5.1.1 (see also Young, 2002). It is controlled by a noise-variance ratio (NVR) parameter that is also optimised during calibration.

Finally the model also takes account of the fact that the forecast errors tend to increase with the magnitude of the forecast (i.e. the errors are **heteroscedastic**). Heteroscedastic errors might be expected for good physical reasons. The prediction of water levels in a flood for example would be expected to be more uncertain as flood waters go overbank in high-magnitude events. A simple function form for the heteroscedasticity is used, such that

$$\sigma_t^2 = \lambda_o + \lambda_1 y_t^2$$

where σ_t^2 is the variance of the stochastic error term, ε_t, which now varies with the magnitude of the predicted output, y, and λ_o and λ_1 are hyper-parameters optimised during calibration. Some typical results of the forecasting system for a test event are shown in Figures 5.5 and 5.6. In Figure 5.5, five hour ahead forecasts are shown for an upstream site that involves only a rainfall to river level forecast. In Figure 5.6, 35 hour ahead forecasts are shown for the downstream site at Buildwas that is the result of the cascade of rainfall-river level and flood wave routing models. Romanowicz and Beven (1998) showed how this form of transfer function forecasting methodology could also be used to constrain the predictions of a hydraulic model of inundation on the River Severn within the GLUE methodology of Chapter 4. In principle, this approach could be used to provide inundation predictions in real time.

5.3.4 The Extended Kalman filter

The Kalman filter can be extended to deal also with parameter and state updating in (at least mildly) nonlinear models. To do so requires a linearisation of the forecast variance (and covariance in the multivariate case), around the nonlinear model predictions between the times t and $t - 1$. In this case we can write (to first order) that the desired target set of states at time t, S^*, in the model $M(S)$ will be a perturbation of the current estimated states of the form:

$$M_t(S^*) = M_{t-1}(S + \delta S) = M_{t-1}(S) + L_{t-1} \delta S \qquad [5.12]$$

where L_{t-1} is a *Jacobian* gradient matrix, evaluated at time $t - 1$, which has elements

$$L_{t-1} = \partial M_{t-1}(S)/\partial S_i \qquad I = 1,2,\ldots N_s \qquad [5.13]$$

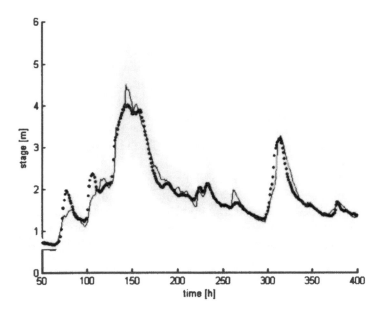

Figure 5.5 Observed (dotted line) and predicted water levels from rainfall-water level modelling at Abermule on the River Severn. This is a test event not used in calibration. The forecasts are five hour ahead forecasts, with state updating using the Kalman filter and gain updating using a recursive least squares algorithm. The shaded area represents 95% prediction limits around the model predictions

Source: Romanowicz et al., 2006, Copyright ©2006 American Geophysical Union, reproduced with permission

where N_S is the number of states. The matrix L extrapolates any perturbation of a state in the model to the target state. The transpose, L^T, can be used to extrapolate a target state at time $t - 1$ back to its equivalent in the model at time t. L^T is also known as the *adjoint* matrix. It is assumed that the same operator can be applied in both forward and backward extrapolation. The Extended Kalman filter (EKF) is then developed by again assuming that the observations have random errors with zero mean and known covariance.

The data assimilation process is then set up as an optimisation problem to minimise a cost function of the squared perturbations (see Box 5.1 for more details). The solution is a set of states, as far as possible consistent with the model dynamics, with the advantage that this should minimise any non-physical instabilities in the solution as a result purely of the data assimilation when it is propagated forward over the next forecasting period. In practice, additional constraints are often added to the optimisation problem to satisfy global balance equations in the model target solution.

This also, however, gives rise to a problem in that all the different potential sources of error might interact in nonlinear ways. Thus, the Extended Kalman filter has some limitations for practical applications. Because it depends on linear extrapolations in any particular forecast step, it is limited to models of mild nonlinearity. If the nonlinearity is too strong there will be a tendency to generate numerical instabilities in the EKF algorithm. It is also limited to cases where the number of states and parameters

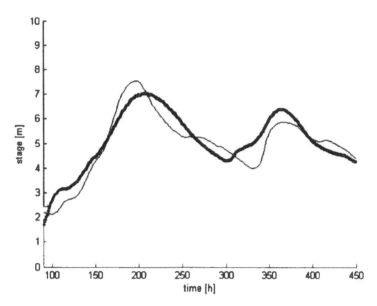

Figure 5.6 Observed (black line) and predicted water levels at Buildwas, downstream of Shrewsbury on the River Severn. This is a test event not used in calibration. The forecasts are 35 hour ahead forecasts, resulting from a cascade of rainfall-water level and water level to water level models in the reaches of the River Severn, with state updating using the Kalman filter and gain updating using a recursive least squares algorithm. The shaded area represents 95% prediction limits around the model predictions

that are allowed to vary is not too great. Otherwise the Jacobian and adjoint matrices become expensive to compute and updating the covariance matrix also becomes computationally demanding. Stability problems will also increase. That is why it has been more popular in hydraulic, atmospheric and oceanographic models, which may have very large numbers of states, to use either variational methods (see Section 5.3) or, more recently, Ensemble Kalman filter methods.

More details of the technical background to the Extended Kalman filter can be found in Box 5.1, including how the error variances of the forecast are calculated in the updating. Applications include updating of soil moisture data (Walker et al., 2001) and parameter estimation in complex aquifer systems (Yeh and Huang, 2005).

5.4 The Ensemble Kalman filter

The Ensemble Kalman filter (EnKF) has been developed to cope with more highly nonlinear models by allowing different sources of error to propagate through the model by using a Monte Carlo sampling strategy. In this way, any effect of the nonlinearities on the way in which the errors affect the forecasts is treated directly in the outputs from the sample of models. At each updating step a form of the Kalman filter prediction correction equations is used to update the distribution estimates for states and/or parameters that are being allowed to vary. This provides the basis for a new Monte Carlo sample for the next forecasting step. In most applications of the EnKF the distributions are assumed to be multivariate Gaussian, requiring the estimation and updating of a full variance–covariance matrix. The concept was first outlined for a weather forecasting example by Evensen (1994), and has since been used in weather forecasting by Burgers et al. (1998), Houtekamer and Mitchell (2001) and Houtekamer et al. (2005); in hydrological modelling applications by Moradkhani et al. (2005a), Vrugt et al. (2006) and Vrugt and Robinson (2007b); in hydraulic models by Madsen and Cañizares (1999), Sørensen et al. (2004) and Weerts and El Serafy (2006); and in land surface to atmosphere flux estimation by Margulis et al. (2002), Reichle et al. (2002) and Crow and Wood (2003). Vrugt and Robinson (2007b) compare the EnKF approach (using updating of an ensemble of a single model structure) with Bayesian model averaging (BMA, which uses several different model structures but does not update the likelihood distributions for each model over time) and conclude that for the rainfall-runoff example they considered, the EnKF performs better than the BMA method.

Details of the extension of the Kalman filter to the ensemble case are also given in Box 5.1. The success of the EnKF methodology depends on the success of the sampling methodology and the number of samples in the ensemble. Recent work has suggested that significant improvements in the data assimilation can be achieved by changes to the sampling strategy (Evensen, 2004, 2006), at least where model errors are small. Clearly there are computational advantages in using as small an ensemble as possible, particularly where computational constraints may be important in actual real-time forecasting. A small ensemble, however, may not adequately represent the range of variability that should be explored in a nonlinear modelling problem.

The results of the ensemble propagation are also analysed under multivariate Gaussian assumptions. This will not always be a good set of assumptions and the more general technique of Particle Filtering (see Section 5.5.1 below) has been developed to allow the Gaussian assumptions to be relaxed. Although this allows greater flexibility, it will normally require a much greater number of ensemble members and it appears as if it is not always advantageous in making operational forecasts for environmental problems (see the Case Study of Section 5.5.1).

5.4.1 Case study: Application of the Ensemble Kalman filter to the Leaf River Basin

An application of the Ensemble Kalman filter to rainfall-runoff modelling in the Leaf River catchment, Mississippi, was reported by Moradkhani et al. (2005) using the hydrological model HYMOD. Figure 5.7 shows a schematic of the model structure used, together with the parameters and storages. They propose the EnKF as a consistent framework for dealing with the limitations of both model structures and input data in hydrological forecasting, and the possibility that the parameters representative of a catchment area might change over time. They implement the EnKF as a dual filter, one to update the five model parameter values followed by one to update the state variables of the model which are here five storage variables. The model is implemented on a daily time step and it is assumed that the current flow in the river is known at each

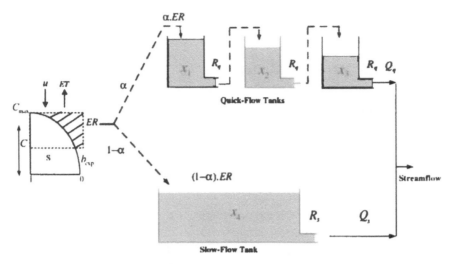

Figure 5.7 Schematic of the HYMOD catchment model

Source: Moradkhani et al. (2005), reproduced with permission of Elsevier

day in predicting the flow at the next day. They are consequently proposing a one time step ahead prediction system rather than a simulation model of the catchment watershed (when the uncertainties in prediction would be expected to be much larger).

Figure 5.8 shows the evolution of the parameter values over three years of daily updating. The confidence bands in this figure are not directly from a single ensemble prediction run but are the distributions of the mean parameter values for ensembles of 50, taken over 500 different realisations.

Figure 5.9 shows the predicted uncertainty for the storage, S, in the nonlinear runoff component of the model. Figure 5.10 shows the one day ahead predictions, with uncertainty bounds.

5.4.2 The Ensemble Kalman smoother

A further extension of the EnKF is the Ensemble Kalman smoother (EnKS) introduced by Evensen and van Leeuwen [2000] as an improvement over an earlier ensemble smoother due to van Leeuwen and Evensen (1998). The EnKS aims to update not only the current states in the model but also states at past time steps. It uses the EnKF as a first step and should never provide results that are worse (in terms of accuracy and error variance of the forecasts) than the EnKF. In practice, the EnKS is not generally used to update all past states in a modelling problem but only states over some fixed lag into the past associated with the decorrelation time of the system of interest, since smoothers will only be of benefit if there is correlation in the effect of perturbations to the system model in making the forecasts.

The EnKS has been used in a variety of environmental applications, including ecology (Gronnevik and Evenson, 2001); oceanography (Brusdal et al., 2003) and the use of radar remote sensing to update soil moisture estimates in improving predictions of land surface to atmosphere fluxes (Dunne and Entekhabi, 2005, 2006).

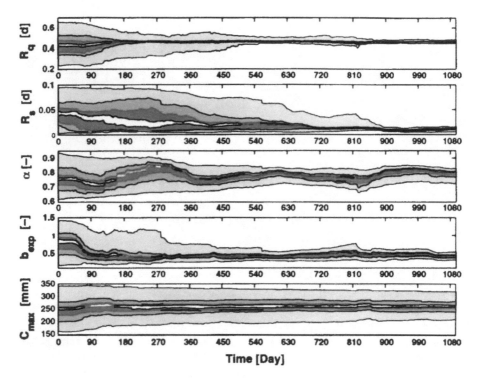

Figure 5.8 Time evolution of the values of the HYMOD parameters as applied to the Leaf River
catchment after three years of daily EnKF updating. Shaded areas correspond to 95%, 75%,
68% and 10% distribution limits calculated over 500 realisations of 50 member ensembles

Source: Moradkhani et al. 2005, reproduced with permission of Elsevier

5.5 The Particle filter

Over the same period an alternative Monte Carlo methodology called Particle filtering (PF) was being developed. Particle filtering is a form of Bayesian learning process
in which the propagation of all uncertainties is carried out by a suitable selection of
randomly generated ensemble members, but the statistics of the resulting distributions
of errors are based purely on the sample statistics without any assumptions about the
nature of the distributions. This method does not therefore require any assumptions
about the prior distributions of the states and/or parameters; the specified prior is
simply sampled and the resulting model realisations (particles) used to create a distribution of forecasts. With a very large sample of realisations the resulting distribution
of forecasts should approximate the posterior distribution well. Each sample is associated with a weight. At the first time step the sampling can be carried out so that the
weights of all the realisations are equal. As new observations are made available at the
next forecasting realisation, the weights can be updated, essentially using Bayes equation (see Box 4.1), to reflect how well each sample was modelling each observation.
The models are then propagated to the next forecast step and the weights re-evaluated
again.

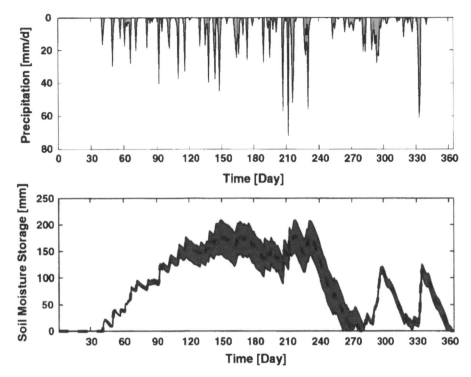

Figure 5.9 Daily rainfalls (upper plot) and evolution of storage variable S after daily EnKF updating in the HYMOD application to the Leaf River catchment after daily updating (lower plot). Shaded area represents 95% prediction limits, dotted line the mean of the ensemble

Source: Moradkhani et al. 2005, reproduced with permission of Elsevier

In principle, this is a very flexible technique. In practice, it suffers from the difficulty that, given the computing constraints of the number of particles that can be used in a sample of the prior, very often the weights of most of the particles becomes very low and the posterior distribution is dominated by one or a very small number of the original large sample of particles. This will happen especially where the specified uncertainty on the observations is small relative to the uncertainty on the prior distribution of particles at any step so that the constraints imposed are very strong. Thus, the PF method has been modified to allow a re-sampling step at each forecast updating. Sequential importance re-sampling was introduced by Gordon et al. (1993); Residual re-sampling by Liu and Chen (1998); and re-sampling using Monte Carlo Markov Chain methods (see Box 4.2) by Doucet et al. (2001). The aim in each case is to improve the representation of the posterior distribution of model states and/or parameters at each time step, when propagating forward to the next forecast. A review of different Particle filter methods is provided by Arulampalam et al. (2002).

Environmental applications of the Particle filter include Moradkhani et al. (2005b), Weertz and El Sarafy (2006) and Smith et al. (2008) in applications to rainfall-runoff modelling problems.

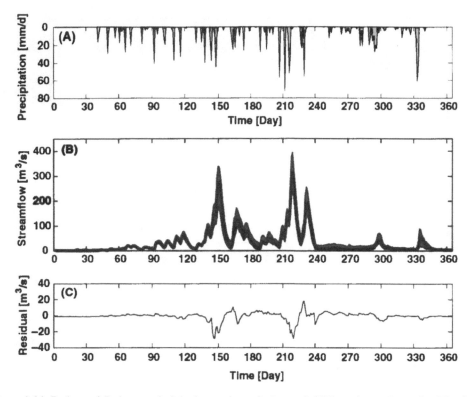

Figure 5.10 Daily rainfalls (upper plot), hydrograph prediction with 95% prediction limits (middle plot) and model error (ensemble mean – observed) (lower plot) after daily EnKF updating in the HYMOD application to the Leaf River catchment after daily updating (lower plot)

Source: Moradkhani et al. 2005, reproduced with permission of Elsevier

5.5.1 Case study: Comparison of EnKF and PF methods on the River Rhine

An application of the two variants on the PF method to flow forecasting in different sub-catchments of the River Rhine basin has been presented by Weerts and El Serafy (2006) in a comparison with the EnKF. The two PF methods were Sequential Importance re-sampling (SIR) and Residual re-sampling (RR). The hydrological model used was the HBV-96 model using an hourly time step, including a snowmelt component so that precipitation, evaporation and temperature data are all required as inputs. The model had previously been calibrated on this catchment for flood forecasting purposes.

The three methods were applied to both a synthetic data set that was generated using the same model structure, with random perturbations to the precipitation inputs, evapotranspiration estimates, temperature and model generated discharges, and a real data set for the Nahe 1 sub-catchment down to the gauging station at Martinstein. In the HBV-96 model the snowmelt routine uses different calculations for different elevation zones. Two vegetation types were also used in setting up the model.

This gives a total of 58 state variables to be updated in running the model. The calibrated parameter values were assumed fixed. The synthetic experiment allows the performance of the various updating techniques to be assessed, in particular to see how far the 'true' values of the inputs and discharge vary for different numbers of ensemble members.

Most of the results described in the paper refer to the synthetic experiment where model structural error is not an issue. Even so, it was found that both the PF methods would often collapse to a single particle, particularly for a small number of initial ensemble members and small variances of the random perturbations. This does not happen with the multi-Gaussian assumptions of the EnKF. Identification of the true discharge is relatively successful with all the methods; identification of the true precipitation less successful until the catchment is wet and there is a direct response to new inputs. Identification of the true evapotranspiration input was less successful, while identification of the true temperature was only possible when the accumulation and melt of snow was sensitive to temperature at around zero degrees. Even then the ability to estimate the true temperature was lost when the error variance on the discharge 'observations' was high. The potential to identify the true states in the model also varied with the hydrological state of the catchment.

Figure 5.11 shows the results of using the three different updating techniques in applying the HBV-96 model to the real data set. Only the ensemble mean predictions for updating after each hourly time step are shown (the article does not show the uncertainties associated with the predictions for this real case). In this case, using 32 particles in each case, the root mean square error is lower in the case of the EnKF, but the article points out that to get a better forecast the EnKF allows the updated input variables to drift beyond the range of the variances specified in the application of the other methods.

The conclusion of this study was that for a small number of model samples the EnKF generally outperformed the PF techniques which were less robust to the choice of prior assumptions about the forms of the distributions of errors in model states and inputs.

5.6 Variational methods

It is well known that the earliest experiments in numerical weather forecasting were carried out by Lewis Fry Richardson (1881–1953) early in the 20th century (e.g. Richardson, 1922). Richardson used a large hall full of computers (who, at that time, were real people doing computations by hand) to carry out the calculations necessary at each point of a gridded discretisation of the atmosphere. Initial conditions for every calculation node were estimated from a limited amount of observations. It is also well known that, although Richardson's computers were using discrete forms of the dynamic equations that were not so different to those used now, the calculations rapidly became unstable. The equations were then, as they are now, highly sensitive to having dynamically correct initial conditions and any small errors in the calculations.

The same applies to similar dynamic predictions of the atmosphere and oceans today. The computers are now digital, the calculations are much more accurate, there are many more observations on which to base the initial conditions for each forecast so that the times scales over which forecasts can be useful have been greatly extended ... but still only to a few days. And it is still valuable to use a data assimilation process

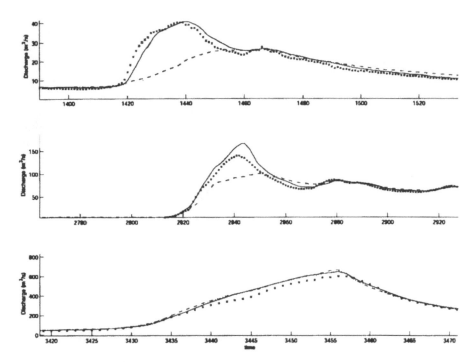

Figure 5.11 Results of applying the HBV-96 model to the Nahe sub-catchment of the Rhine basin for the period 1 Sep 1994 to 31 Jan 1995. Observed discharge (crosses) is compared with ensemble mean prediction using SIR updating (dot-dash line), RR updating (dotted line), EnKF updating (solid line). Updating applied at each hourly time step for 13 hour ahead predictions

Source: Weerts and El Safary, 2006, Copyright ©2006 American Geophysical Union, reproduced with permission

to reinitialise the states of the atmosphere every time a new set of observations becomes available. In the ECMWF, for example, the weather forecasting ensembles are reinitialised every 12 hours before a new forecast is made.

The problem that then arises is that there are many more nodal calculation points (of the order of 10^7 points in space in current numerical weather prediction models) than there are observations (still large but an order of magnitude or so less). Yet the initial conditions must be consistent with the dynamic equations so as not to introduce numerical instabilities as a result of misspecification of the initial state of the atmosphere. This is again, therefore, a process of learning from the discrepancies between the prior observations and the observations (at the large, but limited, number of sites where they are available), but this time in a very high-dimensional state space indeed. The challenge therefore is to produce a data assimilation algorithm that is consistent (as far as possible) with the dynamic equations.

A number of techniques have been developed to address this data assimilation problem, including the use of the Ensemble Kalman filter discussed earlier. Reviews of different types of data assimilation techniques for such large scale problems may be found in Ghil and Manalotte-Rizzoli (1991), Courtier et al. (1993) and Talnay (2003).

The technique used in most operational weather forecasting centres is that of *variational data assimilation*. To understand variational data assimilation methods it is necessary to go back to the concept of a cost function used in the Kalman filter techniques in Box 5.1. Given an estimate of the state of the atmosphere, together with a set of observations, we want to minimise some function of the differences between predicted state and observations. One of the difficulties that arises at this point is that, as well as having more model states than observations, the available observations may be only indirect estimates of variables that occur in the model (for example, there may be *commensurability errors* as described in Section 1.3.3, often called *representational errors* in atmospheric modelling). This may be because they represent different scales, or because they have been taken at different times, or because, like many observations based on satellite remote sensing, they need some additional interpretative model to be useful in conditioning the states to be updated.

The approach taken is similar to the Extended Kalman filter but in the variational approach a different cost function is minimised. This recognises that there is a difference between what is observed and the state variables of the model. A linear operator is defined that will transform a model variable into an observed variable. This is normally defined in terms of the gradient of the dynamic equations, linearising around the background (generally model-predicted) values of the variables of interest. This is implicitly assuming that the model is close to the true state of the atmosphere so that only minor (linear) perturbations will be necessary in the data assimilation.

Three-dimensional variational data assimilation (3DVar) treats the assimilation as if all the observations were available at the same time as model predictions so that it is a problem only of interpolating the innovations in the state variables in space (e.g. Lorenc, 1986). A modification of the approach (4DVar) allows for the fact that the observations may not always be taken at the same time as a model state is estimated. The cost function then includes a component allowing for the time distribution of observations. To do this most simply requires that the model be assumed true (see Box 5.2).

4DVar assimilation is now used in many numerical weather prediction (NWP) schemes, with an ever-increasing number of ground and rawinsonde data. The incorporation of the observation error component in 3DVar and 4DVar schemes also allows the use of data that do not directly estimate model state variables, and, most importantly, the use of satellite remote sensing data and ground based radar. These images generally require pre-processing before the digital numbers are useful and may require an interpretative model before an estimate of the difference from a model state variable can be given. The introduction of remotely sensed data has, however, been extremely important in improving the accuracy of NWP forecasts. More detail on variational data assimilation will be found in Box 5.2. A comparison of 4DVar with the EnKF is demonstrated by Caya et al. (2005) in an application to a cloud resolving atmospheric model over short time scales during the development of a storm. Data for assimilation came from Doppler radar imaging of the storm, available every 10 s. They showed that over short modelled times up to 10 minutes 4DVar gave the best analysis results but for longer times, the way in which the EnKF allowed the error covariance matrices to evolve made it advantageous.

3D and 4DVar have also been used in other areas of environmental modelling. Boni et al. (2001) and Reichle et al. (2001), for example, have used it in assimilating remotely sensed brightness temperatures in updating land surface soil moisture to improve predictions of latent and sensible heat fluxes to the atmosphere; Seo et al. (2003) in flood forecasting, and Yang and Hamrick (2002) for a distributed tidal transport model of salinity variations in an estuary.

5.7 Ensemble methods in weather forecasting

A recent innovation to NWP has been the introduction of ensemble forecasts. In ensemble NWP a number of models are run in parallel (Figure 5.12; see also Figure 5.1 above). Each member of the ensemble is reinitialised following the data assimilation step, but they differ in the initial conditions in a way that is consistent with the estimated covariance of the model states following data assimilation. The runs will therefore all make different predictions into the future. For some states of the atmosphere, the different ensemble members might give quite similar results, but under many conditions they will evolve quite different forecasts after one, two or more days into the future (as in Figure 5.1). In particular, different ensemble members might give quite different predictions of extreme winds, rainfalls or other variables of interest. This gives the forecasters some idea of how confident they should be in the forecasts, even if it is not possible to associate the individual ensemble members with any estimate of probability once they evolve away from the initial states.

In actual forecasting practice, somewhat more sophisticated perturbations to the initial conditions are used, since early studies showed that simple random perturbations did not diverge away from a control simulation as quickly as the real atmosphere (e.g. Toth and Kalnay, 1993). Two methods are currently used operationally to improve the spread of the ensemble within a relatively small number of runs. These are 'breeding' the fastest-growing deviations from the control, and singular vector methods that identify the maximum energy growth in regions of interest over some control period prior to the forecast lead time (see Kalnay, 2003, for more details).

A recent extension to the ensemble forecasting concepts has been the use of perturbations of the parameterisations in an atmospheric model as well as the perturbations of the initial conditions assuming the prediction model is true. This was pioneered by Houtekamer et al. (1996). At the ECMWF such model uncertainty has been implemented as a 'stochastic physics' ensemble technique by Buizza et al. (1999). This increases the spread of the ensemble but, by taking at least some account of model error, seems to lead to somewhat greater forecasting skill particularly for variables such as precipitation. Initial implementation was somewhat crude, with multipliers on parameterised components uniformly chosen between 50% and 150% of the control prediction.

5.8 Summary of Chapter 5

This chapter has looked specifically at the issues that arise in making forecasts in real time where it possible to learn about and correct for forecast errors using observations available in real time. Important examples include weather forecasting and flood forecasting. This has very specific requirements of trying to minimise the uncertainty in the forecasts at the lead time required by decision makers who will use the forecasts.

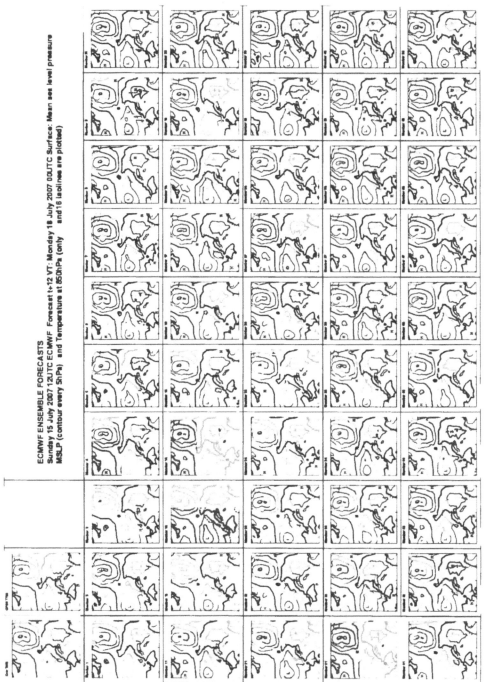

ECMWF ENSEMBLE FORECASTS
Sunday 15 July 2007 12UTC ECMWF Forecast t+12 VT: Monday 16 July 2007 00UTC Surface: Mean sea level pressure
MSLP (contour every 5hPa) and Temperature at 850hPa (only and 16 isolines are plotted)

Figure 5.12 ECMWF control and ensemble forecasts of mean sea level pressure fields for western Europe and North Atlantic. Forecasts for 0000 UTC on 16th July 2007 made at 1200 UTC on 15th July 2007 (reproduced by authorisation of ECMWF)

- Classical techniques of the Kalman filter and Extended Kalman filter can be used to update the states and/or parameters for relatively simple forecasting systems, such as modifying the gain on transfer function models.
- Modern computing power has allowed the development of the Ensemble Kalman filter and Particle filters that can be applied to quite nonlinear models.
- Forecasting with highly distributed models, such as atmosphere and ocean models, have generally made use of the local linearisations of 3D and 4D variational data assimilation.
- This is an area of current interesting developments, including the routine assimilation of satellite remote sensing images into global modelling systems. Such images do not always provide the data required by a model directly, however, and may require an (uncertain) interpretative model to be useful.

Box 5.1 Kalman filter methods for data assimilation [1]

B5.1.1 Recursive least squares
In understanding the background to the Kalman filter it is helpful to look first at the simpler recursive least squares algorithm as a way of implementing the least squares error correction of Section 5.2 for a simple scalar case. Here we will show how least squares estimation can be generalised to the multivariate case for a model expressed in the discrete *state space* form[2]:

$$\underline{S}_t^* = M[\underline{S}_{t-1}, \underline{I}_{t-1}, \underline{\Theta}] + \underline{\varepsilon}_t \qquad \text{[B5.1.1]}$$

where the vector of true states at the lead time of interest $\underline{S}_t^*$ is predicted by the model $M[\ ]$ on the basis of the states S at time $t - 1$, a vector of inputs $\underline{I}$ up to time $t - 1$ and a parameter vector $\underline{\Theta}$, assumed constant. The vector $\underline{\varepsilon}_t$ represents a dynamic model error term that represents the difference between the model estimated forecast and the true state at time t.

In the general case, of course, the model in [B5.1.1] could be nonlinear, but initially we will consider only the case of a linear model, analogous to a linear regression equation, such that the prediction of some observable, y_t, is the sum of a number of perfectly known, linear independent variables, $\underline{h}$, multiplied by coefficients (the vector of coefficients, $\underline{x}$) and corrupted by some random noise, e_t. Thus,

$$y_t = x_1 h + x_2 h_{t-1} + \dots x_n h_{t-n-1} + e_t$$

and in vector form we can write

---

1 The material of Box 5.1 depends heavily on the writings of Peter Young (1984, 2002) who is also thanked for checking the presentation here.
2 In the equations of Box 5.1 and Box 5.2 an underlined variable or the use of curly braces, { }, represents a single-dimensional vector, and a bold capital letter a two-dimensional matrix (see Matrix Algebra Appendix). The notation E< > represents an expected value over variables within the angled brackets.

$$y_t = \underline{h}_t^T \underline{x} + e_t \qquad\qquad \text{[B5.1.2]}$$

where T indicates a vector/matrix transpose and

$$\underline{h}_t^T = [h_t, h_{t-1}, \ldots h_{t-n-1}]; \quad \underline{x} = \left\{ \begin{matrix} x_1 \\ x_2 \\ \vdots \\ x_n \end{matrix} \right\}_t \qquad\qquad \text{[B5.1.3]}$$

We wish to solve this stochastic problem to obtain values of the (stationary) parameter vector, $\underline{x}$. We will assume that the parameters are linearly independent and uncorrelated with the error term. This can then be achieved by minimising a cost function, J, equal to the sum over $k = 1,2,\ldots,t$ samples of the squared deviations of the prediction, $\underline{h}_t^T \underline{x}$ from the observed value $\underline{S}_t$

Thus

$$J = \sum_{k=1}^{t} \left[y_k - \underline{h}_k^T \underline{x} \right]^2 = \sum_{k=1}^{t} e_k^2 \qquad\qquad \text{[B5.1.4]}$$

J will be minimised when its partial derivatives with respect to each of the $\underline{x}$ parameters are equal to zero or when

$$-2 \sum_{k=1}^{t} \underline{h}_k y_{kt} + \left\{ 2 \sum_{kt=1}^{t} \underline{h}_k \underline{h}_k^T \right\} \underline{x} = 0$$

or

$$\underline{x} = \sum_{k=1}^{t} \underline{h}_k y_k \left[\sum_{k=1}^{t} \underline{h}_k \underline{h}_k^T \right]^{-1} \qquad\qquad \text{[B5.1.5]}$$

This solution can be implemented recursively (i.e. updating the solution as each new observation becomes available at a new time t) in the following form. Let $\hat{\underline{x}}_t$ be the estimate of the "true" parameters, $\underline{x}$, at time t. Then [B5.1.5] can be written in the form

$$\hat{\underline{x}}_t = P_t \underline{b}_t \qquad\qquad \text{[B5.1.6]}$$

where $P_t = \left[\sum_{k=1}^{t} \underline{h}_k \underline{h}_k^T \right]^{-1}$ and $\underline{b}_t = \sum_{k=1}^{t} \underline{h}_k y_k.$

If the further assumption is made that the noise, e_t, has a Gaussian distribution, then because of the assumed linearity of the system, the estimates of the parameters, $\hat{\underline{x}}_t$, will also have a Gaussian distribution and P_t is then a covariance matrix, defined as the expected value of the squared estimation errors for the elements of $\hat{\underline{x}}_t$, $P_t = E\langle (x_i - \hat{x}_i)(x_j - \hat{x}_j)\rangle$ i.e. the expected value taken over all possible values of the joint errors between the true and estimated values of parameters i and j at time t. Values along the diagonal of the matrix for $i = j$ represent the variance of the ith parameter.

The covariance matrix P is always symmetric and positive definite. It can be expressed in terms of the variances of the individual parameters and the cross-correlations between the errors in the states as:

$$P = D^{0.5}CD^{0.5} \tag{B5.1.7}$$

where:

$$D = \begin{bmatrix} \sigma_1^2 & 0 & 0 & \cdots & 0 \\ 0 & \sigma_2^2 & 0 & \cdots & 0 \\ \vdots & \vdots & & & \vdots \\ 0 & 0 & 0 & \cdots & \sigma_{N_s}^2 \end{bmatrix} \tag{B5.1.8}$$

and:

$$C = \begin{bmatrix} 1 & \rho_{12} & \rho_{13} & \cdots & \rho_{1N_s} \\ \rho_{12} & 1 & \rho_{23} & \cdots & \rho_{2N_s} \\ \vdots & \vdots & & & \vdots \\ \rho_{1N_s} & \rho_{2N_s} & \rho_{3N_s} & \cdots & 1 \end{bmatrix} \tag{B5.1.9}$$

As a new observation, $\tilde{y}_{t+1}$, is made available at time $t + 1$, P_t and $\underline{b}_t$ can be recursively updated in the form:

$$P_{t+1}^{-1} = P_t^{-1} + \underline{b}_{t+1}\underline{b}_{t+1}^T; \quad \underline{b}_{t+1} = \underline{b}_t + \underline{b}_{t+1}\tilde{y}_{t+1} \tag{B5.1.10}$$

The parameter vector can then be updated as:

$$\hat{\underline{x}}_{t+1} = \hat{\underline{x}}_t + P_t\,\underline{b}_{t+1}[1 + \underline{b}_{t+1}^T P\underline{b}_{t+1}]^{-1}\{\tilde{y}_{t+1} - \underline{b}_{t+1}^T\hat{\underline{x}}_t\} \tag{B5.1.11}$$

Note that this is of the discrete step error correction form

$$\hat{\underline{x}}_{t+1} = \hat{\underline{x}}_t + W_t\{S_t - \underline{b}_{t+1}^T\hat{\underline{x}}_t\} \tag{B5.1.12}$$

where here the innovation weight is given by

$$W_t = P_t\,\underline{b}_{t+1}[1 + \underline{b}_{t+1}^T P\underline{b}_{t+1}]^{-1} \tag{B5.1.13}$$

Young (1984, Appendix 2) notes that this form of recursive least squares error correction was originally presented in algebraic form by Carl Friedrich Gauss (1777–1855) and compares the matrix formulation with Gauss's original derivation.

B5.1.2 The Kalman filter (KF)

The Kalman filter was first introduced by Rudolf Kalman in 1960. It provides a method for the updating of system states as new data become available at successive forecasting time steps. It differs from the recursive least squares algorithm outlined above in that it does not assume that the vector of parameters is time-invariant (so that given sufficient data the parameter values should converge to constant values) but rather that the structure linking parameters and states to the prediction of a measured variable is fixed. The Kalman filter extends the simple least squares algorithm to allow that the states in the model might themselves be stochastic variables that could also be linear functions of a set of independently specified input variables, $\underline{I}$, up to time t such that:

$$\underline{S}_{t+1} = F\underline{S}_t + G\underline{I}_t + \underline{\eta}_t \qquad [\text{B5.1.14}]$$

where F is a state transition matrix that may be dependent on model parameters (assumed known) and defines the dynamic change in the states and the vector $\underline{\eta}_t$ is a zero mean white noise random variable with covariance matrix Q.

The observation equation can also now be generalised to the case of a variable state vector. In doing so we can assume that the "true" value of the predicted variable, estimated by the model as $\underline{S}_t$, is given by

$$\underline{S}_t^* = H^T \underline{S}_t + \underline{e}_t \qquad [\text{B5.1.15}]$$

where H is a matrix representing a fixed linear relationship between all states of the model and the predicted variables. The Kalman filter can then be written in a predictor-corrector form (see, for example, Young, 1984). Given values for the states and inputs at time t the prediction step is given by

$$\underline{S}_{t+1|t} = F\underline{S}_t + G\underline{I}_t \qquad [\text{B5.1.16}]$$

with covariance matrix

$$P_{t+1|t} = FP_t F^T + Q \qquad [\text{B5.1.17}]$$

The correction or updating step, given a new vector of observations, $\underline{\tilde{S}}_t$, is then

$$\hat{S}_{t+1} = \underline{S}_{t+1|t} + P_{t+1|t}H[Q + H^T P_{t+1|t}H]^{-1}\{\underline{\tilde{S}}_t - H^T \underline{S}_{t+1|t}\} \qquad [\text{B5.1.18}]$$

$$\underline{S}_{t+1} = H^T \hat{\underline{S}}_{t+1} \qquad [\text{B5.1.19}]$$

Note that [B5.1.18] again has a discrete error correction or innovation form

$$\hat{\underline{S}}_{t+1} = \underline{S}_{t+1|t} + K_t \{\underline{S}_t - H^T S_{t+1|t}\} \qquad [\text{B5.1.20}]$$

where K_t is called the Kalman gain matrix, equal to $P_{t+1|t}H[Q + H^T P_{t+1|t}H]^{-1}$. The covariance matrix of the estimation, P, is then updated as:

$$P_{t+1} = [I - K_t H^T]P_{t+1|t} \qquad [\text{B5.1.21}]$$

where I is the identity matrix with 1 along the diagonals and zeros elsewhere. The values of the elements in the Kalman gain matrix will depend on the magnitude of the prediction error in [B5.1.20].

Where multi-time step ahead predictions are required up to a lead time L, the prediction step equations [B5.1.16] and [B5.1.17] are used for $\underline{S}_{t+1|t}$ and $P_{t+1|t}$ at successive time steps but without correction since no new observations are available over that time. Note, however, that [B5.1.16] requires some assumptions about the inputs at each successive time step. Since this is also in the future, measurements of the inputs may also not be available, but assuming that the inputs are zero may not be the best strategy (see discussion in main text).

B5.1.3 The Extended Kalman filter (EKF)

The Kalman filter can be extended to deal also with parameter and state updating in (at least mildly) nonlinear models. To do so requires an extension of the state vector to include any varying parameters together with a linearisation of the forecast variance (and covariance in the multivariate case) around the non-linear model predictions between the times t and $t-1$. Now, in the state updating and observations equations, the F, G, and H matrices may not now be constant, but may be functions of the variable parameters and states themselves because of the nonlinearities in the model. By linearising at each time step we can then write to first order that

$$H_{t+1}^T \underline{S}_{t+1} \approx H_t^T \{\underline{S}_t + \underline{\delta S}_t\} \approx H_t^T \underline{S}_t + L^T \underline{\delta S}_t + O(\underline{\delta S}_t^2) \qquad [B5.1.22]$$

where terms in $\underline{\delta S}_t$ of second order and above, $O(\underline{\delta S}_t^2)$, will be assumed negligible and L_t is a *Jacobian* gradient matrix, evaluated at time t which has elements

$$L_t = \frac{dH_t}{dS_i} \quad ; \quad i = 1,2,\ldots,N \qquad [B5.1.23]$$

where N is the number of parameters and states that are allowed to vary. The matrix L extrapolates any perturbation of a parameter or state in the model to the target state. The transpose, L^T, can be used to extrapolate a target state at time $t-1$ back to its equivalent in the model at time t. L^T is also known as the *adjoint* matrix.

The extended Kalman filter (EKF) is then developed by again assuming that the observations have random errors with zero mean and known covariance as in [B5.1.15]. The equivalent prediction and corrector equations to those of the classic Kalman filter [B5.1.16] to [B5.1.21] are now

$$\underline{S}_{t+1|t} = F\underline{S}_t + G_t\underline{I}_t \qquad [B5.1.24]$$

F_t and G_t will now vary in time because of the need to linearise at each time step. Updating of the covariance matrix P_t now makes use of the Jacobian and adjoint gradient matrices in extrapolating the error perturbations from time t to time $t+1$

$$P_{t+1|t} = L_t P_t L_t^T + Q_t \qquad [B5.1.25]$$

The correction or updating step, given a new vector of observations, $\underline{S}_{t+1}$, is then

$$\hat{\underline{S}}_{t+1} = S_{t+1|t} + \mathbf{K}_t \{\tilde{\underline{S}}_t - \mathbf{H}_t^T \underline{S}_{t+1|t}\} \qquad \text{[B5.1.26]}$$

with

$$\mathbf{K}_t = \mathbf{P}_{t+1|t}\mathbf{H}_t[\mathbf{Q}_t + \mathbf{H}_t^T\mathbf{P}_{t+1|t}\mathbf{H}_t]^{-1} \qquad \text{[B5.1.27]}$$

and

$$\underline{S}_{t+1} = \mathbf{H}_t^T\hat{\underline{S}}_{t+1} \qquad \text{[B5.1.28]}$$

The covariance matrix of the estimation, P, is then updated as:

$$\mathbf{P}_{t+1} = [\mathbf{I} - \mathbf{K}_t\mathbf{H}_t^T]\mathbf{P}_{t+1|t} \qquad \text{[B5.1.29]}$$

Again, when multi-time step ahead predictions are required the prediction step is calculated for successive time steps without correction. Every time a new set of observations is available, however, the correction step can be calculated and the predictions reinitialised.

Both the Kalman filter and Extended Kalman filter can also be derived on the basis of an updating of posterior distributions for the states using Bayes equation [see Box 4.2]. This formulation, however, requires that distributional assumptions be made about the nature of the errors (normally that they can be represented by a Gaussian white noise process). The updating step at each time interval can then be interpreted as moving from some prior estimate of the state vector at time t to the posterior at time $t + 1$. Given the Gaussian assumption about the errors, the posterior distribution of the states will also then be multivariate Gaussian. The application of the KF and EKF in the form given above, however, is robust to deviations from Gaussian assumptions, because the vector $\underline{S}_{t+1}$ will still be the minimum covariance unbiased estimator of the true state vector $\underline{S}_{t+1}^*$ (see, for example, Norton, 1986).

B5.1.4 The Ensemble Kalman filter (EnKF)

Further developments of the Kalman filter have been introduced in the last decade to allow data assimilation with highly nonlinear models involving a large number of variables. The concept is similar to the "classical" KF and EKF forms, but rather than using linearisation using a gradient of the model outputs with respect to each state variable included in the filter, the model itself is used to propagate the distribution of uncertainties. To do so requires using an ensemble of model runs over the forecast period, sampled so as to reflect the current error covariance matrix at the start of the period. This gives a new estimate of the covariance matrix at the end of the forecast period based on the ensemble sample. The advantages of this approach are that it can be used with complex nonlinear models for which it is not simple to calculate the Jacobian matrix or the adjoint as required in the EKF.

It was noted earlier that the Kalman filter was developed using orthogonal projections and did not make any particular distributional assumptions about the errors, other than that the distribution should be symmetric. Because the Ensemble Kalman filter requires random sampling of models (the ensemble members), however, it is now necessary to make additional distributional assumptions. In general it is assumed that the error terms are multi-Gaussian in structure (see Section 5.4 in the main text for a discussion of Particle filtering methods that do not require such strong assumptions). The ensemble of models is then run for the required forecast period.

The mean of the states (equivalent to [B5.1.24]) is then calculated for the jth state simply as

$$\bar{S}_{j,t} = \frac{1}{N} \sum_{i=1}^{N} S_{i,j,t}$$

[B5.1.30]

where N is the number of ensemble members. The covariance matrix for the state estimates (equivalent to [B5.1.25]) is calculated from the deviations from the true state over all ensemble members as

$$P_{t+1|t} = \frac{1}{N-1} \overline{E_t E_t^T}$$

[B5.1.31]

with $E_t = [S_{1,t} - S_{1,t}^*, S_{2,t} - S_{2,t}^*, \ldots, S_{N,t} - S_{N,t}^*]$ and, as above, $S_{i,t}^*$ is the true value of the state I at time t. The true values $S_{i,t}^*$ are, however, not known and are replaced by the mean values over the ensemble so that the approximation $E_t = [S_{1,t} - \bar{S}_{1,t}, S_{2,t} - \bar{S}_{2,t}, \ldots, S_{N,t} - \bar{S}_{N,t}]$ is used in [B5.1.31].

Given the observations at the new time step, t, the states are then updated individually for each ensemble member (equivalent to [B5.1.26]) as

$$\hat{S}_{i,t+1} = S_{i,t+1|t} + K_t \{\tilde{S}_t - H_t^T S_{i,t+1|t}\}$$

[B5.1.32]

and the Kalman gain (equivalent to [B5.1.27]) is calculated as

$$K_t = P_{t+1|t} H_t [Q_t + H_t^T P_{t+1|t} H_t]^{-1}$$

[B5.1.33]

where Q is the covariance of the measurement noise as before. Finally, P is then updated as before (equivalent to [B5.1.29]) as:

$$P_{t+1} = [I - K_t H_t^T] P_{t+1|t}$$

[B5.1.34]

A new forecast is then made by generating a new set of ensembles from the covariance matrix and propagating the ensemble of models over the new forecasting period for whatever lead time is required. As soon as new observations are available the correction step can be implemented, and the Kalman gain and covariance matrices updated.

In assuming that the true states can be estimated by the mean over the ensemble, there is an implicit assumption that the model is unbiased. This is not necessarily the case in environmental models and there have been a number of techniques proposed for bias correction in applying the EnKF (see, for example, Dee and Da Silva, 1998; Dee, 2005; DeLannoy et al., 2006; Sørensen et al., 2004). A recent review of Kalman filtering and Ensemble Kalman smoothing techniques is provided by Evensen (2006).

Box 5.2 Variational methods for data assimilation [3]

Variational methods for data assimilation are mostly used in Numerical Weather Prediction (NWP) and ocean modelling, although as noted in the main text there have also been applications in the assimilation of satellite imagery for different purposes. More detail on the different types of variational data assimilation may be found in Kalnay (2003, p.184). Rabier (2005) gives a review of the use of data assimilation schemes in different weather prediction centres, while Rawlins et al. (2007) summarise the 4DVar scheme currently in use at the UK Met Office.

B5.2.1 Three-dimensional variational data assimilation (3DVar)
In data assimilation for large-scale models such as numerical weather prediction models we can distinguish between a number of different time increments. There are the increments at which forecasts are issued, there are the time steps at which the model is run, there are the times at which observations are made, and there are the times at which a model analysis is carried out, updating the states and variables of the models on the basis of assimilating the available observations. In particular, the times at which different observations are available might be different from the model analysis times and, for computational reasons, the analysis increments will be much longer than the model time step.

Variational data assimilation aims to find a model analysis that minimises a cost function with terms for both the deviations from the background model predictions (in numerical weather prediction this is usually a short-term forecast from the previous analysis) and the deviations of the model analysis estimates of observed quantities from the observations themselves. The deviations are weighted by estimates of the inverse of the background model forecast error covariance and by the inverse observation error covariance respectively. In 3DVar, the observations are assumed to be available instantaneously at the same time as the model analysis, so that the corrections or innovations to the states in the model are a function only of the three space dimensions and not of time. Thus, using similar notation to Box 5.1,[4] the cost function, J, is defined as:

$$J = \frac{1}{2}\left[\{\underline{S} - \underline{S}_b\}^T \mathbf{B}^{-1} \{\underline{S} - \underline{S}_b\} + \{\tilde{\underline{S}} - \mathbf{H}^T \underline{S}_b\}^T \mathbf{Q}^{-1} \{\tilde{\underline{S}} - \mathbf{H}^T \underline{S}_b\} \right] \qquad [B5.2.1]$$

3 Thanks are due to Sue Ballard of the UK Met Office for checking the presentation of Box 5.2.

4 There is a "standard" notation for variational data assimilation that has been presented in Ide et al. (1997), though this is not followed by Talnay (2003), and has been simplified for 4DVar by Lorenc (2003b) and others. Here, the presentation mostly follows Talnay (2003) with modifications to make the links to the various Kalman filters of Box 5.1 clearer.

where **B** is the covariance matrix of the errors in representing the true system states by the vector of model background states, $\underline{S}_b$; **Q** is the covariance matrix of the observations, which is sometimes split into two components as **Q** = (**E** + **R**) where **E** is a measurement variance and **R** is a commensurability or representational error to allow for the fact that measured and modelled variables may not be represented at the same scale (see Section 1.4.5); $\underline{\tilde{S}}$ is a vector of observations of system states, and **H** is a matrix that transforms vectors in model space to vectors in observation space (see [B5.1.15]).

This cost function can be derived from either maximum likelihood or Bayesian approaches assuming that the errors in representing the true states with a model analysis and the errors in the observations relative to the true states are normally distributed. An analytical solution to [B5.2.1] can be found by taking the differential of the cost function with respect to S and setting the differential to zero at the minimum.

The solution for the new vector of analysis states, $\underline{S}_a$, is then given by:

$$\underline{S}_a = \underline{S}_b + (\mathbf{B}^{-1} + \mathbf{H}^T\mathbf{Q}^{-1}\mathbf{h})^{-1}\,\mathbf{H}^T\mathbf{Q}^{-1}\,\{\underline{\tilde{S}} - \mathbf{H}^T\underline{S}\} \qquad [\text{B5.2.2}]$$

This has the same form as the innovation equations in Box 5.1 in correcting the background states $\underline{S}_b$ to the states $\underline{S}_a$ with an innovation weight of:

$$W = (\mathbf{B}^{-1} + \mathbf{H}^T\mathbf{Q}^{-1}\mathbf{H})^{-1}\,\mathbf{H}^T\mathbf{Q}^{-1} \qquad [\text{B5.2.3}]$$

For models with a very large number of states or variables (as is the case with the atmospheric models that use variational data assimilation) the evaluation of all the matrices in [B5.2.3] is computationally expensive and [B5.2.1] is usually minimised by an iterative optimisation technique to find the optimal model states after assimilation of the observations. In this way, it is ensured that the data assimilation is consistent with the model dynamics.

Note that in 3DVar, the error covariance matrices must be assumed known. The data assimilation does not update these matrices, although they can be modified externally to the assimilation process according to the state of the system. A modification was introduced by Da Silva et al. (1995) which works on observation space rather than model space that is much more efficient if the number of observations is very much less than the number of model states.

Most current operational NWP systems use an incremental formulation that is computationally cheaper by using reduced resolution analysis increments rather than full analysis fields (Lorenc et al., 2000). There has been an extension to 3DVar called First Guess at Time of observation (FGAT) which allows for observations to occur within some time window surrounding the analysis time. This is based on using the forecast innovation at the time of each observation but applying it at the time of the analysis (see Lorenc et al., 2000). For a more complete treatment of the actual time of observations it is necessary to use the computationally more expensive 4DVar.

B5.2.2   Four-dimensional variational data assimilation (4DVar)
The 3D analysis presented above will be most accurate when all the observations to be assimilated coincide with a time step of the model. This is rather unlikely to

be the case in practice. Thus, the analysis has been extended to allow for the observations to be distributed within a time interval that may extend over multiple model time steps. The cost function is then extended to incorporate a measure of deviation from the background solution integrated from the beginning of the time interval being considered to the time at which an observation is actually available. Thus, it is effectively being assumed that the model dynamics are a true representation of the real dynamics.

The cost function to correct the background states to some new states reflecting the information in the observations can then be expressed as

$$J(\underline{S}_{t_o}) = \frac{1}{2}\Big[\{\underline{S}_{t_o} - \underline{S}_{b,t_o}\}^T \mathbf{B}^{-1}\{\underline{S}_{t_o} - \underline{S}_{b,t_o}\} + \sum_{k=0}^{N}\{\tilde{\underline{S}}_k - \mathbf{H}_k^T\underline{S}_k\}^T \mathbf{Q}_k^{-1}$$

$$\{\tilde{\underline{S}}_k - \mathbf{H}_k^T\underline{S}_k\}\Big] \qquad [\text{B5.2.4}]$$

where $k = 0,1,\ldots N$ is an index of the time increments at which observations are available within the observation window around the analysis time t_o. The optimisation then seeks to find the solution at time t_o, $\underline{S}_{t_o}$, that best satisfies the cost function, including the effect of the model integrations to the time of an observation. The optimisation makes use of the gradient of the cost function with respect to the states required. This makes use of a linearisation similar to that used in the Extended Kalman filter (Equation [B5.1.22] in Box 5.1) as:

$$\mathbf{H}_{k-1}^T\underline{S}_{k-1} \approx \mathbf{H}_k^T\{\underline{S}_k + \delta\underline{S}_k\} \approx \mathbf{H}_k^T\underline{S}_k + \mathbf{L}_k^T\delta\underline{S}_k + O(\delta\underline{S}_k^2)$$

where $\mathbf{L}_k^T = [d\mathbf{H}_k/d\underline{S}_k]^T$; $k = 1,2,\ldots,N$ is the *adjoint* gradient matrix which here is being used to transform any (small) perturbation backwards in time from observation interval k to interval $k-1$ and then successfully to time t_o when the model states will be updated to initialise a new simulation. Lorenc (2003a) shows how 4DVar is related to the EKF in much greater detail.

Making use of this linearisation, the gradient of the cost function with extrapolation back from time t_k to time t_o using the adjoint matrix $\mathbf{L}^T$ may be written as

$$\left[\frac{\partial J(\underline{S}_{to})}{\partial \underline{S}_{to}}\right] = \left[\sum_{k=0}^{N}[\mathbf{L}(t_k,t_o)]^T \mathbf{H}_k^T \mathbf{Q}_k^{-1}\{\tilde{\underline{S}}_k - \mathbf{H}_k^T\underline{S}_k\}\right] \qquad [\text{B5.2.5}]$$

It is evident in [B5.2.5] that for multiple observation increments, the extrapolation of the effects of errors from increment k to time t_o involves k applications of the adjoint matrix $\mathbf{L}^T$ which, because of the system dynamics, will change over time. The gradients of both the Jacobian and adjoint matrices might vary in both sign and magnitude at different points in space and time. The approach is therefore most appropriate where analytical expressions for $\mathbf{L}_t$ and $\mathbf{L}_t^T$ can be developed from the original equations, at the expense of increasing the size of the model code by a factor of the order of two.

Optimisation of the 4DVar cost function remains computationally challenging and attempts have been made both to implement [B5.2.5] in a form that is incremented at each observation interval during the solution and to speed the process by "pre-conditioning" the response surface by transformation of the control variables (Lorenc, 1997).

In fact, Lorenc (1986, 2003) has shown that *if* the model can be assumed perfect and the initial estimate of the covariance matrix B at the start of the period is also correct, then the updating of the model states to the end of the time interval is exactly equivalent to the EKF of Box 5.1. However, 4DVar provides only an approximate update of the error covariance matrix and the analysis covariance matrix. These must, therefore, as in 3DVar, continue to be assumed known at the start of each data assimilation period.

A further variant on 4DVar is called the *Method of Representers*. This takes some account of the model error in integrating from the start of the time period to the time of an observation (Bennett and Thorburn, 1992; Egbert et al., 1994). The model forecast errors are assumed to be random, which results in an extended cost function.

Recent work in NWP aims to deal more directly with the question of model errors (e.g. Dee, 2005; Trémolet, 2006); different distributional forms for the errors (Fletcher and Zupanski, 2006) and the implementation of full Ensemble Kalman filter methods (Lorenc, 2003b; Houtekamer et al., 2005). As computer power increases, the use of the EnKF is likely to become more popular, at least for limited-area problems because of the way it allows the co-variance matrix to evolve over time. However, the problem of dealing with model structural error or bias remains a problem in both methods.

Chapter 6

Decision making when faced with uncertainty

The import of Info-Gap decision theory is that we need neither accept Arrow's pessimistic suggestion that Knight's "true uncertainty" cannot be quantified, nor rely exclusively on Bayesian methods and subjective probabilities. Rather our response to Knight's concerns about the diversity of uncertain phenomena must be to diversify our quantifications of uncertainty and the consequent decision theories which they engender. Each theory will reflect different features of the boundless dimensions of human doubt and ignorance.

Yakov Ben-Haim, 2006

Or sympathies are on the side of the normative camp. There seems to be little point in spending time and effort to make a mathematical model that imitates the "imperfections" in human decision making when the humans are perfectly capable of making imperfect decisions in the first place.

Tim Bedford and Roger Cooke, 2001

6.1 Uncertainty and risk in decision making

Way back in Chapter 1, it was noted that the use of models to predict the response of an environmental system was normally done for a purpose, and in many cases it was done to inform a decision in managing the environment, whether at local, national or global scale. That is why assessing uncertainty in such predictions is important. Uncertainty in prediction implies a risk of being wrong in assessing the response to some change or management strategy. In situations where a decision must be made, a risk of being wrong may imply important serious economic consequences. At the time of writing, the United States and some other nations had not ratified the Kyoto agreement on reducing greenhouse gas emissions. This decision was widely reported as being based on an argument that, since the scientific predictions of the effects of greenhouse gas emissions on future climate are still so uncertain, there is no justification for taking a decision to reduce emissions when it will have such a serious impact on the economy of these countries. At the time of Kyoto their politicians concluded, perhaps dangerously, that the risk of short-term economic recession is greater than the risk of possible long-term climate impacts.

The Intergovernmental Panel on Climate Change does not agree with this assessment (IPCC, 2007). Their considered opinion, despite the uncertainties associated with climate model predictions, is that greenhouse warming is already having an impact on climate, that it is now "very likely"[1] that this is the result of anthropogenic factors, and that the impact will continue if no action is taken to reduce emissions. Over 160 governments have been prepared to accept this advice and ratify the Kyoto Protocol, but the United States is (as yet) by far the greatest producer of greenhouse gases and could potentially have the greatest impact on global emissions.

This is an example of a case where it is, in fact, impossible to make a full quantitative assessment of the risks involved (see discussions, for example by Young, 2001; Pielke, 2002, Pahl-Wostl, 2002, and Hall, 2007). Both sets of opinions are, therefore, political in nature. We cannot properly quantify the uncertainties of climate predictions, nor of the resulting impacts on the global economic system (though some analysts have argued that the need to deal with a significant reduction in greenhouse gas emissions could act as an important *stimulus* to the world economy, in an analogous way to past wars; see Stern, 2007). There are many situations, however, where a more restricted problem allows a quantitative assessment of the risks and costs in making a decision. In these cases methodologies for risk-based decision making have been developed and will be described in the later sections of this chapter.

6.2 Uncertainty in framing the decision context

The most important question at this point is whether taking account of uncertainties will change the nature of the decision that is made. The answer is undoubtedly yes. This has been shown to be important in decisions regarding the uncertainties in the nuclear industry, where decision support that takes account of uncertainty has been normal practice for three decades because of the long time scales of some of the regulatory requirements and the need to evaluate risk in complex systems (e.g. US Nuclear Regulatory Commission, 2004; Reinert and Apostolakis, 2006). The way in which uncertainty might affect a decision, however, will depend on the context of a decision. In risk assessment in the nuclear industry the context has been primarily within the context of the science and engineering of the industry itself, and attempts have been made to quantify the different sources of uncertainty in purely probabilistic terms. In other environmental modelling problems, including global climate change, the context of a decision must be drawn more widely to include more qualitative and epistemic uncertainties and a wider range of stakeholders.

This issue of context has been at the heart of critiques of science-based environmental decision making from a cultural and social perspective (e.g. Funtowicz and Ravetz, 1990, 1999; van Asselt and Rotmans, 2002; Pahl-Wostl, 2002). The argument is that the normal scientific context cannot take account of all the uncertainties involved in environmental decision making, particularly when decisions will have impacts on stakeholders, including the general public, beyond the set of scientists who are attempting to evaluate the risks. Following Funtowitz and Ravetz (1990) these have been called post-normal science problems (see also, for example, Haag and Kaupenjohann, 2001; van der Sluijs, 2002). A particular point that is relevant here is that different groups of stakeholders in a decision might have quite different objectives and quite different responses to the types of uncertainty estimations that have been

---

1 "Very likely" is defined by the IPCC as having a >90% probability of being correct.

outlined earlier in this book. Decision making is then about conflict resolution as much as about trying to maximise the benefits of a decision in some way under uncertainty.

This does not mean to say that the process need be totally political in nature. A formal approach to decision making can very much help to structure the process of making a decision (as implied by the quotation from Bedford and Cooke, 2001, at the head of this chapter).[2] A starting point in this process is the formulation of a decision tree or influence diagram, setting out the structure of the options to be considered and their hierarchical dependencies.

6.3 Decision trees, influence diagrams and belief networks

Decision trees and influence diagrams are basic tools in decision-making theory. They provide a simple way of structuring the different stages and options in a decision problem. They can also be a way of formalising the context of a problem by providing a way of expressing the views of groups of different stakeholders about the different potential options. Thus, they can be used at an early stage of problem formulation without any requirement for expressing preferences or quantifying risks and uncertainties, benefits and disbenefits of particular options.

A decision tree is a structured graph that expresses the hierarchical dependencies involved in different potential outcomes. This may involve one or more decision points; it may involve deterministic or probabilistic dependencies; it may involve one of more value attributes for expressing the preferences between different outcomes (e.g. Figure 6.1). The net benefits for each branch of the decision tree are quantified as a way of deciding on the preferences between different options.

A particular form of decision tree is that of a Cost-Benefit Analysis (CBA), a technique that is widely used in environmental management (including for example deciding on priorities for flood defence expenditure in the UK) where the value of different options on each branch of the tree is evaluated as the simple net balance of costs and benefits. The CBA approach depends heavily on a common agreement about the pricing of costs and benefits that is usually only possible within a limited stakeholder community (and is not always so simple when some of the costs and benefits are difficult to evaluate in pure monetary terms).

Influence diagrams are a different way of structuring the same problem. A simple example, analogous to Figure 6.1, is shown in Figure 6.2. An influence diagram can have a number of different types of nodes including:

- Decision nodes (indicating different alternatives)
- Chance nodes (indicating probabilistic relationships)
- Deterministic nodes (indicating deterministic relationships)
- A final value node (indicating the relative utility of an outcome).

2 A normative decision-making theory is based on what an individual would do if he/she followed certain rational rules of consistency in their beliefs. Normative decision theory is, however, not easily applied to groups of individuals who might individually hold different sets of equally rational beliefs.

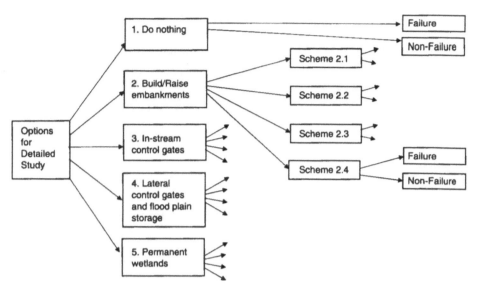

Figure 6.1 Decision tree for evaluation of flood defence schemes before choice of schemes for detailed study. Different schemes under the options for study might include different sites or different heights/areas with consequent different costs and benefits. Each branch might be associated with prior probabilities, or probabilities (such as the failure/non-failure) to be evaluated as part of the study

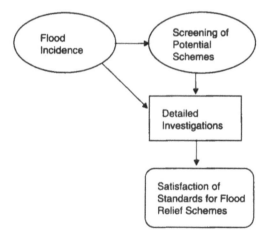

Figure 6.2 Influence diagram for screening of flood defence scheme options. Chance nodes as ovals, decision nodes as rectangles, deterministic nodes as rounded rectangles. The decision tree of Figure 6.1 could be embedded in the node for screening of potential schemes

The nodes are connected by arrows indicating dependencies. Decision trees and influence diagrams are forms of acyclic directed graph, meaning that in following the dependency structure the graph should not return to its own starting point. In the general case, of course, it may be necessary to take account of feedbacks in the system in assessing potential outcomes (e.g. Pielke, 2002, in the context of climate change).

An important concept in the use of influence diagrams is that of conditional independence. A node is conditionally independent of all other nodes that are not its descendent nodes, and dependent only on its parents. Given conditional independence, then all nodes in the diagram can be assigned a probability (see Bedford and Cooke, 2001, Chapter 14 for more details of assigning and verifying such probabilities). A special form of influence diagram is the Bayesian Belief Network, which can be used to express purely qualitative preferences in decision making and will be discussed in Section 6.7.

Influence diagrams can easily be converted to decision trees by ordering the graph such that parent nodes always appear higher in the tree than dependent nodes. Some decision support software systems (such as DPL, see Software Appendix at the end of this book) can switch between these different representations.

6.4 Methods of risk assessment in decision making

The assessment of risk for different decision options requires an estimation of two components: an estimate of how likely it is that an event will occur and an estimate of the consequences of that event (e.g. Kaplan and Garrick, 1981). Quantitative risk analysis requires that both components be given numerical values. In probabilistic risk analysis, how likely it is that an event will occur is expressed in terms of a probability, such that the sum of the probabilities over all possible events envisaged is unity (see Bedford and Cooke, 2001). The consequences of an event will often be expressed as a cost (e.g. expected damages), but might also be assessed in terms of expected number of cancers, loss of life expectancy, or some other severity measure.

Risk assessment methods have been primarily applied in areas where there might be important consequences for human life (e.g. the aircraft industry, chemical industry, nuclear power generation, radioactive waste disposal). In these cases, the systems under consideration are highly complex, include many failure pathways with components of potential human error and uncertain natural hazards as well as technological components that it might be possible to test for failure (though, as the failure of the O-ring seals on the Challenger Space Shuttle launch vehicle showed, this does not necessarily prevent disaster outside the range of available data; see Kirwan, 1994; Draper, 1995). In such complex systems it is generally impossible to quantify properly the probabilities and potential consequences of all possible event scenarios and resort is often made to expert opinion in assessing risk. Such assessments must, therefore, be subjective in nature and therefore associated with some level of uncertainty. Since, if we tried to assess the uncertainties in either probabilities or consequences in such a risk assessment, any estimates would be necessarily uncertain, there is a danger of an infinite regress of uncertainties in assessing uncertainties.

Bedford and Cooke (2001) address this problem by reference to De Finetti's representation theorem. They suggest that we can only assess the probabilities of quantities (variables, events) that are, in principle, observable even if our initial assessments are only subjectively determined. De Finetti's theorem then shows that for any quantity for which the probability does not depend on the ordering of occurrences, then this is equivalent to a belief that the occurrences are independent and identically distributed. Thus, as an increasing number of occurrences are observed, then the estimated probabilities should converge on the true frequencies of occurrence that are not then uncertain. The argument applies to both Bayesian and frequentist approaches to probability. Thus, even if probability is interpreted as a subjective degree of belief, they argue that it is not necessary to invoke any concept of uncertainty about probabilities, as the belief will converge on the true probability as more information about occurrences is obtained.

This is, however, an asymptotic argument, i.e. it depends on the possibility of observing long sequences of occurrences. This is rarely the case in practice, particularly for rare events such as floods, landslides and earthquakes, and the responses of environmental systems will often depend on the ordering of events (see, for example, Beven, 1981, and Newson, 1980, for hydrological/geomorphological examples). Thus, even for quantities that are, in principle, observable, it may be desirable to associate estimates of probabilities with a degree of uncertainty, rather than relying on the degree of belief resulting from observations of only a few events (or even no events in the case of beliefs based purely on expert opinion). There will clearly also often be uncertainty associated with the consequences of an event. This then implies an associated uncertainty in the estimation of the risk associated with any particular event. Thus, the application of risk assessments in practice may be difficult, but there are theories of imprecise probabilities and probability bounds analysis that are being developed and that might be useful in these types of situations (see Box 3.1 and Walley, 1991, 2000; Ferson and Ginzburg, 1996; Hall, 2003; Hall et al., 2006; Sander et al., 2006; Zadeh, 2004, 2005; Klir, 2006). That does not mean to say, however, that the exercise of assessing risk is not worthwhile.

6.5 Risk-based decision-making methodologies

Risk is clearly an issue in decision making (e.g. Morgan and Henrion,1990; Keeny and Raiffa, 1993; Bedford and Cooke, 2001; Young, 2001; Aven, 2003). It would be desirable to deal with risk in decision making in a rational way, even if the levels of risk might themselves be uncertain. This is not a simple problem because it is one for which there may be no "correct" answer since it will depend on the aims of the individual making the decision, and different individuals may have different aims in mind. Decision makers can be risk-averse, risk-neutral, risk-accepting or merely have specific vested interests to consider. The case of greenhouse gas emissions and ratifying the Kyoto agreement is an obvious example of different governments and different non-governmental organisations having quite different aims in mind.

In this section we will follow the *normative* approach to rational risk-based decision making. In actual decisions, of course, individuals do not always act rationally and there is another set of decision methodologies based on trying to assess the actual aims, beliefs and preferences of individuals (see also the InfoGap concept of *robust satisficing* decisions in Section 6.9 below). If, however, we limit the scope of our study here to the role of scientific prediction and uncertainty in the decision-making process, the normative approach can serve as a methodology for bringing science into the decision process that might also ultimately include political, sociological and personal elements. As such, it can act as an aid to the decision maker in taking rational account of the environmental science in so far as that is helpful. This is not to say that the science itself is always free of such political, sociological and personal elements (see

Chapter 2), but we can perhaps expect such elements to have less effect as the science progresses.

Normative decision-making theory stems from the work by von Neumann and Morgenstern (1944), Savage (1972), and others who attempted to formalise a theory of the rational individual through concepts of utility. In effect, the preferences of the decision maker are represented as a set of probabilities of expected outcomes and a *utility function* on the set of possible decisions (actions). The utility function must have real values and is often represented in monetary terms as a cost function (such as the damages likely to be incurred by different actions to control flood levels in the Lake Como Case Study in Section 6.5.4 later in this chapter). The utility function should be chosen so as to reflect the preferences of a decision maker. One definition of making a rational decision is then to maximise the utility functional over all possible actions given the probabilities of potential outcomes.

How does this process relate to the risk assessment issues raised in the previous section? Risk assessment involves the estimation of both the probability of an event and the consequences of that event. It will be expected that the probabilities will be directly related to the probabilities of outcomes in decision making while the assessment of the consequences might also be directly related to the utility function. Where the risk assessment takes account of the potential decisions, through, for example, the use of a decision tree or influence diagram, then the probabilities and consequences, as conditioned on a decision being made, will be very closely related to the probabilities of outcomes and the utility function for different actions. Hence the phrase *risk-based decision making* (see Box 6.1 for more detail).

6.5.1 Assessing the preferences of the decision maker

The actual decisions, however, will depend on the preferences of the decision maker. A decision maker can choose to be risk-accepting or risk-averse. One advantage of the normative decision-making framework, however, is that the two components of the process, probabilities of outcomes and utility function, can be separated. Thus, the preferences of the decision maker, as expressed in the utility function, can therefore be changed (in a form of sensitivity analysis) without changing the probabilities.

This starts, however, to get a little circular. The decision maker could decide to define the utility function so that it reflects his beliefs in what the correct decision should be, regardless of the real rationality of that decision. This can certainly happen where the utility function is difficult to represent in purely monetary terms, for example in the aesthetic costs and ecological cost of building a flood defence system. Evaluations of apparent utility can also be affected by prioritisation issues, for example where two different flood defence schemes are competing to have higher priority in the national flood defence plans. In practice, however, the ordering of the process is generally reversed. Having estimated the probabilities for the set of outcomes, a rational utility value for each outcome is assessed prior to any decision being considered. The utility function then serves to define the preferences for the decision maker. Maximisation over the utility function then provides a basis for a rational decision. There is still some potential for circular reasoning, but the need to have an explicit common basis in assessing utility across outcomes in deciding on preferences can provide a check on the rationality of the choices made.

6.5.2 Indifference between actions

It is worth noting at this point that the risk (as probability * utility) serves to *rank* the potential outcomes in some order of priority. Given the uncertainty in evaluating both probabilities and utility, the process will not necessarily lead to a clear decision. Quite often we will find that there are sets of actions for which the assessed risk, as defined by the decision maker, is essentially equal. There is an analogy here with the equifinality issue discussed earlier in model calibration; in decision theory this is a matter of the decision maker being *indifferent* to different sets of actions. There may, of course, be several sets of indifferent actions at different levels of utility. The decision maker may then choose to seek additional information if maximising the utility results in a set of actions of similar utility value.

One solution to indifference is to add one or more additional utility functions. This can also result, however, in a lack of dominance of one set of actions over another in that different decisions might be optimal for different utility functions. This is analogous to the concept of the Pareto optimal sets of parameter values in model calibration discussed in Section 4.4. More detail on the use of multiple utility functions is given in Box 6.1.

6.5.3 Adding uncertainty and more information

Both components of the decision making process, probabilities of outcomes and the utility of actions, can be subject to uncertainty. In principle, there is no difficulty in extending the analysis to uncertainties in either component by integrating over the utility function for each (uncertain) outcome to obtain marginal utilities for any action. While this may be difficult to calculate analytically, depending on the nature of the appropriate distributions and functions, such marginal utility values can be determined numerically, for example by Monte Carlo or Latin Hypercube sampling (see Box 3.2).

One of the interesting aspects of decision theory is then the possibility of evaluation of the costs and value of collecting more information in constraining the uncertainty and making a decision. This can be done by a form of analysis in which the availability of certain types of information is evaluated before it is collected and the sensitivity of the resulting decision tested *as if* that information were known perfectly (see Box 6.1). If the costs of obtaining the additional information can be assessed, then a decision can be made as to whether it is worth the investment to go ahead with getting the information with a view to making a better decision. In many cases the costs will be small relative to the expected benefits, but not in all.

An example of this approach in a groundwater contamination problem is provided by Freeze et al. (1992). Their approach is firmly within a Bayesian framework. In setting up a groundwater model there will be *prior* estimates of model parameters and boundary conditions. Taking account of how well the model fits the available observations will lead to *posterior* distributions of parameters (and perhaps boundary conditions) and estimates of uncertainty in the model predictions. It is then possible, however, to ask the question as to which of a number of different types of measurements, or measurement sites, might be most useful in constraining the prediction uncertainties. They call this a *pre-posterior prior* analysis since by taking prior estimates at the potential observation sites from the existing model and allowing for measurement error, it is possible to check the potential effect of a new observation on the posterior distributions and, indeed, whether the cost of taking new measurements is less than the benefit that might accrue to the decision being made. The real observation might turn out to be different, of course, but this can be a useful way of assigning priorities in designing a measurement program within an assigned budget. In their groundwater case, each new observation site requires drilling an expensive new borehole and carrying out experimental tests to determine values of transmissivity and storage parameters.

6.5.4 Case studies: Decisions for flood warning and control in Lake Como, Italy and the Red River, N. Dakota

In Chapter 5 we considered some examples of flood forecasting as an application of data assimilation methods. This is clearly not just an exercise in getting better models; flood forecasts are made operationally in many countries of the world to inform decisions about whether to issue flood warnings or use flood control reservoirs or controllable flood defences with a view to saving life and mitigating the costs of flood damages. This is a classical environmental decision-making problem that can be used to illustrate the principles involved in trying to quantify the relevant probabilities and risks.

The example is taken from Todini (1999, 2004) and involves flood control and warning for the towns on the edge of Lake Como in Italy. The lake outlet is controlled by a dam at Olginate for multipurpose use. The lake is storing water for agricultural use; it is storing water for electricity generation, and gates on the dam can also be used to release water to reduce the effects of flooding. The risk of flooding seems to be increasing over time, partly due to subsidence in the centre of the town of Como. The decision problem is made more difficult by the fact that controllable storage in the lake is only about 5% of the annual inputs and that the control gates, when fully open, can release only half the peak inflows into the lake. This means that in large flood events the lake will fill and overflow in three to five days.

The components of the decision in this case are the current storage in the lake (which is known rather well), the potential future inputs to the lake in the next one to ten days, and the expected damages that might occur if the lake overflows. There is also a requirement, however, to maximise the benefits of the water stored for both irrigation and electricity generation purposes over the year. We therefore have two conflicting cost functions, both of which involve uncertainties. The first is used to minimise the loss of water, conditional on future input scenarios developed from the historical record, relative to the optimal release. This is implemented on a ten-day time step. Then, in times when a flood is possible, a second cost function is considered that is used to minimise the deviations from increasing the optimal ten-day release strategy and preventing the lake reaching the permitted minimum level and minimum river flows while also minimising the expected damages due to flooding.

The Lake Como decision support system has been a success. Flooding of intermediate events has been significantly reduced (though it remains difficult to have an impact on the largest events), and water deficits have also been significantly reduced. Todini (2004), however, stresses how taking account of uncertainty in making the forecasts of predicted behaviour is important. Without it, the results may lead to poor decisions (see also the discussion of the Red River example later in this section).

This can be illustrated in Figure 6.3 (from Todini, 2004). This shows how, in a flood control problem, the uncertainty in the predicted future water levels starts to impact on the cost function well before the predicted level actually starts to exceed the level of protection afforded by the flood defences. If the uncertainty is neglected, there is no apparent cost and no reason to issue a flood warning.

A particular example of this is provided by the Red River floods in 1997 (Pielke, 1999). The US National Weather Service (NWS), which has responsibility for flood forecasting in the US, issued a warning for Grand Forks, North Dakota, that the flood would peak at about 15.1 m. On 22nd April, the actual peak was 16.6 m and resulted in extensive flooding and an estimated \$2 bn in damages. After the flood, inaccurate flood warnings were cited as a cause of the disaster. While it is clear that the flood peak was underestimated, the forecast performance was in line with the accuracy of past forecasts but (back in 1997) there had been no attempt by either the NWS or the authorities dependent on the forecast to take responsibility for the associated uncertainty in the forecasts and act accordingly. This was, in fact, another case of a flood caused by too much rain and for such extreme events there is usually limited scope for damage mitigation (though adequate warnings and evacuation procedures can save lives). Pielke and Conant (2003) point out that we cannot now say how the local authorities might have reacted differently if the forecast had been higher. The lesson, however, is that in making decisions it is well advised to be aware of uncertainty in the information provided.

6.6 The use of expert opinion in decision making

There are many situations where, even though it may be possible to assess the uncertainty in model predictions under different potential scenarios (for example

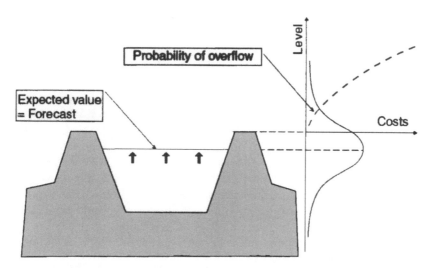

Figure 6.3 Taking account of uncertainty in assessing potential costs of flooding

Source: Todini, 2004, Copyright ©2003 John Wiley and Sons Limited, reproduced with permission

using the techniques discussed in Chapters 3 and 4), there may be no rigorous scientific way of assessing the probability of the different scenarios. I once had a paper on the impacts of climate change on flood frequency rejected because a referee suggested that we had not done an adequate uncertainty analysis because we had not taken account of the probability of the different climatic scenarios, but this is an example of a case where even expert climate modellers would be reluctant to provide estimates of such probabilities. Thus, any estimate of a probability for different scenarios in such cases will be necessarily subjective, based on the reasoned opinions of one or more experts.

There is an extensive literature on the use of expert opinion in decision making (e.g. Morgan and Henrion, 1990; Cooke, 1991; O'Hagan et al., 2006). Expert opinion is useful whenever it is impossible to collect the data or make the model runs necessary to estimate the expected probability of an outcome. Expert opinion has also been widely used to estimate prior probabilities of model parameter values that may be difficult or impossible to measure directly in a particular application. In both cases, the need is for the expert or panel of experts to define some form of distribution function that can then be used in the decision-making process.

The first problem is then to choose the experts. The choice will influence the outcome. Should the panel of experts include the complete range of scientific opinions available on a particular topic; or should some scientists with widely divergent views be excluded from the analysis? Can each expert be considered independent in their assessments (when often they will have similar training and working experience)? Can each expert be considered to be unbiased? Should stakeholders from outside the scientific domain but who might have important insights into a problem be included (see the identification of threats to ecosystems by stakeholders in Carey et al., 2007)? These questions are major issues in the use of expert opinion.

6.7 Combining the opinions of experts: Bayesian Belief Networks

In a Bayesian Belief Network or Decision Tree, the initial estimates of the likelihoods associated with each branch will normally be prior likelihoods elicited from experts. These might be likelihoods of a particular future event, or estimated success or failure of a particular outcome as on the branches of Figure 6.1. A general introduction to the use of Bayesian Belief Networks to decision making is provided by Jensen (2001). We will consider here only the very simplest case of combining the opinions of experts, assumed to be independent, within a Bayesian context.

A general formulation of Belief Networks for application to environmental systems has been provided by Varis (1995, 1997), extending earlier work by Pearl (1988) in the area of artificial intelligence. A recent review of some of the issues involved, and software available, is provided by Uusitalo (2007). Other methods for combining expert assessments are given in Cooke (1991), O'Hagan et al. (2006) and Wang et al. (2005) (see also Section 6.8 below). The decision maker starts with a prior distribution for the quantity of interest and asks the panel of experts to make their own assessment of that quantity. Formally, Bayes equation can then be used to create a posterior distribution of belief for the decision maker, conditional on the opinions of the experts. Bayes equation, in this context, calculates a posterior on the basis of multiplying a prior belief in a choice (usually expressed as a probability) by a likelihood expressing the belief in that choice conditional on the expert opinion, scaled by a constant such that the sum of the probabilities in each choice is unity. The critical term therefore is the likelihood, p(E|C) where E is the evidence provided by the experts, for each of the choices, C. O'Hagan et al. (2006) point out that the construction of such a likelihood function is not a simple matter because what is really strictly required is for the decision maker to make a rather sophisticated evaluation of his prior expectations of the information that the experts will provide. With multiple experts and multiple choices, the complexity of doing so mounts rapidly. This will perhaps only be a real problem if the decision maker has information on how the available experts have performed in past studies; otherwise he or she might choose to use a non-informative prior.

Two approaches are commonly used to formulate a likelihood in this context, assuming either additive or multiplicative errors. For additive errors, the value given by the expert is assumed to be the sum of the true value and a normally distributed error. The mean and variance of the error term can be chosen by the decision maker on the basis of some subjective evaluation of the expected bias and accuracy of each expert (zero bias and constant variance assumptions will often be made for simplicity). A multiplicative error is easily treated in exactly the same way by taking logarithms of the estimated values such that the error again becomes additive and normality is again assumed with respect to the log values. Cooke (1991) recommends carrying out a preliminary evaluation exercise so that the accuracy of the experts can be judged and weighted accordingly – this requires constructing an example that is both realistic but for which the answers are known to the decision maker but not to the experts. This can also raise a question about the correlation between expert opinions – if two experts provide identical results should their joint evidence reinforce their opinions, or is one of them just redundant? This is a value of information issue similar to that discussed in model conditioning in Chapter 4.

The independence assumption can be relaxed, but at the expense of the decision maker being required to estimate a covariance matrix for the expected errors across all experts. Since there may be no strong basis for estimating the dependence in opinions between the different experts, this is often neglected.

6.7.1 Adding empirical evidence to a belief network

An important feature of Bayesian Belief Networks is the possibility of refining those initial estimates as new empirical information becomes available from more quantitative data sources that might support one or more of the expert opinions. In this case we can again apply Bayes equation to combine the different sources of evidence. The evidence might be of quite different types, e.g linguistic, class interval or fully quantified (if uncertain) model predictions or observations (see the Case Study below). Treating the evidence in terms of likelihoods in Bayes equation, however, gives a common metric (likelihood) in refining the uncertainty associated with each option (as used in Chapter 4).

However, in applying Bayes equation to update a posterior likelihood for different potential options it has to be remembered that there is an implicit assumption that all the potential outcomes have been enumerated so that the sum of the posterior likelihoods will always be unity (this assumption can be relaxed using imprecise probability theory at the expense of making the analysis much more difficult).

There have been a number of interesting applications of Belief Networks in environmental applications. Reckhow (1999) and Borsuk et al. (2004) have applied the method to water quality predictions; Varis and Kuikka (1997) and Borsuk et al. (2006) have looked at the problem of estimating and managing fish stocks; while Wooldridge and Done (2004) have implemented a prediction system for coral reef bleaching conditioned on a variety of data sources. Crome et al. (1996) showed how expert opinion could provide useful prior information in assessing the effects of rain forest logging on bird and animal populations. Wolfson et al. (1996) report some practical experience of eliciting expert opinion in two case studies, one concerned with the possibility that radioactive waste was leaking from a landfill site, the other with the remediation of soil lead contamination associated with the operation of battery recycling factories (a site designated as a "Superfund" site in the US). They demonstrate how a Bayesian approach to decision making can be an iterative process, involving negotiation between stakeholders to agree on prior estimates and utility measures. It can also involve agreement on the collection of additional data to inform the decision.

6.7.2 A case study

The use of Bayesian Belief Networks involving both expert opinions and quantitative data will be illustrated by a study of the impact of grazing management on bird populations in sub-tropical Australia, reported by Martin et al. (2005). In this application, the aim was to combine surveys of bird populations with expert opinion to estimate the persistence of different species under different grazing conditions. Thirty-two experts were asked about the impacts of grazing on 31 different bird species. Each expert gave an opinion as to whether each species would increase, decrease or remain constant under low, moderate or high grazing conditions. They were asked only to give opinions for those species for which they felt confident in their opinion.

Field surveys for the different species were also available. However, the field data contained a much higher number of zero sightings than would be expected assuming either Poisson or Negative Binomial distributions to describe frequencies of occurrence. Some of these zeros may be a true ecological effect; others may be spurious due to the difficulties of observing low numbers of a particular species. The study used a two-component model to represent this, one for the occurrence of the species at a particular site, the other for abundance conditional on occurrence. Model parameters for both components included variance components for species, grazing level and within grazing level for each species. These random effects were assumed to be unbiased with gamma-distributed variances.

This two-component model was then extended by combining with the information from the experts. The unweighted mean expert scores were also assumed to be subject to random effects for species, grazing level and within grazing level variability with normally distributed means and gamma-distributed variances. There are thus a large number of mean and variance parameters to be estimated. This was done using the type of Metropolis–Hastings Monte Carlo Markov Chain algorithm described in Box 4.3.

It was found that the precision of the expert opinions was smallest under low and moderate grazing levels but only slightly higher for the highest grazing level. Figure 6.4A shows the unweighted average scores for a selection of the bird species. Figure 6.4b shows the equivalent field data with varying degrees of tolerance to grazing. Over all the species, there was agreement between the experts and field data for ten of the species, and for another eight the experts predicted a greater impact on a species than was observed in the field. Agreement was less good when either the expert precision (agreement) was low or where the field data were poor (showed few sightings). When expert precision is low then the effect on the posterior is minimal; but when the field data were poor but the agreement among the experts was good, then the prior judgements were useful in constraining the posterior predictions of impact.

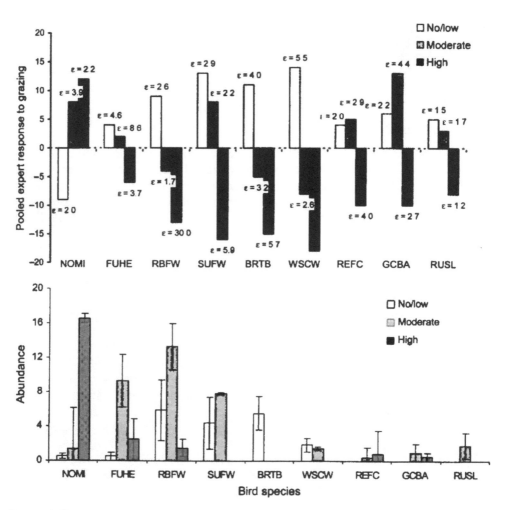

Figure 6.4 Comparison of pooled expert opinion (top) and field survey (bottom) of the impact of grazing intensity (no/low, moderate, high) on selected bird species in sub-tropical woodland in Australia. Values of ε represent precision (inverse of variance) of the expert opinions

Source: Martin et al., 2005, reproduced by permission of the Ecological Society of America

In this case, the study did not go as far as to base any management decisions on the results of the modelling results, but clearly this would be possible. Only one of the 31 species studied (the Noisy Miner, code NOMI) was predicted as benefiting from high grazing levels (a conclusion supported by data from elsewhere that many of the experts would have been aware of). Several species appeared to benefit from moderate grazing levels within the low-density woodland of the study area. These were species in decline elsewhere as a result of woodland clearance.

6.8 Evidential Reasoning methods

Not all expert opinions can be expressed in the quantitative terms required for a risk-based utility function decision theory. Much useful information from experts may be expressed in the form of multi-dimensional "soft" data, for example, on interval scales (e.g a value or even a measure of belief in a particular value might be high, medium, low or negligible). These types of data are not easily expressed in terms of continuous probability scales or utility functions.

One approach to making use of such data is in the Evidential Reasoning methods of Wang et al. (2005, 2006). Evidential reasoning is based on the Dempster–Shafer (D–S) theory of evidence, one of a number of different forms of taking a non-probabilistic approach to uncertainty (see Klir, 2006). Dempster–Shafer theory was originally designed for deterministic evidence but there have been more recent attempts to extend the theory to uncertain evidence, represented by fuzzy variables or vague variables defined only over intervals in a way that can be used for aggregating multiple attributes within a decision tree or belief network.

Shafer (1976) suggests that Bayesian probabilities are a special case of the theory of evidence that can be used when all beliefs can be expressed in the form of probabilities, and when all outcomes are known *a priori* (there is no ignorance of potential outcomes). The D–S theory of evidence is more general in that it can allow for ignorance in using two measures associated with any probability assignment: a *belief measure* and a *plausibility* measure.

For any mutually exclusive set of propositions $H = \{H_1, \ldots, H_N\}$ we can define a belief structure in terms of a basic probability assignment (bpa), $m(A) = [0,1]$ where A is any subset of H and

$$\sum_{A \subseteq H} m(A) = 1.$$

Given the set of m(A) assignments defined in this way, a belief measure (Bel) and a plausibility measure (Pl) are defined as

$$Bel(A) = \sum_{B \subseteq A} m(B)$$

$$Pl(A) = \sum_{A \cap B \neq \Phi} m(B)$$

where $\subseteq$ indicates that B must be included in A, $\cap$ indicates the union of two sets, and Φ is the null set, i.e. $A \cap B \neq \Phi$ means that A and B must have some elements in common. These measures are connected by the relationship

$$Pl(A) = 1 - Bel(\bar{A})$$

where $\bar{A}$ represents the complement of A. In this framework *ignorance* is defined as the difference between the belief and plausibility measures. As plausibility and belief converge, the decision should become clearer. The heart of the Dempster–Shafer theory is then a method for combining the bpa assignments from different sources of error in a way that means that evidence can be combined in any order.

6.8.1 Case study: Use of Evidential Reasoning in assessing management options for Rupa Tal Lake, Nepal

Wang et al. (2006) have applied the ER approach to environmental impact assessment. They identify four elements in this approach: the identification of the relevant factors expected to have an impact on the environmental system; the ER framework for the identified factors; the ER aggregation of the evidence regarding these factors; and a utility interval based ER ranking method. They show how the approach can be applied even when the impact associated with each factor can only be assessed in categorical terms resulting from elicited expert opinion (e.g. major positive impact, significant positive impact, . . ., no impact, . . ., major negative impact). They apply the methodology to an assessment of alternative methods to conserve Rupa Tal Lake, Nepal, which was subject to both eutrophication and rapid sedimentation. The assessment considered physical factors (such as change in volume, change in sedimentation, change in crop and grazing areas); biological ecological factors (change in lake fisheries, biodiversity, primary productivity, macrophytes and disease vector populations); sociological and economic factors (e.g. housing impacts, tourism impacts, disease impacts . . .); and economic/operational factors (e.g. crop incomes, fishery incomes, costs of maintenance, . . .). These factors were assessed using 11 impact grades, from major positive impact (E) to major negative impact (–E) for each of four different management options. For each option belief measures can be calculated for each of the interval assessment grades. These can then be combined with an expression of relative utility that can be modified to account for whether a decision maker, is risk-accepting, risk-neutral or risk-averse (e.g. Figure 6.5). This gives minimum, maximum and average utilities for each alternative, taking account of the uncertain nature of the assessment criteria. The result is a relative ranking of the different options, where overlap of the ranges of utility expresses more or less indifference between the options. This is a relatively simple way of applying such a multi-attribute decision framework in real-world problems where much of the information or evidence about impacts and utility might be highly uncertain or the result of subjective expert opinion (see also the Info-Gap approach outlined in Section 6.10 and Box 6.2).

A

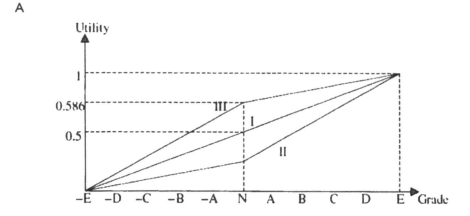

B

Options	Expected Utility			Rank
	Minimum	Maximum	Average	
Option 1: Do nothing	0.3489	0.3953	0.3721	4
Option 2: Build high retaining dam	0.5291	0.5755	0.5523	1
Option 3: Build lower retaining dam	0.4950	0.5182	0.5066	3
Option 4: Build upstream sedimentation basin	0.5192	0.5248	0.5219	2

Figure 6.5 A. Utility functions used in the application of Evidential Reasoning by Wang et al. (2006) with curves representing risk-averse (I), risk-neutral (II) and risk-accepting (III) decision-making strategies. B. Table of expected utilities for the four options considered in the Rupa Tal Lake example (reproduced with permission of Elsevier)

6.9 Decision support systems

The types of utility function-based decision frameworks outlined above have been incorporated into a wide range of computer-based decision support systems (DSS) for environmental decision making. Such systems can be designed to provide decision guidance or to allow the user to explore many different "what-if" situations rapidly, with user-specified changes in implementation strategy, utility function etc. Environmental DSS often include spatial databases and output their results in the form of maps. They do not always take account of uncertainty in the inputs on which the decisions depend. Some indeed purport to find an optimal (maximum utility) solution without any consideration of uncertainty at all.

The degree to which uncertainty is incorporated into an environmental DSS does vary, but there are some in which at least some prior uncertainty estimates can be incorporated. An important example of an environmental DSS in the USA is the FRAMES-3MRA system that is supported by the US Environmental Protection Agency (EPA). FRAMES-3MRA is the Framework for Risk Analysis in Multimedia Environmental Systems, Multimedia, Multi-pathway and Multi-receptor Risk Assessment, for the evaluation of hazardous waste management facilities (Marin et al., 2003). The FRAMES-3MRA has 17 different scientific modules to model the release, transport and risk associated with different hazardous pollutants. It requires 966 multi-dimensional input variables and parameters, of which 185 are specified as stochastic distributions. Databases for 28 different organic contaminants and 15 metal contaminants are specified within the DSS, together with a wide range of other data on receptors (including different ages of people) and fate processes (chemical reactions, linear and nonlinear partitioning, volatilisation etc). The system is characteristically run with 1,000 Monte Carlo realisations of variables specified as distributions in order to test for compliance with regulations on a receptor by receptor basis.

The potential for such a large number of possible combinations of waste site configurations, parameters and input variables, and the need for Monte Carlo simulations creates an important computational problem. Babendreier and Castleton (2005) show how this can be solved practically for FRAMES-3MRA using a network of standard PCs running in parallel, even if it is still only possible to make a limited search of such a large dimensional space. The results can be output as distributions for various receptors in space and time showing the potential for exceedences of critical levels for individual contaminants and hazardous degradation products.

This is, however, only a forward uncertainty or sensitivity analysis, similar to those discussed in Chapter 3, without conditioning on site data (which would make the problem even more computationally demanding). It is, like very many DSS of this type that are designed to estimate "risks", a very deterministic form of stochastic risk analysis wherein the results depend directly on the prior assumptions used. Its value in terms of *decision support* is then more in providing a complete and coherent framework with which to structure each new application while making access to common forms of data required easy to the user. The disadvantage is that some assumptions (such as reaction and partition coefficients) may be fixed as constants in the system and its databases, without the user having easy access to them and without any exploration of uncertainty in those values in real applications. Other DSS of this type include those included in the UK Risk Assessment for Strategic Planning (RASP) for fluvial and coastal flooding (Sayers et al., 2002). Such systems represent at least a start in taking account of uncertainty in decision making but are still based on assumptions that the underlying model structure is correct and that the many input and parameter values can be adequately quantified in any application.

A rather important point is assessing the uncertainties for different alternatives in decision making is the question of dependence between parameter values in making predictions for each scenario. This is often ignored by assuming that parameter values used in different scenarios are either independent or have identical marginal distributions (are fully dependent). Reichart and Borsak (2005) point out that this can lead to misleading results where some parameters might be partially dependent. They demonstrate a technique for dealing with partial dependence by the use of copulae (see Section 3.3.5) in an application looking at policy alternatives to control inputs of phosphorus into a lake. They recognise, however, that providing a technique to allow the effects of partial dependence to be accounted for is actually a lot easier than specifying the nature of that dependence in any particular application when there might only be very limited data about effective parameter values available.

6.10 Info-Gap decision theory

There is another form of theory of decision making under uncertainty. This is the Info-Gap analysis of Yakov Ben-Haim (2006). It is based on quite different principles to the type of probability/utility function-based decision theory outlined in the previous sections. Ben-Haim (2005, 2006) notes that not all uncertainties can be represented in the form of probabilities, but may be subject to quite different forms of unquantifiable uncertainties. As mentioned previously, Ben-Haim refers to Frank Knight's (1921) distinction between risk – which can be assessed in terms of probabilities and therefore insured against – and "true uncertainty" which cannot. Probability/utility function theory deals very effectively with risk. It is subject to being wrong because of the unquantifiable "true uncertainties" of future change, innovation, non-stationarity and the unexpected. Info-Gap theory aims to provide assessments of the relative value of different decisions in the face of such uncertainties. The assessment of risk already assumes that there is a considerable stock of knowledge or information available to the decision maker. Info-Gap theory is for use in situations where there is a lack of knowledge or information.

To allow this, Ben-Haim assumes that some "best estimate" description (or, given the problems of identifying best representations of a system discussed earlier, at least a nominal description) of the system of interest is available (as for any form of decision). That might be a quantitative model of the types that have been discussed earlier in this book; it might be a probabilistic representation; it might be a purely qualitative description. The essential idea that underlies the theory is that as we move away from this nominal description (perhaps because of uncertainty in future input data, or model parameter values, or model structures), then the utility of the outcome of any decision will become more and more uncertain in some consistent way. This process can be defined generally in the following way:

$$U(a,\bar{M}(\Theta)) = \{M(\Theta) : |M(\Theta) - \bar{M}(\Theta)| \leq a\} \quad a \geq 0$$

$U(a,\bar{M}(\Theta))$ is the set of all possible descriptions (or model predictions of a variable of interest in cases of interest here) whose deviation from the nominal model is nowhere greater than a, noting that the model output $M(\Theta)$ may be varying in space and time. The sources of uncertainty might also be quite general in this approach. They might come from different model structures, different parameter sets or different input conditions but unlike Bayesian theory or the GLUE methodology, it is not necessary to try to put any form of likelihood weight on the predictions. In particular, they might include Knightian or epistemic uncertainties. Ben-Haim (2006) suggests that this is a major advantage of the Info-Gap methodology relative to the utility function-based approaches of Box 6.1. Here, we wish to assess only the relative deviation of the outcome away from some nominal case for any value of α.

Given that deviation, however, two functions may be specified, representing *robustness* and *opportuneness* (using the nomenclature of Ben-Haim, 2006). The robustness function expresses the greatest level of uncertainty (in terms of the values of α) for which failure should not occur. The definition of what constitutes a failure, of course, will depend on the particular characteristics of a project. Ben-Haim gives examples involving the extinction of a species, the failure of an engineering structure, epidemiology and drug testing, and portfolio investment and monetary policy. In all these cases there may be future uncertainties that are not easily handled in a probabilistic framework, although he also shows how the method can be used to evaluate the impact of uncertainty in probability distributions when these are required for certain decision frameworks.

The opportuneness function, on the other hand, is the least level of uncertainty at which success is assured. For any decision we would ideally like to maximise the value of α for which the decision is robust, while minimising the value of α at which success is assured. Clearly these requirements are generally in opposition to each other. In particular, an "optimal" deterministic solution that would ensure success (which effectively has $\alpha = 0$) would be very unlikely to be robust in the face of significant uncertainties. What Ben-Haim shows is that robustness and opportuneness functions can provide, in cases of real uncertainty, a rational and consistent framework for assessing the *relative* merits of different decisions. More details of the Info-Gap methodology will be found in Box 6.2.

A particular environmental example, currently pertinent but not discussed in Ben-Haim's book, is (again) the issue of predicting climate change. To date there have been a number of different predictions of climate change in the 21st century, taking account of different emissions scenarios and using different model structures in different research institutes around the world. These models differ in their parameterisations of different processes and solution methodologies for the governing equations. The finer the grid scale of the model, the better the scales of movement of the atmosphere and boundary conditions can be resolved, but the longer the run times of the models. Thus, the finest global model resolutions take months on the largest supercomputers to produce a single deterministic simulation covering historical and future time (generally from about 1850 to 2100). The outcomes from the different simulations have been summarised in a number of reports from the Inter-Governmental Panel on Climate Change (most recently IPCC, 2007).

The differences in the predictions of the different models provide a certain range of uncertainty in future climate that has not changed very much over the four IPCC reports issued between 1990 and 2007. There has also been an experiment[3] to make tens of thousands of runs of a global climate model in the background clock cycles of the personal computers of registered participants around the world. This has produced a wider range of predictions of future temperature change. The model used was, however, simplified and much coarser resolution to allow it to run (albeit slowly) on a PC. Thus, both sets of results, from the few deterministic fine-scale models to the thousands of coarse-scale models, produce predictions of the future that are uncertain in different ways. They are uncertain because the inputs to the models (different emission scenarios) have unknown uncertainty. They are also uncertain in the parameterisations used which are often simplified and difficult to identify with any security for every grid element in the model (see, for example, the study of land surface hydrology by Franks et al., 1999). They are also uncertain in their predictions of historical climate variability when compared against estimates of grid-averaged observational data, particularly in some regions of the world, although there is a commensurability

---

3 See www.climateprediction.net.

issue in making such assessments as to how best to process the limited observations that are available to match the grid-scale predictions of the model. These uncertainties are difficult to quantify in any probabilistic sense, even though some efforts are being made in this direction (e.g. Rougier, 2007; Rougier and Sexton, 2007). The model predictions have therefore generally been treated simply as potential "scenarios" of future change that might or might not happen.

In fact, this situation will become even more interesting in 2008 when the first large-scale set of regional ensemble climate predictions will be released by the UK Hadley Centre. This ensemble is intended to consist of a few hundred runs for which different inputs and parameter values have been varied based on expert judgements of the type of uncertainty that might be expected. Will this prior judgement be sufficient to span the range of possibilities of what might happen in the future? Almost certainly it will not (see Stainforth et al., 2007a). Will the prior judgement be sufficient to be able to interpret the outcomes from the ensemble modelling process in probabilistic terms? Well, yes in terms of the relative probability of an outcome within the ensemble, but certainly not in terms of the probability of actual outcomes in the future when all sorts of unexpected things might happen. For example, both the United States and China might suddenly adopt a policy of drastically reduced emissions as a spur to new technological developments in power generation and transport (I said unexpected, but that is not to say impossible – the start of such a policy is being seen in Australia as a response to extended drought, and the State of California already plans more severe controls on car gas emissions than anywhere else in the world to try to mitigate the occurrence of smog in the major conurbations).

Thus, the prediction of future climate change is subject to the type of Knightian uncertainties that the Info-Gap approach is intended to address. It is interesting to speculate whether, if the ensemble forecast experiment currently being run at the Hadley Centre had been designed using an Info-Gap methodology, the sets of simulations being run would have been chosen quite differently. It will be interesting, once the ensemble simulation results are released, to see how they might be used in different ways in different decision-making frameworks.

While the Info-Gap has a quite different philosophical approach to decision making under uncertainty, essentially avoiding modelling the *nature* of the uncertainty, we can usefully here clarify the issues it raises within a similar set framework to those used by GLUE or fuzzy set approaches. In these methods, the performance of a potential model is evaluated in terms of one or more fuzzy membership or generalised likelihood measures (see, for example, Figures 4.5.2, 4.5.3). In these approaches, in general, models that do not give acceptable predictions are given a membership or likelihood value of zero. This already, however, assumes that adequate information is available to be able to decide on the limits of acceptability. The InfoGap approach is aimed at situations when this might not be possible, but for which it is possible to evaluate the implications of a nested set of predictions (from one of more models) on a potential decision. We can visualise this in a similar way. Figure 6.6 shows schematically the ranges of predictions for different levels of the uncertainty measure, α, around some baseline case at $\alpha = 0$. It can be seen therefore, that in assessing model predictions, the Info-Gap approach does not require an a priori decision about model acceptability, but rather allows the range of possibilities to expand as nested sets for different levels of α. There are parallels here with the issue of deciding acceptability in the face of unknown input errors which were discussed in the context of the GLUE methodology in Section 4.5.3. There, it was suggested that limits of acceptability might need to be expanded to allow for the fact that whether a model was deemed to be acceptable or not might depend on the particular realisation of (non-error free) input data.

Both robustness and opportuneness models depend upon the Info-Gap model $U(a, \tilde{M}(\Theta))$ which has a centrepoint $\tilde{M}(\Theta)$ that is some nominal estimate of the uncertain model of the system of interest. It has been shown above how uncertainty, and consequently robustness and opportuneness, can be expressed in terms of the uncertainty measure a. It is worth stressing that a does not refer at all to model parameters, boundary conditions or structures. It refers directly and only to uncertainty in the variables of interest in the decision, regardless of how those variables have been predicted. The uncertainty could be the result of Monte Carlo experiments given prior distributions of parameters; it could simply be the result of speculation around some nominal estimate based on prior experience, especially where it may be very difficult to quantify future behaviour in any way. Info-Gap decision theory can be used across this range of available information and its concepts of robust-satisficing decision behaviour seem to provide a useful framework for difficult decision-making contexts.

6.10.1 Case study: Info-Gap decision making in designing flood defences

Ben-Haim (2006) discusses in detail the possible forms of robustness and opportuneness functions that might be used in making decisions in different contexts, with a variety of practical examples. To illustrate the approach we will take an example based on that in Hine and Hall (2005) and Hine (2007) that deals with the failure of flood defence embankments. The reader may remember seeing pictures in the news of the embankment breaches at New Orleans that resulted in such widespread flooding,

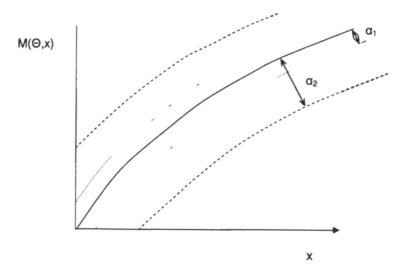

Figure 6.6 Schematic sketch of a-uncertainties associated with some prediction $M(\Theta, x)$ as a function of a control variable x. Solid line is baseline relationship for $a = 0$

damage and loss of life, following the impact of Hurricane Katrina in 2004. Over 1,000 breaches occurred elsewhere along the Mississippi in the major flood event of 1993 that resulted in $15–20 bn damages and the towns of Valmeyer, IL, and Rhineland, MO, being moved to higher ground. Crudely summarising the Natural Disaster Survey Report on the 1993 floods, it was reported to the President that there had been too much rain; and that the event was more extreme than the defences could have been expected to cope with (Josephson, 1994). Post-Hurricane Katrina, however, issues of maintenance of the defences were raised, particularly when the Mississippi delta is sinking (see Horne, 2006; Van Heerden and Bryan, 2006). This is a natural result of the consolidation of the delta sediments but is being exacerbated by reduced sediment supply from upstream and rising sea levels.

Flood defence design is always a matter of compromise between the costs of construction (the higher an embankment the greater the cost of build and maintenance), the degree of protection against floods of different magnitude but poorly known frequency, and the impacts of building defences to protect one site on other sites upstream and downstream. Design standards vary in different countries and for different purposes. Local defences for a river running through a town in the UK are normally designed to protect against a flood of a 100-year return period (that is to say the flood that, on average, has a 0.01 probability of occurring in any particular year). Nuclear power stations close to rivers, on the other hand, might be protected to avoid flooding by a flood with a 10,000-year return period (0.0001 probability). The number of floods for which we have observations, however, is quite small so that estimating the magnitude of the 100-year return period event is quite uncertain. Estimating the magnitude of the 0.00001 probability event will be grossly uncertain.

An interesting example of the problems involved in flood defence design is provided by a recent local flood event in Carlisle, Cumbria, UK (Mayes et al., 2006). Carlisle is a town of 70,000 people, situated where the rivers Eden, Irthing, Caldew and Petteril meet, with a total upstream catchment area of the order of 1,000 km^2. In January 2005, with the ground already near saturated by winter rain, some 200 mm of rain fell over the catchment area, causing the worst floods since 1822. Parts of Carlisle were flooded, including the police station and emergency centre, apparently by a combination of fluvial flooding and the local failure of drainage systems. It just so happened that, at the time, plans for an improved flood defence system were being displayed in the foyer of the regional Environment Agency building for public consultation. These had been designed to cope with the 0.01 probability event, but the improvements had not yet been built. If they had been built, however, parts of Carlisle would still have been flooded because the river discharge in the actual event was greater than that estimated for the 0.01 probability event. The defences are now being redesigned to cope with the 0.005 probability (200-year return period) event, at greater cost but with the purpose of reassuring the local population. The estimate of the 0.005 probability event will be a little less uncertain than previously because the January 2005 event has allowed the flood frequency estimates to be re-evaluated.

Whether reassuring the local population by building flood defences is a good idea, of course, is another matter. In the past, the implementation of flood defence schemes has tended to result in greater investment in infrastructure on the flood plain that has been protected. This is one reason why, even if there is no strong evidence that the frequency of floods is increasing over time (e.g Kundzewicz and Robson, 2000;

Robson, 2002), the economic impact of flooding in the UK (and elsewhere) has been rising dramatically. Because of the compromise between cost and degree of protection, there remains a finite (albeit small) probability that a larger event will come along, as illustrated at Carlisle. Even when the new defences are built at Carlisle there is still a finite (albeit smaller) probability that a still larger event will come along. It is still the case that to avoid increasing the long-term economic impacts of flooding, the best policy is to avoid building on flood prone areas. In countries such as Bangladesh, of course, this is very difficult indeed and it is often the case that the poorest people who live in the most vulnerable locations are the least prepared to cope with flood risk (Brouwer et al., 2007).

The uncertainties in flood defence evaluation arise because we cannot know the sequence of flood events that will occur in the future. We might, on the basis of historical data, be able to make a rough estimate of the expected frequency of events of different magnitude. In making such an analysis we always assume that there is a finite probability that a larger event than those in the historical record will occur but that larger extreme events will be rarer (have lower probability). Since we do not have enough data to evaluate the probabilities by enumeration (we may be trying to estimate the peak river discharge for the 100-year return period event with only 50 years of data or less), we usually resort to fitting a parametric probability distribution to the available data and extrapolating to more extreme conditions. Thus, there is uncertainty in the form of that distribution (the distribution of extreme values is known to be skewed but there is no general consensus on which distribution should be fitted to the data and many different types of skewed distribution have been used including log normal, log Pearson Type III, generalised extreme value, generalised Pareto, and Wakeby distributions), in the parameters of the distribution, in predicting the frequency of more extreme events than the range of the available data, and in the potential for changing in the frequencies during the design lifetime relative to the historical period as a result of climate change or variability. In addition, there are uncertainties in the potential for failure of the flood defences, in the economic data with which to assess damages should a flood occur, and in the interpretation of a peak discharge in terms of actual water levels, especially where it is expected that in mobile bed rivers sediment erosion and deposition during the flood hydrograph might change the cross-section of the channel.

Clearly, however, engineers design and implement flood defences despite these uncertainties. The traditional engineering approach to this has been to design conservatively, by allowing for additional "freeboard" in design to increase the factor of safety. This is a strategy tending to the robust, but increases cost and reduces potential for opportunistic reward if the defence is not overtopped during the design lifetime. Info-Gap decision making allows a more sophisticated assessment of the real uncertainties involved in the design process.

Essentially, each uncertain relationship is represented as an expanding envelope around the nominal relationship, scaled by a value of a. The decision maker must then balance robustness and opportuneness which Hine and Hall (2005) represent for this problem in terms of benefit–cost ratios. In particular, the critical value of benefit–cost ratio for the design to be robust should be no less than 1 (i.e. the project savings in damages over the design lifetime should not be less than the discounted costs of implementation and maintenance). The opportuneness is represented in terms of the possibility that a much higher benefit–cost ratio will be achieved if the sequence of floods is such that a less costly defence will not be overtopped.

Hine and Hall (2005) make the point that over a whole portfolio of schemes, the potential for such opportunistic rewards could be useful for cases where funding is limited (as in the UK, where even schemes with high benefit–cost ratios are not always approved because of budget limitations, even though the spend on the build and maintenance of flood defence schemes currently amounts to £700 m per annum). For a number of schemes the high returns from some might offset a reduction in robustness for each individual scheme. Having said that, a reduction in robustness means that there is a possibility that more homes will be flooded. Thus, a risk-averse decision maker might choose to prefer robustness over opportuneness. An Info-Gap analysis allows this type of investigation. Figure 6.7 shows how a flood defence design process can be formulated within an Info-Gap framework, taking account of the uncertainty in the stage (water level) – discharge relationship, the probabilities of occurrence for different discharges, and specified damage curves. Since both the latter relationships are also uncertain, the analysis could be extended to allow for these additional uncertainties (see Hine, 2007).

Figure 6.8 shows the resulting robustness and opportuneness curves for different flood defence strategies, expressed in terms of benefit–cost ratios and values of a and β. These results are based on assuming uncertainty in the stage–discharge alone, using an energy-loss Info-Gap model. On both plots, when a and β equal 0, the results are those which would be obtained by a purely deterministic calculation or optimisation. Figure 6.8A shows that the benefits for all three strategies decrease within increasing uncertainty, and that the levee+widening strategy, in particular, has minimal robustness to uncertainty (benefit–cost ratio falls below 1 very quickly). Figure 6.8B shows

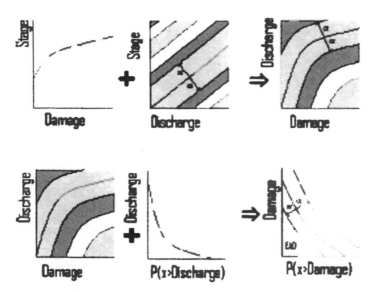

Figure 6.7 Info-Gap α-uncertainties applied to the stage–discharge relationship in assessing uncertainty in potential flood damage

Source: Hine, 2007, reproduced with permission of Daniel Hine

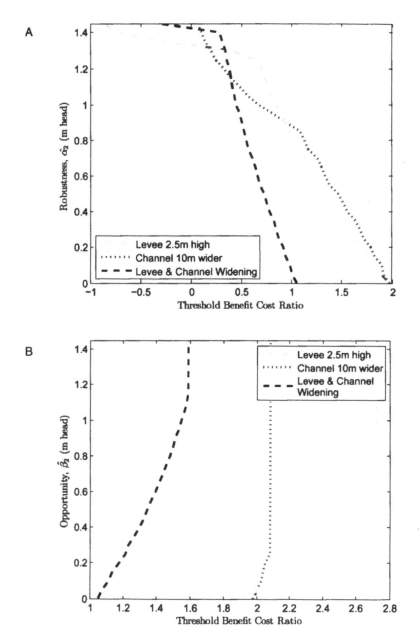

Figure 6.8 Robustness functions (A) and opportuneness functions (B) for different flood defence strategies expressed in terms of cost–benefit ratios

Source: Hine, 2007, reproduced with permission of Daniel Hine

that the levee alone strategy gives the greatest potential for opportune gains as uncertainty increases. Clearly other designs can be rapidly compared in this way, but the comparison becomes more complex as additional sources of uncertainty are included. Hine (2007) also gives examples where uncertainty in flood frequencies is also added to the analysis.

6.11 The issue of ownership of uncertainty in decision making

Decision making in the face of uncertainty involves a variety of stakeholders. There are scientists and consultants who are doing their best to evaluate and quantify the risks associated with a particular decision. There are policy makers who are formulating the framework for the decision-making process. There are the people who will actually make the decision based on all the available information. There are the stakeholders who will be affected, in some cases adversely, in some cases beneficially, by a particular decision. Not all of these groups will be used to methods of assessing and dealing with uncertainty. Indeed, as we have seen in the previous chapters in this book, despite the manifest uncertainties (both epistemic and aleatory) in making decisions about environmental systems, it has largely been neglected. Or rather, none of these groups has been prepared to take ownership of the obvious uncertainties in the decision-making process.

If, in future, decisions are to be made that are robust to uncertainty, then there needs to be an appreciation of the sources and methods of assessment of uncertainty by the different groups, even if not complete understanding. But it is clear that uncertainty means different things to different people. Not all stakeholders and decision makers will have had training in probability statistics or Bayesian methods. They may have had only limited training in soliciting the opinions of experts and other stakeholders in formulating the context of decisions. Even within the scientific community, different scientists have quite different views about the best way to represent and condition uncertainties based on data (see Chapter 4 and Beven, 2006a). In the wider context, there are a wide variety of perceptions and interpretations of risk and uncertainty.

This is an interesting issue of *semiotics*. Semiotics is the study and interpretation of signs. Here the signs we are interested in are the words risk and uncertainty. If different stakeholders have different conceptions of what is meant by these signs, and if they are to properly communicate about concepts of risk and uncertainty and the way they might be used in decision making, then the different stakeholders will need some common basis to do so.

Considering again the problem of flood risk, this has been the subject of a paper by Faulkner et al. (2007) who suggest that there is a need for a *translatory discourse* between the different groups. Typical sets of signs in this area are maps of flood-prone areas at a certain flood return period (in the UK typically for the 100-year return period event, Figure 6.9). Stakeholders in this case range from the Environment Agency who have responsibility for mapping flood risk and issuing of flood warnings in England and Wales, consultants who carry out the modelling work on which to base the flood inundation maps, local authorities and emergency services who have to respond to flood events, and members of the public in the flood risk zone (or outside it)

'1% (1 in 100) or greater chance of happening each year'

'Up to 0.1% (1 in 1000) chance of occurring each year'

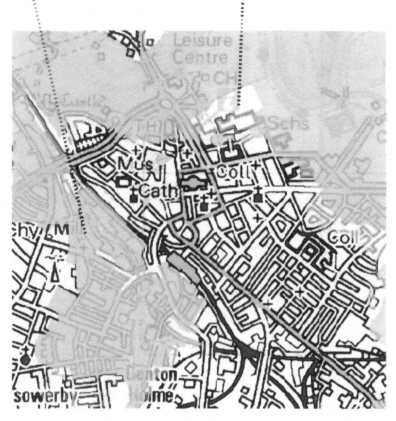

Figure 6.9 UK Environment Agency Indicative Flood Plain Map for part of the town of Carlisle showing different levels of risk. The Carlisle flood of January 2005 was estimated as having an annual probability of exceedence less than that for the 0.01 (1 in 100-year) event map shown

who might have to react to flood warnings and the values of whose houses might be affected by the mapped flood extent.

However, what do these signs mean? The flood inundation map is produced by a hydraulic model driven by an estimate of the 100-year flood discharge. The hydraulic model is uncertain, and may not have been checked against inundation data from past flood events since such data sets are relatively rare. The estimate of the 100-year discharge is also uncertain. It will usually have been made on the basis of much shorter periods with an extrapolation of the tail of some fitted distribution of the extreme events in the record to an annual probability of exceedence of 0.01. There is uncertainty in choice of distribution (and therefore tail behaviour) that could be chosen; there is uncertainty in the extrapolation. Combining these uncertainties, it is clear that the maps of the potential flood extent of the 100-year return period event should also be uncertainty (or fuzzy) in the representations of its outlines (see also

Romanowicz and Beven, 1998, 2003; Pappenberger et al., 2006c,d; Smemoe et al., 2007). But what would a local land use planning committee, or somebody thinking of buying a house in the area make of this as a sign of the risk of flooding?

Indeed, what would either make of the concept of the 100-year event? It is actually already an attempt to translate the technical estimate of the event with an annual probability of exceedence of 0.01 into a more readily understandable form. It does not mean that it is the flood that will happen every 100 years. Only that, given a very long period of record with stationary statistics, it is the best estimate of the flood that would occur *on average* every 100 years. Numerical experiments suggest that about 1,000 years of annual floods would be required to get an estimate of the 100-year event with small uncertainty. But, as already suggested, we have to make that estimate on the basis of much shorter periods of data (even the very longest records of river discharge are rarely over 100 years, and most records are of the order of 10 to 50 years). In fact we have to make that estimate knowing that the distribution of flood magnitudes may not be stationary because of climate and land use changes (see Clarke, 2007 for an example of significant change and Robson, 2002, for an example of the difficulty of identifying change in the face of uncertainty). We may also have to make that estimate not knowing what the correct form of extreme value distribution might be (in the UK opinion about the form of distribution that should be used changed between the Flood Studies Report of 1975, and the Flood Estimation Handbook of 1999; in the US yet another form of distribution is required by law in some states).

Similar issues arise in flood warning. The signs of Figure 6.10 were developed by the UK Environment Agency as a readily understandable way of communicating uncertainty to the public based on forecast flood levels. A number of different levels are used because of the difficulty of predicting flood levels precisely. This is an area where uncertainty in interaction with the public is an important issue in that a definitive statement that there will be a flood will undermine confidence if that flood does not actually happen, leading to a lack of public response to the next warning. This would consequently increase damages and risk of mortalities if there was a lack of public response and a flood did then occur (the "crying wolf" effect of the old fable Roulston and Smith, 2004). In the same way, houses becoming flooded (even if only

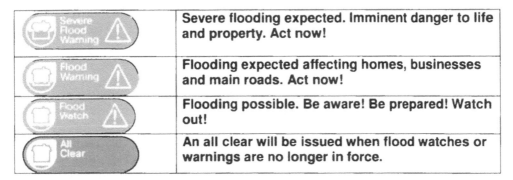

Severe Flood Warning	**Severe flooding expected. Imminent danger to life and property. Act now!**
Flood Warning	**Flooding expected affecting homes, businesses and main roads. Act now!**
Flood Watch	**Flooding possible. Be aware! Be prepared! Watch out!**
All Clear	**An all clear will be issued when flood watches or warnings are no longer in force.**

Figure 6.10 Signs used by the UK Environmental Agency Flood Warning System and their explanation in words (see http://environment-agency.gov.uk/subjects/flood/826674/830330/882451/111627/?version=1&lang=_e)

very locally) when no warning is given also means that the responsible Agency will be criticised. It leaves open, however, what the correct public response should be to the different flood levels given the uncertainty with which flood levels can be predicted.

Faulkner et al. (2007) suggest that one way of developing an appropriate translationary discourse in flood risk management is in the form of the *process* of formulating Codes of Practice for different types of application (see also Pappenberger and Beven, 2006). The final Codes of Practice might define specific methods for eliciting opinions in framing the context of a decision, specific methods that should be used for model predictions in informing the decision-making process, specific methods for assessing the uncertainty in those model predictions, and a specific decision-making framework, but in the process of developing the Code of Practice it would be necessary to involve a wide range of relevant stakeholders and ensure that there was understanding of the issues by all. There would then be a common basis for discussion of the concepts and methods to be used in different application areas. They identify a number of stages in the process to include enhancing understanding by agreeing contested terms, agreeing communication needs and goals, agreeing on relevant tools and methods and agreeing a plan for interaction. This type of interdisciplinary exchange might need new methods of exchange (Mansilla et al., 2005), such as the uncertainty estimation decision tree Wiki pages experiment of Figure 1.4. New ethical standards focused on uncertainty analysis and a sense of shared ownership of uncertainties will need debating within professional bodies that already have professional codes (see, for example, Ersdal and Aven, 2007).

Any Code of Practice that results from this process might well involve suggestions about how to involve different stakeholders, especially professionals at local level, in setting the context and shared ownership of the analysis. This all implies, of course, the application of time and resources and therefore an initial decision about how important an issue it is to take account of uncertainty in different application areas. One result of the practitioners' workshop on Risk and Uncertainty in Flood Risk Management held at Lancaster University in January 2006 was the professionals' suggestion that they would be happy to make some estimates of uncertainty in flood hazard and flood risk if they were provided with some *recipes* to use (and preferably recipes that would not be much more expensive to apply than existing procedures). They also doubted, however, that unless more professional leadership was evident, they would be unlikely to be able to provide estimates of uncertainty to clients, unless clients demanded them. This suggested an initial unwillingness to embrace ownership of uncertainty, or at least a reluctance to be criticised for going out on a limb and deciding which methods to use. It is indicative of the current context for moving uncertainty analysis concepts into practical application. An interesting attempt to overcome some of these barriers and create such a translational discourse in the Netherlands is reviewed in the next section.

The problem has also been addressed in the context of a European Union project called HarmoniRiB. This is one of a number of projects aimed at harmonising practice in the area of water resources management in the EU. The scope and outcomes of HarmoniRiB are outlined in Refsgaard et al. (2005). These included the provision of a set of tools for assessing and describing uncertainty (see Brown and Heuvelink, 2007); a conceptual model for data management; and establishing test data sets for different types of application in river basin management for representative basins across the EU.

6.12 The NUSAP methodology

Some of these issues have, however, already been grasped in the Netherlands where an approach to developing a Code of Practice has been developed using the NUSAP methodology.[4] The NUSAP methodology has developed out the concepts of *post-normal science* expressed by Funtowicz and Ravetz (1990, 1999). Funtowicz and Ravetz have been instrumental in formulating a response to these issues in a context where even some basic scientific issues may be contested but where decisions must be made in a world increasingly connected, where decisions are treated as pressing, where the implications of decisions may have real impact, where uncertainty abounds and the values expressed by different groups of stakeholders are disputed. Examples abound, from global climate change to local resource management. They suggest that in such a context there is a real need for involving all stakeholder groups in a way that is transparent, open and subject to agreed quality control procedures (see also Nowotny et al., 2001; Harremöes et al., 2001; Krayer von Krauss et al., 2005; Janssen et al., 2005).

One of the first major applications of these concepts in the environmental field has an interesting history. In 1999, the Netherlands National Institute for Public Health and the Environment (RIVM) was criticised in the national public press by one of their own employees for grossly underestimating the uncertainties associated with the model predictions on which many reports and decisions were based. The criticism was the start of a major media debate in the Netherlands about the credibility of environmental statistics and models, and the wider issues of the role of science in policy and decision making.

This very public airing of the issues involved in post-normal science resulted in a major review of RIVM's activities and the initiation of a project to develop a set of operational uncertainty assessment tools. This immediately raised issues of the quantification of uncertainty and, in particular, how to treat epistemic uncertainties within a framework that could be common to the wide range of RIVM activities. A project started in 2001 with a wide range of stakeholder consultations. The uncertainty assessment methodology, essentially a Code of Practice in the terminology of the last section, was developed within the NUSAP framework, going "live" in 2005 within the Netherlands Environmental Assessment Agency part of RIVM (RIVM/MNP).

NUSAP is an acronym, an abbreviation of five different ways of assessing uncertainty (Numeral, Unit, Spread, Assessment and Pedigree). It is framed so as to complement quantification of uncertainty assessments (Numeral, Unit, Spread) with expert judgements of reliability (Assessment) and the multi-criteria evaluation of the historical background to the methods and data used (Pedigree). The implementation at RIVM/MNP has developed the Assessment and Pedigree stages into a process of Guidance for Uncertainty Assessment and Communication. The Guidance is intended for use in framing the uncertainty context of a new project, in structuring methods of uncertainty estimation and choosing quantification methods during a project, and in reviewing and communicating results after a project (see for example, van der Sluijs et al., 2003, 2005a,b; Janssen et al., 2005).

4 See www.nusap.net.

The Guidance, in practice, takes the form of a mini-check list and short booklets containing more detailed questionnaires (the Quickscan) backed up by detailed guidance documents. These cover the areas to be addressed in a more qualitative evaluation of the Assessment and Pedigree for particular projects' group into problem framing, involvement of stakeholders, selection of indicators, appraisal of the knowledge base, mapping and assessment of relevant uncertainties, and reporting of uncertainty information. It is recognised that different levels of effort will be justified depending on the importance and resources available to a project, but the minimal effort required would be a simple choice of the wholly, partly, or insufficient categories of the mini-check list so that, in carrying out a project, a framework exists for an individual scientist or decision maker to structure thinking about uncertainty in the wider context of the project.

Experience with this system is still limited (van der Sluijs et al., 2005a,b). Almost certainly it will require some modifications over time as more experience is gained both within RIVM and in the response to RIVM reports by external stakeholders. It is, however, a start on addressing the issues posed by different types of uncertainty within an operational interdisciplinary institutional context.

6.13 Robust adaptive management in the face of uncertainty

Decision makers in all realms of life have always made decisions in the face of uncertainty. The history of the world is a history of decisions made in the face of an uncertain future. Perhaps, as this chapter has attempted to show, we can appreciate the role of uncertainty in the decision-making process and frame it in more rigorous ways in the 21st century, but that does not mean that we can predict all possible outcomes. We should therefore be prepared to expect the unexpected. Studies of actual decision-making histories, as well as experimental studies, show that people are affected in their decision-making behaviour by experiencing the unexpected. Regret becomes a component of decision making (Bell, 1982). They become more risk-averse over time, with a tendency to avoid apparently optimal decisions. This has led to a number of paradoxes in utility-based decision theory that, Ben-Heim (2006) suggests, can be resolved if, in the language of the Info-Gap methodology, decision makers are considered as more *robust-satisficing* than optimisers.

In making decisions about environmental systems we should certainly expect the unexpected. That does not just mean more extreme events than have been experienced before, because we should always be anticipating that in any distribution of events something more extreme may come along. But there are many epistemic uncertainties in environmental systems, and many different examples show that we might experience responses of a different type than we have seen before, responses that were not envisaged beforehand and would therefore not have been taken into account in any formal decision analysis whether by Info-Gap or utility function methodologies. Environmental systems are complex and the element of surprise seems to be widespread in the management of environmental systems and the response of decision and policy makers to surprise becomes part of the dynamics of a managed system. Peterson et al. (2003) give an interesting example, albeit based on simple hypothetical models, about how a lake system subject to eutrophication can be managed in a rational way to collapse.

So we should be generally wary, and this suggests taking an adaptive management strategy in the face of uncertainty (Holling, 1978; Walters, 1986). Adaptive management means being prepared to change strategy as the future unfolds with the results of decisions being monitored and regularly reviewed. It also means trying to be robust in decision making in the sense of ensuring that a decision made now does not preclude a future change in strategy.

There are many examples of adaptive strategies in environmental management. Some of the most important examples are also highly political in nature. The history of setting quotas for fish catches to try to achieve sustainable populations and yields in the face of uncertainty about the population numbers and dynamics is a history of scientific advice being mitigated by political expediency. The Kyoto Protocol on reduction of greenhouse gas emissions was an attempt at adaptive management that seems to have failed as a result of national political considerations, particularly in the United States.

A current interesting example is the Water Framework Directive (WFD) in the European Union. There have been a number of past directives on water issued by the EU, including the bathing waters directive (1978), drinking water directive (1980), the groundwater directive (1980), and urban waste water treatment directive (1991). The WFD of 2000 was the most extensive piece of legislation enacted by the European Commission and now applies to all 25 countries of the EU. The formulation of the WFD was based on scientific advice but was couched in political terms. Thus, it is specified that all designated water bodies in the EU must reach "good ecological and chemical status" for "sustainable use" by the year 2015 (unless a water body could be deemed heavily modified by man in which case it could be derogated until 2027) (EU, 2000).

The definition of "good ecological and chemical status" was not specified in the WFD. The definition of "sustainable use" was not specified in the WFD. They were left for the national water authorities in the countries of the EU to agree on. The science on which to base such definitions is still highly uncertain but the WFD does not refer to uncertainty but rather to a complementary phraseology of "adequate level of confidence and precision" (see the discussions of Newig et al., 2005, and Mysiak and Sigel, 2005). What is clear from the science is that in many cases the time scales for improvement of many water bodies would be much longer than the few years to 2015. In the case of nutrients (especially phosphorus) in rivers and major lakes, for example, the storage in the catchment system and the release times from storage could certainly mean that, even given a drastic reduction in anthropogenic inputs to the system, the rate of decline in concentrations would often be slower than the time frame of 2015. Other factors may be far less subject to control, such as the suggestion that the availability of phosphorous in the Baltic Sea from existing sediment sources totally dominates any current point or diffuse anthropogenic sources. Similar considerations of uncertainty arise in the implementation of the total maximum daily loads approach to water quality management in the US (Chen et al., 2007).

It seems therefore that the full import of the WFD legislation is unlikely to be satisfied by 2015. But, if the requirements of the legislation are approached sensibly, the quality of waters throughout the EU will generally improve, even if, as van der

Brugge and Rotmans (2007) suggest, the legislation is not sufficient to stimulate real change. Improvements in water quality have already been seen as a result of the previous water directives. This is a case where moving the water quality of designated water bodies towards improvement may have a cost but is achievable, whatever good ecological and chemical status and sustainable use might mean in practice. Adaptive management is here a matter of having a monitoring system in place (as required by the WFD) and a learning approach to management (what Pahl-Wostl, 2007, calls "learning to manage by managing to learn") so as to put in place a system that might be robust to change and unexpected events. This strategy can be pro-active, in what van der Brugge and Rotmans (2007) call *transition management*.

The use of adaptive management in the face of uncertain knowledge about complex environmental systems is an attractive concept. However, it is also fraught with pit-falls, a good discussion of which is provided by Gregory et al. (2006). They point out that the approach might appeal to both environmental managers who see it as a way of putting off a difficult decision into the future; and by scientists who see it as a way of justifying the funding experiments to learn more about the complexities of particular systems, often without thought to the impacts of such experiments to the system or other environmental or societal considerations. The result is that many applications of adaptive management are poorly thought through and unsuccessful. The most suc-cessful applications seem to be on small-scale and relatively simple systems, but Gregory et al. suggest that it may be on the large and complex and messy systems that it may actually be needed most. In this context, structured decision-making frame-works such as the Bayesian Decision Networks or Info-Gap analysis discussed above may be helpful in providing a formal structure within which different stakeholders can interact in deciding on adaptive management strategies (e.g. Failing et al., 2004; Andersson, 2004; Olsson and Andersson, 2007; Croke et al., 2007). For readers who are interested in pursuing this topic further, a good starting point is the discussion of criteria for deciding whether adaptive management is an appropriate strategy in Gregory et al. (2006).

6.14 Uncertainty and the precautionary principle in decision making

There is one aspect of uncertainty in relation to environmental policy formulation and decision making that has generated an enormous amount of discussion and literature. This is the question of when, in the face of scientific and other uncertainties about what might happen in the future, the precautionary principle should be invoked. Krayer von Krauss et al. (2005) discuss this issue and point out that proponents of activities that might potentially have detrimental effects often use scientific uncertainty as an argument for postponing or relaxing regulation. Environmental activists, on the other hand, often use scientific uncertainty as a reason for invoking the precautionary principle that no change should be allowed if there might be unforeseen environmental impacts. At this rather superficial level, the issues are quite clear. The practical applica-tion of a precautionary principle, however, appears to be much more difficult (Harremoës et al., 2001). This is, in part, because there is a linguistic uncertainty about what is meant by a precautionary principle, and also, in part, because there is no general agreement about how uncertainty in impacts might be assessed in relation to its implementation when there might be potential benefits of allowing a particular development (Weiss, 2006) and where the impact might involve extreme, low-frequency events (Basili, 2006). A full discussion of this issue is beyond the scope of this book but the interested reader might find that looking at the special issue of *Water Science and Technology* that contains the Krayer von Krauss et al. (2005) paper will be a good starting point, since there are a variety of different viewpoints on the issue and a number of case studies presented.

6.15 Summary of Chapter 6

Many decisions about environmental systems are still made without adequate account being taken of the uncertainties inherent in assessing the response of those systems. The following points summarise the issues addressed in this chapter.

- Decisions require assessments of relative risk, and formal risk-based decision-making methodologies are available to provide a framework for structuring the decision-making process including the use of decision trees and utility functions. These methods can be extended to include non-probabilistic uncertain information that might be important to a particular decision.
- But not all uncertainties are easily quantified in a formal decision-making framework. In particular, there may be Knightian or epistemic uncertainties that cannot be easily formulated in terms of utility functions. The Info-Gap methodology provides a method for decision making in this context, in a way that leads to *robust-satisficing* decisions, that seems to explain "sub-optimal" (in terms of utility) decisions made in a variety of different real applications.
- Framing the context of a decision is important in revealing the range of stakeholders that should be involved. Explaining uncertainty in decision making to different groups of stakeholders might require a form of *translationary discourse* to facilitate communication. A first attempt at implementing such a system has been based on the NUSAP concepts in the Netherlands.
- Uncertainty, and particularly epistemic uncertainty about environmental systems and the future inputs to which they might be subjected, implies taking robust decisions that allow adaptive management, while being aware of the limitations of applying the concepts of adaptive management to complex environmental systems. Monitoring and review of decisions should be an essential part of environmental management.

Box 6.1 Basic risk-based decision theory

This necessarily brief exposition of risk-based decision theory follows largely the normative approach to decision making set out in Bedford and Cooke (2001) where much more detail will be found (see also French, 1986; Keeny and Raiffa, 1993). It will be assumed that the decision maker, in discussion with relevant stakeholders, has formalised the context of the problem in terms of a decision tree (Figure 6.1 in the main text) that represents different sets of actions and their consequences.

B6.1.1 Evaluating preferences for different sets of actions

The fundamental action in rational decision making is to rank different sets of actions (or pathways through a decision tree). Thus, given N sets of actions A_1, A_2, A_3 . . . . A_N, we wish to obtain a ranking such that the value of $A_1 \geq A_2 \geq A_3 \geq$ . . . . A_N, where the subscripts now represent the order of preference rather than any original numbering of options. This requires that the value of all the sets of actions be *commensurate*, and a commonly used scale is that of net monetary value. Sets of different options for which the utilities are not comparable (are incommensurate) cannot be ranked in any rational way. There may, of course, be more than one type of utility measure that could be assigned to the sets of actions, to some of which the decision maker might wish to assign different weights. Multiple utility measures are considered further in B6.1.2 below.

Given the uncertainties in the potential outcomes of a set of actions, the decision maker must formulate preferences in an uncertain decision space representing all possible sets of actions. A theory for doing this was provided by Savage (1972) who showed that a rational decision can be represented in terms of probability for any point (set of actions) in the decision space, and expected utility for that point, such that the product of probability and utility orders the actions in the same order as a rational decision about preferences. The estimates of probability and uncertainty might themselves be uncertain, of course, in which case an integral product over some form of representation of that uncertainty would be required.

Application of this theory therefore requires the definition of the probabilities associated with different pathways in the decision tree and a utility value for every end point on the tree, taking account of uncertainties as necessary. The theory is then employed to define the set of preferences. This will not always lead to a clear preference for one set of actions over another (there may be *indifference* to different sets of actions), particularly when there is more than one measure of utility applied to the different sets of actions. Indifference might be resolved by the decision maker weighting the different measures of utility (Section B6.1.2), or by investing in obtaining more information to clarify the decision (Section B6.1.3).

There are some other ways of formulating preferences. The *maximin criterion* ranks alternatives on the basis of their worst possible potential outcomes. The decision maker would then choose the action for which the worst potential outcome is better than any other worst potential outcome. An equivalent *minimax criterion* can be used with measures of disutility, so that the decision maker would choose the action which minimises the maximum potential loss.

Savage (1951) also suggested the *minimax regret criterion* which aims to minimise the difference between the best that could happen and what might actually happen (there is a certain similarity here with the robustness and opportuneness functions of the Info-Gap theory described in Box 6.2). The concept therefore is to take an action that is associated with the least regret in the future. Another way of expressing this is that it minimises the future loss if what actually happens turns out to be as bad as it could be. This would be one way of trying to minimise the risk of making a real mistake.

B6.1.2 Multiple utility measures

It will often be the case that a decision must take account of multiple benefits and disbenefits of a decision, i.e. multiple utility functions. Different utility functions may then be competing in the sense that any solution must present a trade-off between different types of benefits. We are then effectively faced with a multi-criteria utility optimisation problem, for which there may be no single solution (see Chapter 4 where the same problem arises in the context of model parameter calibration). Such trade-offs can lead to indifference that results from balancing different criteria (in the same way as the Pareto optimal set in the calibration problem). What is then needed is a means of assessing the indifference sets over all the criteria, and of ranking those indifference sets in order of preference to maximise the utility of the decision made.

This can be difficult, since the different criteria may have different scales (they may not all be expressed as monetary benefits) and the decision maker may wish to give them different weights. There may also be dependencies between utility measures that may need to be taken into account. The problem therefore needs to be simplified. One set of possible assumptions is that of Multi-attribute Utility Theory (see Keeny and Raiffa, 1993).

Multi-attribute Utility Theory is a means of assessing multiple utility measures given uncertain outcomes. A basic assumption is that of independence of different measures of utility. This assumption greatly simplifies the mathematics to the extent that non-independent measures are often transformed to try to ensure independence. A multi-attribute utility function is then a linear function of the individual marginal utility functions contributing to the decision process. Different types of independence assumptions are possible. *Additive independence* means that the multi-attribute function can be written as a simple weighted sum of the marginal utility functions. An interesting application to a watershed reclamation problem is provided by Elshorbagy (2006). On the basis of multi-year simulations a multi-criteria decision analysis is used to estimate the dominance of one strategy over others in terms of a "probability of making the right decision (PRMD)". A confidence index is also estimated to assess the robustness of the PRMD.

B6.1.3 Informing the decision: pre-posterior prior analysis and the value of additional data in Bayesian decision making

In many situations there is the potential to gain additional information to help guide a decision but at additional cost. This might be particularly valuable, for example, where there is indifference to a set of actions that maximise the utility function. Not all additional information will be cost-effective, however. In some cases the cost of obtaining the information may be greater than the expected benefits. In other cases, there may be a chance that, having commissioned an additional study, the information gained may not prove to be sufficient to decide between competing alternatives. Thus, it would be useful to be able to evaluate the potential value of obtaining additional information.

Additional research can inform both components of the decision analysis: the probability of potential outcomes and the utility function. Refining the estimates of probabilities is easily done by the application of Bayes equation (see Section

4.3). Thus, given some prior probability distribution of outcomes $p(\Omega)$, and a further assessment of the likelihood of Ω given the additional evidence E as $p(E|\Omega)$, then the posterior distribution is calculated using [4.1] as

$$p(\Omega |E) = p(\Omega) \, p(E| \Omega) \, / \, C \qquad \qquad [B6.1.1]$$

The additional costs of obtaining the new information should be accounted for in assessing the utility function. This is evidently most easily done if the utility function is expressed directly in monetary terms.

Refining the utility function associated with the potential outcomes will also generally incur a cost in obtaining additional information. Here, it may be difficult to assess *a priori* what impact on utility the additional information will have. It is readily apparent, however, that any additional study will not be justified if the costs will be greater than any potential benefits to be gained in maximising the utility function.

In both these cases the concept of the *value of perfect information* is useful. This is defined as:

$$V_{PI} = \sum_{i=1}^{\Omega} p(i) U_{max} (i) \qquad \qquad [B6.1.2]$$

where the summation is over all possible outcomes $i = 1, 2, \ldots \Omega$, $p(i)$ is the prior probability of i and $U_{max}(i)$ is the utility of the optimal decision given an outcome. V_{PI} therefore represents a mean optimal utility assessed *as if* we had perfect additional information. We can then assess the cost of gaining additional information relative to the utility of assuming perfect information to decide whether to go ahead with refining our estimates and the associated costs, and whether it will actually make any difference to the decision. Clearly if the additional information makes no difference to the preferences of the decision maker, then it might not be worth collecting the additional information even if the estimates of $U_{max}(i)$ might be refined by doing so.

In adaptive management strategies it is normal to continue to monitor the system of interest in deciding about future management actions. In this case the monitoring programme might be set beforehand to observe a particular state of the system or to meet the demands of legislation rather than being optimal in informing a decision. The information collected can be used to refine/update a model over time as well as determining the utility of different management strategies. An ecological example of the use of Bayesian decision making in the management of grasslands is given by Dorazio and Johnson (2003).

B6.1.4 The expected value of including uncertainty in making decisions
It has been argued in Chapter 6 that it is important to allow for uncertainty in decision making in that it might actually improve the decision that is chosen. It would be advantageous to be able to quantify this improvement to make a convincing case for all the extra work involved. Morgan and Henrion (1990) suggest a measure for the expected value of including uncertainty within a Bayesian decision framework.

The value of including uncertainty can be calculated as:

$$V_{IU} = \sum_{i=1}^{\Omega} p(i)[L_d(i) - L_B(i)] \qquad [\text{B6.1.3}]$$

where L_d is the loss associated with the deterministic optimum decision that minimises loss ignoring uncertainty and L_B is the decision that expected loss over the prior probability density $p(i)$ of the $i = 1, 2, \ldots \Omega$ options.

Box 6.2 Info-Gap decision theory

The Info-Gap methodology of Yakov Ben-Haim (2006) assumes that some "best estimate" description of the system of interest is available (as for any form of decision). As noted in the main text, that description might be a quantitative model of the types that have been discussed earlier in this book; it might be a probabilistic representation; it might be a purely qualitative description. This is the nominal description of the system. The assessment of uncertainty in the Info-Gap methodology is based on the very general idea that, as we move away from this nominal description, then the utility of the outcome of any decision will become more and more uncertain in some consistent way.

As before, let $M(\Theta)$ be the set of predicted variables for a model run with set of parameters Θ. Let $\tilde{M}(\Theta)$ be the outputs from nominal run of the model. Then, over a large number of models (different parameter sets, different boundary conditions, different model structures), we can evaluate for every set of outputs the InfoGap function (using similar symbolism to Ben-Haim, 2006)

$$U\left(a, \tilde{M}(\Theta)\right) = \left\{ M(\Theta) : |M(\Theta) - \tilde{M}(\Theta)| \le a \right\} \quad a \ge 0 \qquad [\text{B6.2.1}]$$

$U(a, \tilde{M}(\Theta))$ is the set of all possible models whose deviation from the nominal model is nowhere greater than a, noting that the model output $M(\Theta)$ may be varying in space and time. The greater the value of a, the greater the range of model outputs that will be included in the set. Thus, the sets for different values of a will be nested and, generally, will be convex (a convex set is one in which the line segment joining any two members of the set is always within the set). Note that this refers to functions of the model *outputs*; it does not imply that the associated sets of individual model parameters will be nested or convex because of the complex nonlinear interactions that might be involved in a particular model structure. Note also that there is no explicit representation of model error in this representation, only of deviation of a predicted variable away from some nominal value (which might itself differ significantly from the actual response of the system in cases of real uncertainty and lack of information about the nature of the system responses).

Given the deviation, a, Ben-Haim frames the decision-making process in terms of two functions representing robustness and opportuneness. The *robustness function* expresses the greatest level of uncertainty (in terms of the values of α) for which failure should not occur. The *opportuneness function* is the least level of uncertainty at which success is assured. The decision maker will then wish to maximise the value of α for which the decision is robust, in the sense of satisfying the basic requirements of success, while minimising the value of α at which success is assured. Ben-Haim (2006) discusses in detail the possible forms of robustness and opportuneness functions that might be used in making decisions in different contexts, with a variety of practical examples.

Equation [B6.2.1] is only one possible relationship that might be used to describe the deviations from the nominal model run. Ben-Haim (2006) provides a number of alternatives, including an energy-bound model for a vector of model outputs $M(\Theta,t)$ (such as a time series of model output variables):

$$U\left(a,\tilde{M}(\Theta,t)\right) = \left\{M(\Theta,t): \int_0^\infty \left(M(\Theta,t) - \tilde{M}(\Theta,t)\right)^2 \leq a^2\right\} \quad a \geq 0 \qquad \text{[B6.2.2]}$$

A further representation is a form envelope-bound model that is useful when the deviation of different variables might best be evaluated on a fractional scale, relative to the nominal value. Thus:

$$U\left(a,\tilde{M}(\Theta,t)\right) = \left\{M(\Theta,t): \left|\frac{M(\Theta,t) - \tilde{M}(\Theta,t)}{\tilde{M}(\Theta,t)}\right| \leq a\right\} \quad a \geq 0 \qquad \text{[B6.2.3]}$$

More generally, envelope-bound models can have a general functional form:

$$U(a,\tilde{M}(\Theta,t)) = \{M(\Theta,t): |M(\Theta,t) - \tilde{M}(\Theta,t)| \leq a\varphi(t)\} \quad a \geq 0 \qquad \text{[B6.2.4]}$$

where $\varphi(t)$ is a function defining the envelope (such as that shown in Figure 6.6 in the main text).

This general envelope-bound model can also be extended to multiple vectors of different outputs. For the case of N output variables, for example, a possible representation of the deviations is:

$$U(a,\tilde{M}(\Theta,t)) = \{M(\Theta,t): |M_n(\Theta,t) - \tilde{M}_n(\Theta,t)| \leq a\varphi_n(t): n = 1,\ldots,N\}$$

$$a \geq 0 \quad \text{[B6.2.5]}$$

Thus, with an appropriate choice of envelope function $\varphi_n(t)$ for each of N output variables, the deviations may be compared directly in terms of a consistent value of uncertainty measure a.

When evaluating models as simulators of the real system, the nominal model output can be replaced by observed values, in which case the deviation becomes the total model error. Thus, if $\varepsilon_n = (M_n(\Theta,t) - O_n(t)): n = 1,\ldots,N$, we can rewrite [B6.2.5] in the form:

$$U(a,M(\Theta,t)) = \{M(\Theta,t): -a\varphi_n(t) \leq \varepsilon_n \leq a\varphi_n(t): n = 1,\ldots,N\} \quad a \geq 0 \quad \text{[B6.2.6]}$$

There is an analogy here between both fuzzy measure and generalised likelihood approaches to uncertainty. Note, however, that unlike both fuzzy and GLUE approaches, the value of a does not have to be fixed in evaluating which models will have more weight in prediction. The value of a simply expands to include a wider and wider range of predictions.

Ben-Haim (2006) also gives details of Fourier-bound models, Info-Gap representations of uncertain probabilities, discrete option (rather than continuous a) models, and non-convex Info-Gap models (including linear systems with uncertain coefficients). The general envelope model [B6.2.6] will serve our purpose to explain the step to decision making here. The discrete model is often of interest in environmental applications since it represents a situation in which only discrete scenarios of the future can be evaluated. In the discrete model, there may be J possible scenarios and we are interested in the deviation of each scenario from some base case. We can represent this in terms of some preference vector π within which the elements represent the integer rank of each scenario in terms of deviation from the base case $\pi^{(o)}$. It is possible that two scenarios j_1 and j_2 might have equal rank $\pi(j_1) = \pi(j_2)$. Then, if Π^k is the set of preference ranks that are no more k single preference changes away from the base case, then an Info-Gap uncertainty model can be defined in the form:

$$U(a,\pi^{(o)}) = \left\{ \bigcup_{k=0}^{a} \Pi^k \right\} \quad a = 0,1,2,\ldots \qquad \text{[B6.2.7]}$$

such that $U(a, \pi^{(o)})$ is the set of all scenarios that differ from the base case by no more than a changes in rank. As a increases, this structure provides nested sets of scenarios of increasing deviation from the base case.

Every model run will be associated with some consequence relative to the decision that is being considered. Since, for any given value of a, we would normally expect to be encapsulating a range of potential models, all with scaled deviations less than a, then as a increases, this implies a wider range of consequences. However, it would be expected that, since models with similar values of a by definition have similar outputs, the consequences associated with models with similar values of a will be similar (if we make proper allowance for the fact that the uncertainty scale does not differentiate between positive and negative deviations but that in some problems the consequences of positive deviations might be quite different from the consequences associated with negative deviations). However, we expect the range of possible consequences to change broadly with the uncertainty measure. In particular, as a gets larger, the potential for a decision based on the model predictions to be wrong will also get larger.

This is where the concept of robustness becomes important. In particular, we are interested in the maximum value of a at which all the minimum requirements for the outcome of a decision are satisfied. If we are happy that we can estimate the potential outcomes to within this level of uncertainty a then we will have a decision that is robust to uncertainty and unlikely to fail. Ben-Haim expresses robustness in the form that given a decision vector (q) then

$$\hat{a}(q) = \max \{a : \text{minimal requirements always satisfied}\} \qquad \text{[B6.2.8a]}$$

Opportuneness, on the other hand, is defined in a complementary way as the minimum value of a for which the decisions (q) will lead to total success:

$$\hat{b}(q) = \min \{a : \text{sweeping success is possible}\} \qquad \text{[B6.2.8b]}$$

Small uncertainty would generally give a high possibility of success but with a high possibility of making the wrong decision in the face of uncertainty. Together, Ben-Haim calls robustness and opportuneness, the *immunity functions* in the decision-making process.

Interpreted in environmental modelling terms, we could say that, if we based a decision on a single deterministic model outcome (perhaps the "optimal" model), then *if the model was correct* we would have a high chance of total success in our decision. Given the uncertainties in environmental models, however, this would be very unlikely to be a robust decision. We will often be in this type of situation of trying to balance the potential for total success and robustness to uncertainty.

It is therefore quite important how these immunity functions are defined in relation to the consequences of a decision. Consequences are usually defined in terms of some form of reward (monetary or otherwise). In the context of environmental modelling let this be defined as $R(q,u)$ where u is an outcome within the set $U(a,\bar{M}(\Theta))$ (see Figure B6.2.1). Robustness can then be defined as the minimal satisfactory value of $R(q,u)$, r_c. Opportuneness can be defined in terms of some significant outcome level of $R(q,u)$, r_w, where we would expect $r_w \gg r_c$. Equations B6.2.8 can then be expressed, for any level of uncertainty a, as:

$$\hat{a}(q,r_c) = \max \{a : (\min R(q,u)) \geq r_c\} \qquad \text{[B6.2.9a]}$$

$$\hat{b}(q,r_w) = \min \{a : (\max R(q,u)) \geq r_w\} \qquad \text{[B6.2.9b]}$$

These definitions can easily be extended to the cases of multi-variate measures of reward. Satisfying these conditions of both robustness and opportuneness implies that the sets of outcomes consistent with the specified levels of r_c and r_w are not empty. If there are no outcomes consistent with r_c then any decision will be vulnerable to uncertainty and there is no possibility of a robust set of decisions. If there are no outcomes consistent with r_w then there is no opportunity for a highly beneficial outcome. In either case it may be necessary to revisit the problem formulation to see if some of the uncertainties can be reduced, or consider changing the values of r_c or r_w.

Ben-Haim makes the point that the decision maker need not just accept the uncertainties associated with a particular problem, particularly when faced with a situation where it appears that realistic levels of uncertainty do not allow a robust solution. By collecting more data or clarifying certain contentious issues it might be possible to reduce the uncertainties in representing the system to allow the *preferences* between different sets of decisions to be made clearer.

Preference is an important word within the Info-Gap framework. Essentially, we are aiming to establish preference of one set of decisions over another, given

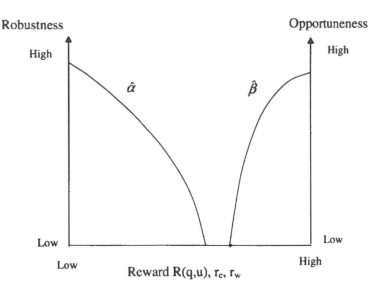

Figure B6.2.1 Robustness and Opportuneness Curves for different levels of Reward

the uncertainties in representing the system. Given the uncertainty, these preferences cannot necessarily be uniquely established (in the same way that we cannot easily differentiate models as hypotheses of how the system is working). Nor can we necessarily quantify the degree of reward by which one set of decisions should be preferred over another (the basis of decisions in utility function theory in Box 6.1), but the robustness and opportuneness functions provide a framework within which to try and express the *relative* preference of different potential decisions. We can illustrate this as follows.

In the flood defence example of Info-Gap decision analysis in the main text, the reward for protecting a certain place can be expressed in (approximate) monetary terms in terms of the damages expected to be saved over the design lifetime of the defences given the uncertain occurrences of future floods. The higher the defences, the greater the damages that will be saved, but the greater the costs (and, often, the more intrusive the defences are for the local community). The windfall reward in this case is the saving in construction and maintenance costs by minimising the height and extent of the defences. We can then (for a simple case) represent the robustness in terms of savings in damages, and the opportuneness in terms of savings in costs. Since these are both on the same (monetary) scale, they can be represented graphically by plotting the uncertainty measure a against the critical values r_c and r_w. The robustness curve $\hat{a}(q,r_c)$ and opportuneness curve $\hat{b}(q,r_w)$ are then trajectories in this space (Figure B6.2.1).

This figure may be interpreted as follows: as the minimal required gain in protection r_c is increased, robustness to uncertainty necessarily decreases. As the critical required saving in costs (the windfall reward, r_w) is increased, the opportuneness increases. When a is high then the opportunities for savings in costs will be high (r_w is high), but the risks of failure will be high (r_c is low). As uncertainty decreases, then the possibility for windfall reward will get smaller since the costs of the necessary design will be more certain (r_w is low), but the risks of failure will decrease (r_c is high). The trade-offs evident in the decision are apparent. In stating a difference in preference between different decisions, a decision maker must weigh the balance of robustness and opportuneness (or try to reduce the uncertainty).

Given two sets of decisions q and q', a decision maker would normally prefer an option for which $\hat{a}(q,r_c) > \hat{a}(q',r_c)$, or more generally for a set of all feasible decisions:

$$\hat{q}_c(r_c) = \arg \max \hat{a}(q,r_c)   q \in Q \qquad [B6.2.10]$$

This choice $\hat{q}_c(r_c)$ is called the *robust-satisficing* decision, given a choice of r_c. It might depend on the choice of r_c, depending on the shape of different $\hat{a}(q,r_c)$ curves for different q in the type of plot shown in Figure B6.2.1 (since different $\hat{a}(q,r_c)$ curves might intersect).

This is clearly not the only way of preferring one decision over another, however. Where cost savings are important, a decision maker might be prepared to take a greater risk and state preferences on the basis of opportuneness. In this case the preference, for a chosen reward of r_w, would be of the form:

$$\hat{q}_w(r_w) = \arg \min \hat{b}(q,r_w)   q \in Q \qquad [B6.2.11]$$

It is very likely that these two preferences $\hat{q}_c(r_c)$ and $\hat{q}_w(r_w)$ will be different (and might vary with the choice of r_c and r_w). In most cases, however, a robust-satisficing decision will be preferred.

It is worth noting that the robustness curve can also be looked at in another way. The trajectory of $\hat{a}(q,r_c)$ shows that robustness decreases as we reduce the uncertainty level allowed for. In particular, it goes to zero when we are prepared to rely on a single representation of the future behaviour of the system. This might be an optimal model, or "best estimate". The critical level of reward in this case will be maximised, but the protection against failure of that decision will be minimal. For more details of the flood defence example, see the main text in Section 6.10.1. Another example from the ecological literature is a study of the management of the Sumatran rhino in Regan et al. (2005). They show that a previous risk-based analysis that assumed that probabilities and utilities were known should have been expected to fail in the face of uncertainties.

Chapter 7

An uncertain future?

> It is quite conceivable that our society will tire of devoting its wealth to science, especially if the implied promises held out when big projects are launched do not materialise ... It is as much out of a concern for their own survival that scientists must acquire the habit of scrutinizing what they do from a broader point of view than has been their custom.
>
> Alvin Weinberg, 1967

7.1 So what should the practitioner do in the face of so many uncertainty estimation methods?

In what has gone before, this book has attempted to clarify the different notions of uncertainty associated with environmental modelling. In particular, we have looked at forward uncertainty analysis in Chapter 3, the learning process associated with having historical observations in Chapter 4, the case of real-time forecasting in Chapter 5, and how uncertainty enters into decision making in Chapter 6. In all of these cases, the predictions made will be uncertain, but will often have been made for a purpose, and some of the different ways in which uncertainties can be taken into account in the decision-making process have been outlined in Chapter 6. Throughout, there has been a recognition that achieving "*the*" model of an environmental system may be fraught with difficulty. Instead, there may be many different model structures, and parameter sets within model structures, that are consistent in some sense with the uncertainties in the available data, and many different ways of estimating uncertainty in the predictions.

From a practitioner's point of view this would appear to give rise to a certain difficulty. Just how should a practitioner decide on what models and uncertainty estimation methods to use? The choice is not always that clear but I hope that this book has at least provided a guide to such decisions for different purposes and in different circumstances, starting with the decision tree for choosing different uncertainty methods in Chapter 1 and expanding on the various available techniques in the later chapters.

This problem of deciding which technique to use may be, to some extent, a transitional problem. There may be a future consensus about which techniques to use as we learn more about how to estimate the uncertainties associated with environmental systems. There is, after all, a certain commonality amongst all the available techniques. Most uncertainty analyses involve the following steps:

1   The choice of model(s) to be considered.
2   The choice of prior assumptions about uncertainties in the driving variables and model parameters.
3   The choice of how to evaluate model predictions based on the prior assumptions against observations, with either implicit or explicit assumptions about the nature of different types of errors, to derive posterior uncertainty estimates of inputs and/ or model outputs.
4   The choice of how to use the posterior uncertainties to make decisions based on the predicted future behaviour of the system under study.

These steps are common whether the assessment is being made within a formal statistical framework with prior and posterior probability distributions, or within a fuzzy framework, or within a scenario framework (including dealing with epistemic or "true" uncertainty), or for simulation or real-time forecasting purposes.

Step 3 above depends, of course, on the availability of observations that can be used to evaluate model predictions (either historical data in evaluating simulation models, or real-time data in forecasting). In the absence of such observations, only a forward uncertainty or scenario analysis is possible, based on the expert judgement of the modeller. This is the crucial step, however, and even where observations are not yet available in a particular study, then the practitioner should consider whether there might be cost-effective ways of making either observations or "soft" information available that can be use to condition the prior model predictions and constrain the prediction uncertainties.

The effect to which this is possible depends critically on the information content of the observations or other conditioning data. In effect, the different methods that have been presented in earlier chapters differ primarily in the assumptions that they make about the information content of observations (including model inputs) and how that information interacts with the predictions of (generally) nonlinear models. In particular, the choice of a suitable likelihood measure in making use of different types of observations in conditioning, remains an open question despite the belief of some that a suitable likelihood for the formal Bayes statistical methodology of Chapter 4 can *always* be found!

Thus, while it would be wonderful to be able to close this book by making some recommendations or providing sets of guidelines about which methods are best to use for different types of cases, we are not yet at that stage (see discussion of this issue in Section 4.7). We simply do not know enough about the content and limitations of observations to be able to provide such guidelines. This applies to guidance on the choice of likelihoods and to guidance on the value of different types of observation in constraining the prediction uncertainties.

Why is this all so difficult? Because of the fact that, when we evaluate a model prediction relative to an observation, we can quantify the residual error, but we cannot disaggregate all the different sources of uncertainty (model structure error, input data error, parameter estimation error, commensurability error and observation error) that contribute to that residual. This is simply not a well-posed problem. There will always be multiple solutions that will not be easily differentiated to allow consensus. But perhaps, as we learn more about the real value of different types of data, such a consensus will begin to develop (as is beginning to happen, for example, in methods for using data assimilation in generating ensemble weather predictions, even if these do not fully evaluate all the uncertainties involved).

This leads on directly to the issue of using uncertainty estimates in informed decision making. That there are well established techniques for decision making under uncertainty, as outlined in Chapter 6, does not seem to have been widely accepted in practical environmental management (see Beven, 2006b). Decision makers are not often interested in uncertain predictions of the impacts of different strategies. They want to know what *will* happen, despite the fact that this may be beyond the current capabilities of the science. Here we enter the realm of modelling in relation to research ethics (e.g. Aven, 2007; Ersdal and Aven, 2007; Weiss, 2006). It seems that this widespread reluctance to deal with uncertainty is partly a matter of accepting responsibility. If there is no model uncertainty, then responsibility for the prediction (and by implication the outcome) is partly transferred to the modeller. The modeller, having chosen a model that is "fit for purpose" and implemented it according to accepted criteria, will, in turn, pass the responsibility to the model. If the predictions turn out to be wrong, the decision maker can blame the modeller and the modeller can blame the model. Everyone has done their best within the limitations of the science, but there remains an issue about the ethics of the use of models beyond the range for which predictions can be considered as reliable (see also Beven, 2002a; Frame et al., 2007).

Incorporation of uncertainty into this process brings out some interesting additional issues in this context. In one sense, for the modeller (and model) it is a protection against being wrong, since a wider range of outcomes is predicted. The responsibility is therefore diminished. For the decision maker, however, there is now a real decision to be made about the acceptable level of risk of an outcome. The responsibility is consequently increased. Is this why it has taken so long for risk-based decision making to permeate more widely, or is it just a lack of training in risk concepts in the various stages of the education process? Everyone makes risk-based decisions in everyday life (driving a car, crossing the road, taking a flight in a plane, personal investments, choosing a partner), even though the uncertainties are often great and the perceptions of risk may be purely qualitative and poorly defined (Adams, 1995). Why not, therefore, in more formal quantitative management contexts?

Morgan (1994) stated that, while considerable progress has been made in uncertainty analysis for environmental modelling, it cannot be denied that civilisation has advanced by simply muddling along and not explicitly acknowledging uncertainty. However, he makes some good practical arguments for making uncertainty analysis a component of any project:

- it makes one think about the processes involved and the decisions based on our model results;
- it makes predictions of different experts more comparable and leads to a transparent science;
- it allows a more fundamental retrospective analysis and allows new or revised decisions to be based on the full understanding of the problem and not on only a partial snapshot;
- decision-makers and the public have the right to know all limitations in order to make up their own minds and lobby for their individual causes.

7.2 The problem of future histories – unknowability and uncertainty

Some things will, of course, remain unknowable, particularly about the boundary conditions for predicting system responses into the future. There is no doubt that there is an increasing market for environmental models across the range of scales from local management and remediation tools to global change. The initial conditions for such predictions will be known in broad scope, although the detailed characteristics of even local systems may be difficult to define precisely. Future boundary conditions, however, are clearly the stuff of speculation even in the relatively short term. Longer-term climate predictions using GCM models might be adequate with which to draw "very likely" inferences about the effects of man on the global climate (IPCC, 2007) but cannot be considered reliable, providing highly uncertain predictions of current climate at regional scales even for mean monthly predictions. Despite widespread recent speculation about the impact of climate change on extreme events (floods, droughts), there will be even more uncertainty about the changing probabilities of extremes under changing climatic conditions.

Perhaps, as recognised more than 40 years ago by Alvin Weinberg in the quotation at the head of this chapter, as scientists we need to be circumspect about overselling our predictive capabilities. "To prophesy is extremely difficult – especially with respect to the future" is an old proverb that expresses the problem well. Beven (1993) refers to potentially false prophecies rather than predictions about environmental responses to change. He notes that if false prophets were still put to the sword or stoned to death (or even if they lost a salary increment), we might be much more careful about some of the predictions that we make. One way of being more circumspect is to take the question of predictive uncertainty, and unknowability of some characteristics of environmental systems (what we have referred to earlier as epistemic or Knightian uncertainty), seriously.

As we have seen, there will be uncertainty arising from different possible models and parameter sets and uncertainty arising from different scenarios of possible boundary conditions. The possibilities are numerous and it may be difficult to assess or assign any probability of occurrence or likelihood to each possibility except in some subjective way. As discussed in Chapter 4, we should be wary of the strong conditioning of models that is the result of the application of "objective" formal likelihood methods in real applications when input errors and model structural errors are important. This is a particular problem for application-critical predictions and it might be necessary to take more explicit account of the subjectivities and epistemic uncertainties inherent in the modelling process in decision analysis (such as in the Info-Gap methods of Chapter 6).

Early in this book we saw how any environmental modelling problem could be formulated as a learning process, and how this learning process was analogous to a mapping of the system of interest into a model space. Limitations in knowledge, data, and lack of experimental techniques for discriminating between model scenarios should be expected to lead to a degree of irreducibility of the set of feasible models and consequently to uncertainty in inference and prediction, but we can still pose the learning processes as a scientific methodology.

7.3 But is the uncertainty problem simply a result of using poor models?

If we are honest, then it has to be said that in many types of environmental system the answer must be yes. However, we can qualify that answer in a number of ways. The fact that the necessary scientific understanding or easily applied measurement systems do not exist at the current time does not take away the demand for predictions. That demand will be met by scientists extending the range of predictability of their models. This may be done in good faith, and using the best available models at the time, but may sometimes be beyond the limits of the science. To cite Alvin Weinberg again, he has suggested that many problems of this type may indeed be *trans-scientific* (Weinberg, 1972; see also the discussion of Philip, 1980, with reference to soil science). Examples of this abound, and include most assessments of the impacts of the effects of possible climate change. There is certainly a demand for such predictions and funding available for the necessary research, despite the fact that it may be impossible to assess the inherent uncertainties involved (and therefore we have no idea if the results produced within the current modelling limitations are really meaningful). Given the funding, however, there is then the benefit of scientists extending their science by attempting to meet that demand and, as a consequence, improving the available models. The current generation of coupled global ocean–atmosphere models are considered to be a major advance on previous global atmosphere models in which the oceans appeared primarily as a "flux correction" to the global energy balance. The questions may currently be trans-scientific, but the benefits to improving the modelling capabilities in different areas of environmental science may be tangible. But we make such improvements primarily by a conditioning process, and in particular by rejecting current models as hypotheses where it can be shown that they are failing with respect to the available observations.

There is an issue, however, as to whether there are environmental modelling problems that are truly trans-scientific. Are there indeed limits to predictability for environmental systems? These are, for the most part, systems that exhibit nonlinear responses to input variables and that often have complex positive and negative feedbacks. The theory of nonlinear dynamics suggests that such systems might show chaotic behaviour and therefore be very difficult to predict, if not inherently unpredictable (see for example Lorenz, 1993; Prigogine, 1997; Smith, 2007). Chaotic systems are highly sensitive to the specification of initial conditions. One simple definition of chaos is that predictions originating from initial conditions that are very close will diverge exponentially. Certainly, it is known that many models of environmental systems are sensitive to the specification of initial conditions. This is why current weather forecasting models are generally run in an ensemble forecasting mode, with multiple runs using different initial conditions and why the divergence of the resulting predictions after a few days means that the predictions have utility only for relatively short term forecasts (Kalnay, 2003).

One way of exploring some of these issues is using modelling as a deductive tool to explore the performance of a system of interest *as if* it had the features corresponding to a certain set of assumptions. That is essentially what is being done in global change simulations, since it is known that there are limitations in how far the models reflect even current conditions. The results of such model runs are sometimes called *projections*, rather than predictions of what might actually happen to the system in the future. Although such deductive models are often used, with specified assumptions, boundary conditions and parameter sets, to represent real places (or even the global circulation system) there is no guarantee that the behaviour of the real system will actually correspond to the model. We should still expect some element of surprise in the real system (Smith, 2007).

This then implies that it is important to continue to monitor the system response, and continue to test models as hypotheses of how the system works (at least where the model predictions are important enough to justify the expense). In this way, models can continue to be checked for fitness-for-purpose and, where limitations are found, model structures or parameters can be improved. As we have seen in this book, the many different sources of uncertainty mean that it is very difficult to differentiate model structural error and data limitations in making such assessments. Hypothesis testing and assessing whether a model is fit-for-purpose are consequently also difficult, but it is likely that future developments in modelling technology will concentrate attention on these aspects of the modelling process that, in the past, have been treated in an ad hoc way in most areas of environmental modelling.

7.4 Accepting an uncertain future

New computer technologies seem likely to change the way that environmental models are constructed and used. There will be new hardware and software solutions based on distributed high-performance parallel computers, linked by fast network connections that, to the user, should appear as a single machine (the concept of the GRID). The user should not have to worry about where the data necessary for a project are stored, nor where any computational tasks are run. The possibility of using high-speed computer networking to link together distributed database and computational engines means that it will become possible to couple together models of many more environmental systems across disciplinary boundaries and across national administrative boundaries. This is, in fact, already possible and is already happening on a limited basis as demonstrated, for example, in the regional water resources models under construction in Denmark (Henriksen et al., 2003), in the national environmental management models being used in the Netherlands (van der Giessen, 2005) and in the Europe-wide Flood Alert System at the European Union Joint Research Centre in Ispra, Italy.[1]

These new possibilities raise some really interesting questions about how this type of interdisciplinary prediction, such as that required for implementing the EU Water Framework Directive, or the US Total Maximum Daily Loads framework (TMDL), might be best achieved. In the past, comprehensive modelling systems have been constructed as large complex computer programs. These programs were intended to be general, but have proven to be expensive to develop, difficult to maintain and difficult to apply because of their data demands and needs for parameter estimation or calibration. With new computing technology it will be possible to continue in the same vein, but with more coupled processes and finer spatial and temporal resolutions for the

1 http://efas.jrc.it/.

predictions. It is not clear, however, whether this will result in a real improvement in model accuracy and utility because the problems inherent in the current generation of distributed environmental model do not necessarily easily go away with improvements in space and time resolutions of the component models.

Those problems include the possibility of Type III modelling errors, i.e. the neglect of processes because of lack of understanding of how the system works, that is an ultimate constraint on how well that system can be predicted. We will not, however, learn much about such model structural limitations when we are prepared to accept existing "calibrated" models in making predictions rather than treating models as hypotheses about how the system works that might be rejected (see discussion in Beven, 2002a,b, 2006a).

In the future, one of the features of having the possibility of large catchment, regional or national scale models is that *everywhere* is represented. This is actually already true of course in global climate models and the Earth Simulator in Japan,[2] but the current level of local detail is inadequate to be able to relate predictions to local scales of interest. In the future, however, we will have, *de facto*, environmental models of everywhere at scales that local decision makers can relate to directly. This will be an important issue in relation to the uncertainties associated with predictions. Once all places are represented, *data* may assume a greater importance than model structures as a means to refine the representation of each place within a learning framework. The result may be a new way of looking at environmental modelling that transcends the traditional goal of incorporating all our understanding of the complexity of coupled environmental systems into a single mathematical framework with a multitude of parameters that cannot easily be identified for any particular place (Beven, 2000, 2002b; Young, 2003; Young et al., 2004). This approach does not imply defining a single model of everywhere. That is too inflexible to allow for multiple representations across scales for a given application or for future innovations. Rather it implies an open architecture framework within which the place, rather than any particular model, is the focus of attention.

Consider, for example, the possibility of modelling the subtle (and interdisciplinary) coupling between atmospheric forcing, catchment response, river runoff and coastal interaction with tidally-dominated sea level for flood prediction purposes. Model codes for all these different components are available from different institutes and on different computing systems. Data for the model components might reside in different types of databases on other distributed computing systems. Predictions made with the full system will require the dynamical coupling of many processes and components to capture these subtleties. Built on the fluxes within those models, air and water pollutant transport models and biogeochemical models could additionally be implemented locally within the regional scale domain. Each component should be able to assimilate data transmitted from field sites and assess the uncertainty in the predictions. Such an integrated system should operate both in real time, assimilating data and boundary conditions from larger scale models, and displaying the "current state of the environment", as well as providing the potential to update model predictions into the future under different scenarios.

2 http://www.es.jamstec.go.jp/esc/eng/.

The components could share 4-D/5-D visualisation tools with appropriate inter-active user interfaces. Users will be able to access the current data, visualise predictions for particular locations and play what-if scenario games over different time scales. The structure of the system would be such as to facilitate and even stimulate improvements to the representation of different components and the constraint of predictive uncertainty by field data collection. The potential capabilities of the new computing technology being developed for grid-scale computing underlie all these components, though much could already be achieved using the Web technology of today. Examples of steps towards this type of integrated system (albeit essentially raster-based) include the US Inter-Agency Object Modelling System (OMS, Leavesley et al., 2002) and the UK Coastal Observatory System.[3]

7.4.1 Modelling as a learning process about places

Even if the very best graphical user interface was available, however, all environmental modellers recognise (or, at least, should recognise) that their predictive capability is limited, and that something is learned about the limitations of the modelling process in every application to a new site (though they virtually never say so in published papers or reports to clients – there are clearly strong incentives to be positive, even if this results in making predictions with models that have not actually provided very good simulations of calibration data). In the past, the learning process has tended to be treated as learning about the parameter values required in the landscape to model space-mapping, or, more rarely, about the adequacy of particular model structures (see, for example, the discussion in Beven, 2002a,b). Once everywhere is represented this emphasis will change to a process of learning about the idiosyncrasies of particu-lar *places*, albeit that initially this will most probably be treated as a problem of learning about parameter values appropriate to a place (Beven, 2007). Treating model-ling more explicitly as a learning process will allow a new approach to be taken to this problem based on a methodology that will match scale-dependent model objects, databases and spatial objects in applications within the areas of interest. One of the most exciting benefits of the possibilities provided by the possibility of representing everywhere in environmental modelling is the potential to implement models available from different institutions as a process of learning about specific places. As suggested earlier, it will be possible, in fact, to have models of *all* places of interest. However, as argued by Beven (2000, 2001, 2002a,b), as a result of scale, nonlinearity and incom-mensurability issues, the representation of place will be inherently uncertain so that this learning process should be implemented within the types of uncertainty estimation framework discussed in this book.

7.4.2 Learning about model structures

That is not to say that models of places will not require process representations, nor that current process representations are always acceptably accurate. The approach does not resolve the problem of making errors in the choice of models (whether Type I,

---

3 http://coastobs.pol.ac.uk.

Type II or Type III errors) but, by setting the modelling problem in the context of a learning framework for specific places, it does allow a gradual refinement of how places are represented, including (at least in principle) allowing the rejection of models as they are shown to be incompatible with new data. There will always be a real research question about how detailed a process representation is necessary to be useful in predicting the dominant modes of response of a system, given the uncertainties inherent in representing the processes in places that are all unique and the process of learning about that place. This appropriate complexity issue has become obscured by the desire to build more and more scientific understanding into models, including physical, chemical and biological components (but see Young, 2001, 2003; Young et al., 2004). This desire is perfectly understandable; it is a way of demonstrating that we do understand the science of the environment, but, as discussed earlier, it results in models that have lots of parameter values that cannot be easily measured or estimated in applications to real places.

There is always a certain underlying principle in science that, as we add more understanding and eliminate empiricisms, then the application of scientific principles should become simpler and more robust. This does not seem to have been the case in many environmental models where the observations can often be seen to be inconsistent with model predictions, suggesting that the model could be (and perhaps ought to be) rejected as a hypothesis of how the system is working. In some cases this might arise as a result of checking model performance on more than one performance measure (e.g. Gupta et al., 1999; Freer et al., 2003, 2004; Parkin et al., 1996; Choi and Beven, 2007); in other cases it might be the collection of new types of data that reveal inconsistencies (e.g. the geochemical modelling of the Birkenes catchment discussed by Hooper et al., 1986; the modelling of the Plynlimon chloride data set by Kirchner et al., 2001 and Page et al., 2007).

There are two important points to be made about model inconsistencies. The first is that we might learn more from model rejection than acceptance. Rejection of a model hypothesis, where properly justified, is an important stage in model development and improvement. The second is that model rejection might be purely a result of inadequate boundary-condition data (or observations with which the model is being compared). We would wish to avoid the rejection of a good model in such cases, and should therefore take account of the potential for input and observation error in model evaluation (Beven, 2006a).

This view of assessing uncertainty in environmental models suggests two requirements for work in the future. The first is for creative experiment: collecting measurements that will allow for different hypotheses and assumptions to be tested in a way that *eliminates* some of the set of feasible or behavioural models. This is not a simple task, in that failure in a test can be often be avoided by the simple addition or refinement of auxiliary assumptions (such as heterogeneity of parameter values) that allow underlying model structures to be protected and that many of the possible measurements may not have great power in discriminating between different models and parameter sets (Morton, 1993). The second is for continuing monitoring of sites both to inform robust adaptive decision-making processes and so that the likelihoods associated with particular models can be refined as time progresses within a learning process. It probably remains an open question as to whether this strategy, as it evolves in symbiosis with model development and improvement, will increase or decrease the uncertainty in predictions of future change.

7.5 Future-proofing modelling systems: adaptive modelling, adaptive management

An essential element of this strategy will be the need, as far as possible, to "future-proof" the model and database systems used; avoiding, for example, a strict raster-based approach or a commitment to one particular modelling framework. The key will need to be flexibility and adaptive modelling so that improved model structures can be easily made available, or simpler model structures can be used when adequate for particular purposes. Raster databases will continue to be driven by remote-sensing imaging inputs to the modelling process, and, in some cases, by convenient numerical solution schemes for partial differential equations. However, it is often inappropriate to force an environmental problem into a raster straightjacket. Treating places as flexible active objects might be one way around this future-proofing problem. Defining the spatial domain of a prediction problem, i.e. the *place* with its particular scale and characteristics, would then allow that place, as an active object within a modelling system, to search for appropriate methods and data for resolving that problem, and also for appropriate methods and data for providing the boundary conditions for the problem (which might then involve other modelling or data extrapolation techniques).

There are some interesting implications of such an approach. One is that the variety of modelling methods available to solve a prediction problem might be able to be compared more readily, leading to better understanding of issues of appropriate model complexity for different applications. This will especially be the case if, as part of the learning process, simulations are saved to be compared with later observations of the real outcome (as is routine in atmospheric modelling in the evaluation of forecast skill).

The learning framework that underlies this approach is best suited to systems that are not changing or, at least, not changing rapidly. New data should then allow a refinement of the acceptable model representations and reduction in the predictive uncertainty in modelling particular places. However, most prediction problems involve learning about a system as it is changing (even though such non-stationarities are often neglected in traditional model calibration exercises). Predictions of the effects of current or future change under different scenarios will be even more uncertain than the simulation of current conditions, but there has been very limited work on estimating the uncertainties of potential outcomes in future scenario simulations, and still less on the conditioning of those predictions as the changing conditions are monitored. Data assimilation, in this context of changing conditions, then becomes a tool for following drift in system response (within the limitations of data uncertainties and variability) with the possibility of changing management strategies as the prediction of future responses also changes over time.

There is one very important corollary to modelling as a process of learning about places in this way. If such a learning process cannot proceed without data assimilation, and data assimilation requires the continued collection of data, learning about places will imply both the continued monitoring of the systems of interest (particularly in detecting non-stationarities in the future that might indicate model inadequacies) and more directed, cost-effective, local measurement campaigns to learn more about places of particular significance (flood-risk sites, lake and river sites of particular ecological significance; storm water overflow sites etc). This is, however, likely to happen quite naturally as local stakeholders start to relate much more directly to the local predictions about the future of the system that they have to manage or live in by means of networking and visualisations (e.g. Failing et al., 2004; Olsson and Berg, 2005; Olsson and Andersson, 2007).

It is appropriate to end here by stressing that the estimation of prediction uncertainty should not be considered as the end point of a modelling process (though often presented in this way). It is better thought of as only the start of a longer-term process of constraining the uncertainty by improving our knowledge and observations of the environmental system of interest. During that process, it is well to remember that when models are used for a purpose, then it will be necessary to involve the relevant stakeholders in the modelling process, especially once they can start to relate directly to what is being predicted about their local system.

7.6 Summary of Chapter 7

This chapter has looked at some of the implications of uncertainty estimation in environmental models in looking to the future:

- The future is uncertain, both in terms of what will be the boundary conditions in different applications, and in terms of the more widespread evaluation of the prediction uncertainties involved in making predictions.
- This does not mean that we need to abandon a pragmatic realistic approach to environmental modelling. Learning about modelling environmental systems can be formulated as a process of continued evaluation of models as hypotheses of how the system responds to forcing.
- One of the implications of new data sources and computational power in the future will be that models of everywhere will become a reality at scales to which local decision makers can relate. Some of the issues in implementing models of everywhere in a way that will be robust to future innovation are discussed.
- Models of everywhere will change the nature of how prediction problems are approached. It will become much more an adaptive process of learning about places than the application of particular model structures. This should include the use of data assimilation to improve the representation of process and facilities for learning from the successes or failures of past predictions.
- Uncertainty estimation will need to be an intrinsic part of such systems and any decision support systems that the predictions are intended to inform, in a way that recognises the potential for multiple model structures to be evaluated in representing places.
- Uncertainty estimation is not an end point; it is the start of a process of improving knowledge and data with a view to constraining the uncertainty in future.

Appendix I: A (brief) guide to matrix algebra

Several sections of this book, and particular Boxes 4.1, 5.1 and 5.2 have equations expressed in the form of vectors and matrices. Here a brief revision guide to matrix algebra is given for those readers who may have not used it for a long, long time.

One-dimensional vectors

A one-dimensional column vector has the form:

$$\underline{a} = \begin{Bmatrix} a_1 \\ a_2 \\ \vdots \\ a_n \end{Bmatrix}$$

The transpose of a one-dimensional column vector produces a row vector such that:

$$\underline{a}^T = \{a_1, a_2, \ldots, a_n\}$$

The product of a vector and a scalar variable has the form:

$$\underline{ka} = \begin{Bmatrix} ka_1 \\ ka_2 \\ \vdots \\ ka_n \end{Bmatrix} \qquad \underline{kb} = \{kb_1, kb_2, \ldots, kb_n\}$$

The dot product of a column vector and a row vector produces a single value as follows:

$$x = \underline{ab} = \begin{Bmatrix} a_1 \\ a_2 \\ \vdots \\ a_n \end{Bmatrix} \{b_1, b_2, \ldots, b_n\} = \{a_1 b_1 + a_2 b_2 + \ldots + a_n b_n\}$$

The addition of two-column vectors is a vector with elements that are the sums of each row:

$$x = a + b = \begin{Bmatrix} a_1 \\ a_2 \\ \vdots \\ a_n \end{Bmatrix} + \begin{Bmatrix} b_1 \\ b_2 \\ \vdots \\ b_n \end{Bmatrix} = \begin{Bmatrix} a_1 + b_1 \\ a_2 + b_2 \\ \vdots \\ a_n + b_n \end{Bmatrix}$$

Two-dimensional matrices

A two-dimensional n by m matrix has the form:

$$A = \begin{bmatrix} a_{1,1} & a_{1,2} & \cdots & a_{1,m} \\ a_{2,1} & a_{2,2} & \cdots & a_{2,m} \\ \vdots & \vdots & & \vdots \\ a_{n,1} & a_{n,2} & \cdots & a_{n,m} \end{bmatrix}$$

The transpose of a two-dimensional matrix has the form:

$$A^T = \begin{bmatrix} a_{1,1} & a_{2,1} & \cdots & a_{n,1} \\ a_{1,2} & a_{2,2} & \cdots & a_{n,2} \\ \vdots & \vdots & & \vdots \\ a_{1,m} & a_{2,m} & \cdots & a_{n,m} \end{bmatrix}$$

A matrix is symmetric if $A^T = A$.
A special case of a symmetric matrix is the Identity Matrix, I, where

$$I = \begin{bmatrix} 1 & 0 & 0 & \cdots & 0 \\ 0 & 1 & 0 & \cdots & 0 \\ 0 & 0 & 1 & \cdots & 0 \\ \vdots & \vdots & \vdots & & \vdots \\ 0 & 0 & 0 & \cdots & 1 \end{bmatrix}$$

The inverse of a two-dimensional matrix is denoted A^{-1}. It occurs frequently in the solution of sets of n linear simultaneous equations that can be written in algebraic form as

$$a_{1,1}x_1 + a_{1,2}x_2 + \ldots + a_n x_n = c_1$$

$$a_{2,1}x_1 + a_{2,2}x_2 + \ldots + a_{2,n}x_n = c_2$$

$$\vdots$$

$$a_{n,1}x_1 + a_{n,2}x_2 + \ldots + a_{n,n}x_n = c_n$$

And in matrix form as

$$A x = c$$

where A is an (n by n) matrix of equation coefficients and x and c are column vectors. We can then solve the equations for values of the elements of x in the form

$$\underline{x} = A^{-1} \underline{c}$$

The inverse can only be formed when the matrix has a non-zero determinant. When this is not the case, the simultaneous equations will not have a solution for real values of x. The determinant of a matrix (denoted as $|A|$ or $det(A)$) is a scalar that represents a scale factor for volume when A is interpreted as a linear transformation.

The product of a matrix and a scalar variable has the form:

$$kA = \begin{bmatrix} ka_{1,1} & ka_{1,2} & \cdots & ka_{1,m} \\ ka_{2,1} & ka_{2,2} & \cdots & ka_{2,m} \\ \vdots & \vdots & & \vdots \\ ka_{n,1} & ka_{n,2} & \cdots & ka_{n,m} \end{bmatrix}$$

The addition of two matrices of the same dimensions has the form:

$$A + B = \begin{bmatrix} a_{1,1} & a_{1,2} & \cdots & a_{1,m} \\ a_{2,1} & a_{2,2} & \cdots & a_{2,m} \\ \vdots & \vdots & & \vdots \\ a_{n,1} & a_{n,2} & \cdots & a_{n,m} \end{bmatrix} + \begin{bmatrix} b_{1,1} & b_{1,2} & \cdots & b_{1,m} \\ b_{2,1} & b_{2,2} & \cdots & b_{2,m} \\ \vdots & \vdots & & \vdots \\ b_{n,1} & b_{n,2} & \cdots & b_{n,m} \end{bmatrix}$$

$$= \begin{bmatrix} a_{1,1} + b_{1,1} & a_{1,2} + b_{1,2} & \cdots & a_{1,m} + b_{1,m} \\ a_{2,1} + b_{2,1} & a_{2,2} + b_{2,2} & \cdots & a_{2,m} + b_{2,m} \\ \vdots & \vdots & & \vdots \\ a_{n,1} + b_{n,1} & a_{n,2} + b_{n,2} & \cdots & a_{n,m} + b_{n,m} \end{bmatrix}$$

The product of an n by m matrix and a column vector (m by 1) has the form:

$$A\underline{b} = \begin{bmatrix} a_{1,1} & a_{1,2} & \cdots & a_{1,m} \\ a_{2,1} & a_{2,2} & \cdots & a_{2,m} \\ \vdots & \vdots & & \vdots \\ a_{n,1} & a_{n,2} & \cdots & a_{n,m} \end{bmatrix} \begin{Bmatrix} b_1 \\ b_2 \\ \vdots \\ b_m \end{Bmatrix} = \begin{Bmatrix} b_1 a_{1,1} + b_2 a_{1,2} + \ldots + b_m a_{1,m} \\ b_1 a_{2,1} + b_2 a_{2,2} + \ldots + b_m a_{2,m} \\ \vdots \\ b_1 a_{n,1} + b_2 a_{n,2} + \ldots + b_m a_{n,m} \end{Bmatrix}$$

To form the product of two matrices they must be of size (n by m) and (m by p) so that they have the same inner dot product dimension m. The result then has the form:

$$AB = \begin{bmatrix} a_{1,1} & a_{1,2} & \cdots & a_{1,m} \\ a_{2,1} & a_{2,2} & \cdots & a_{2,m} \\ \vdots & \vdots & & \vdots \\ a_{n,1} & a_{n,2} & \cdots & a_{n,m} \end{bmatrix} \begin{bmatrix} b_{1,1} & b_{1,2} & \cdots & b_{1,p} \\ b_{2,1} & b_{2,2} & \cdots & b_{2,p} \\ \vdots & \vdots & & \vdots \\ b_{m,1} & b_{m,2} & \cdots & b_{m,p} \end{bmatrix}$$

$$= \begin{bmatrix} b_{1,1}a_{1,1} + b_{2,1}a_{1,2} + \ldots + b_{m,1}a_{1,m} & b_{1,2}a_{1,1} + b_{2,2}a_{1,2} + \ldots + b_{m,2}a_{1,m} & \cdots \\ b_{1,1}a_{2,1} + b_{2,1}a_{2,2} + \ldots + b_{m,1}a_{2,m} & b_{1,2}a_{2,1} + b_{2,2}a_{2,2} + \ldots + b_{m,2}a_{2,m} & \cdots \\ \vdots & \vdots & \\ b_{1,1}a_{n,1} + b_{2,1}a_{n,2} + \ldots + b_{m,1}a_{n,m} & b_{1,2}a_{n,1} + b_{2,2}a_{n,2} + \ldots + b_{m,2}a_{n,m} & \cdots \end{bmatrix}$$

It can be seen that the form of matrix multiplication is such that $AB \neq BA$.

Example of matrix multiplication

A typical example of the use of matrix multiplication in this book involves the formation of a cost function that involves the product of a column vector of error terms, its transpose and a square weighting matrix (sometimes the inverse of an error covariance matrix). An example is equation [B4.1.2] in Box 4.1 where:

$$J = \{\varepsilon\}^T \mathbf{W} \{\varepsilon\}$$

If we can evaluate m errors, then this will involve the multiplication of a row vector (1 by m), a square matrix (m by m) and a column vector (m by m). The result therefore will be a scalar variable since the outer dimensions of the multiplication are (1 by 1).

Appendix II: A (brief) guide to software

This appendix gives a brief guide to software packages that might be useful for the different techniques described in the book. All the URL addresses have been checked in July 2007 at the time of submitting the manuscript but there is always the possibility that they will become out of date. A website for use with this book will be found at http://www.uncertain-future.org.uk where the links will be kept up to date and any new links will be added. If any links do not work, please check the site and leave a message if necessary. If you would like to recommend any software that is not mentioned here and that you have found useful, please also leave a message.

General mathematics and statistics packages

There are a number of general mathematical and statistics packages that have extensive facilities for random number generation and other forms of analysis. The best-known are Matlab (which also has Statistics and Fuzzy Toolboxes) from The Mathworks (http://www.mathworks.com/); Mathematica from Wolfram Research (http://www.wolfram.com/); and Mathcad from ptc Inc. (http://www.ptc.com/app server/mkt/products/home.jsp?k=3901). Octave is a freeware program that is broadly compatible with Matlab code (http://www.gnu.org/software/octave/).

Packages that are more specialised for statistical programming, including Bayesian methods, are S-PLUS from Insightful (http://www.insightful.com/). R is a freeware version that mimics much of the functionality of S-PLUS (http://www.r-project.org/).

A number of useful routines will also be found in the Numerical Recipes collection (see Press et al., 2007 and http://www.nr.com/ where some past versions of Numerical Recipes can be downloaded).

Methods for forward uncertainty analysis

There are two well-known add-ins for Microsoft Excel that provide facilities for forward uncertainty analysis. These are @RISK (http://www.palisade.com/risk/default.asp) and Crystal Ball (http://www.crystalball.com/).

The Data Uncertainty Engine (DUE) is a stand-alone package that is of interest because it supports a wide variety of distributional forms for different variables as well as routines for uncertainty in spatial rasters in 2D, spatial vectors in 2D and 3D (including positional uncertainty) and time series (Brown and Heuvelink, 2007).

Uncertainty propagation is through Monte Carlo simulation. The DUE program is available as freeware at http://www.iamg.org/CGEditor/index.htm.

UNICORN (uncertainty analysis with correlations) is a stand-alone package supporting various dependencies between different univariate distributions, generated using copulae. A light version can be downloaded from http://ssor.twi.tudelft.nl/~risk/. The tutorial describes more advanced facilities available in a Professional version. UNICORN supports the exercises in Kurowicka and Cooke (2006).

The EU Joint Research Centre (JRC) Simlab package includes a wide variety of distribution generation functions (https://simlab.jrc.it/docs/html/main.html) with a "cook book" for new users.

Random number generators

Pseudo-code for different random number generators can be found in the Numerical Recipes books (http://www.nr.com/).

Pseudo-code for the Mersene Twister and links to various implementations can be found via the entry in Wikipedia (http://en.wikipedia.org/wiki/Mersenne_twister)

Methods for global sensitivity analysis

The Simlab package noted above includes the Sobol' method of generalised sensitivity analysis. JRC also maintains a sensitivity analysis forum with details of some other software packages (http://sensitivity-analysis.jrc.it/forum/default.asp).

Support for the HSY generalised sensitivity analysis is included in the GLUE packages (see below).

Methods for model calibration/conditioning

John Doherty's PEST (http://www.sspa.com/PEST/index.html) is a general nonlinear regression model calibration package that can be used with any model outputs. It includes facilities for regularisation of high-dimensional problems (see Moore and Doherty, 2006).

The methods described in Hill and Tiedeman (2007) are based on the UCODE_2005 package developed by the USGS and therefore available freely (http://water.usgs.gov/software/ucode.html).

Both PEST and UCODE_2005 routines are being included in the new Jupiter project, details of which can be found at http://www.mines.edu/igwmc/freeware/ucode/.

There are many packages for specific applications that include model calibration facilities. Those mentioned in the text include the USGS MODFLOW groundwater modelling package (http://water.usgs.gov/nrp/gwsoftware/modflow.html) and the USDA CXTFIT package for conservative/reactive solute transport problems with mobile/immobile storage (http://www.ars.usda.gov/Services/docs.htm?docid=8910).

Software to implement the GLUE methodology can be found at the Lancaster GLUE site (http://www.glue-uncertainty.org). A Matlab implementation is also available from the EU Joint Research Laboratory (http://eemc.jrc.ec.europa.eu/softwareGLUEWIN.htm). These packages also include routines for HSY generalised sensitivity analysis.

Software for MC^2 methods

There are a number of software packages for MC^2 sampling, including the S-based WinBUGS (http://www.mrc-bsu.cam.ac.uk/bugs) and the open source R-based equivalent OpenBUGS (http://mathstat.helsinki.fi/openbugs/). Congdon (2006) uses WinBUGS routines as well as some Matlab routines. The Numerical Recipes books provide example algorithms for an MC^2 sampling (http://www.nr.com/).

John Doherty's MICA is a stand-alone package that can take output from any model and use it within an MC^2 algorithm. It is used in Gallagher and Doherty (2007) and is available at http://www.sspa.com/pest.

George Kuczera's NLFIT, originally a model calibration package based on nonlinear regression techniques, now includes MC^2 methods within a Bayesian framework (http://www.eng.newcastle.edu.au/~cegak/).

Software for fuzzy methods

A Fuzzy Logic Toolbox is available for Matlab. Mathematica also includes functions for fuzzy methods. There is a wide variety of other packages available, many of which are reviewed at http://www.fuzzysoftware.org/software.htm where there are links to sources of freeware code. C++ code with some useful functions comes with the book by Cox (1994).

Software for real-time forecasting

Matlab includes a Time Series Analysis and Forecasting Toolbox. Another Matlab toolbox that incorporates most of the methods developed by Peter Young is CAPTAIN (http://www.es.lancs.ac.uk/cres/captain/).

WinBUGS also includes support for time series analysis and forecasting using Bayesian methods.

Software for decision analysis

Reviews of software for decision analysis may be found in Uusitalo (2007) and at http://faculty.fuqua.duke.edu/daweb/dasw.htm.

Some particular examples are DPL from Syncopations (http://www.syncopationsoftware.com/), the Excel add-in Decision tools from Palisade (http://www.palisade-europe.com/decisiontools_suite/) and an Excel add-in for Bayesian Belief Networks FCBeNe (Varis, 1997).

Software for parallel computation

CONDOR is a useful piece of software for running Monte Carlo simulations on networks of PCs. More information may be found at http://www.cs.wisc.edu/condor/manual/v6.1/1_Overview.html.

Glossary

Actualism A philosophical position that holds that everything that there is must be actual; contrasts with possibilism that allows that some things in reality that have not been experienced might be possible.

Aleatory Uncertainty The way in which a quantity varies in some random stochastic way in a system. Often used in contrast to *epistemic uncertainty*.

Alpha(α)-cut Used in fuzzy set theory to define a level of uncertainty with respect to the range of a *degree of membership* function (normally [0–1]).

Attractor A concept from nonlinear system dynamics to denote a trajectory of the system to which all other trajectories will tend given sufficient time. Attractors can be both simple (e.g. a single stable point in *state space*) or very complex (e.g. the butterfly attractor of Lorenz, 1963).

Autocorrelated Errors A time series of model residuals that exhibit correlation at successive time steps. In distributed models, errors might be correlated in both space and time.

Auxiliary Conditions The set of variables (and sometimes hypotheses) that need to be specified to run a model for a particular case. Includes initial conditions, boundary conditions and parameters representing system characteristics.

Axiom A proposition accepted as being fundamentally true on the basis of principle, past research or purely hypothetical speculation. Axioms usually specify that something is or is not true and are the foundation for *deductive reasoning*.

Bayes Equation Equation for calculating a posterior probability given a prior probability and a likelihood function. Used in the GLUE methodology to calculate posterior model likelihood weights from subjective prior weights and a likelihood measure chosen for model evaluation.

Bayesianism/Bayes Explanation Approach in which subjective, empirical (inductive) and theoretical (deductive) information can be combined. Fits in well with current ideas about the sociological context of science and the way it is done in practice.

Behavioural Simulation A simulation that gives an acceptable reproduction of any observations available for model evaluation. Simulations that are not acceptable are non-behavioural.

Black Box Model A model that relates only an input to a predicted output by a mathematical function or functions without any attempt to describe the processes controlling the response of the system.

Blind Validation   Evaluation of a model using parameter values estimated before the modeller has seen any output data.

Boundary Conditions   Constraints and values of variables required to run a model for a particular flow domain and time period. May include input variables such as rainfalls and temperatures; or constraints such as specifying a fixed potential (Dirichlet boundary condition) or impermeable boundary (Neumann boundary condition) or specified flux rate (Cauchy boundary condition).

Calibration   The process of adjusting parameter values of a model to obtain a better fit between observed and predicted variables. May be done manually or using an automatic calibration algorithm.

Chaos   The behaviour of a system in which infinitesimally small differences in initial conditions lead to exponentially divergent system trajectories at later times. Both deterministic and stochastic systems can exhibit chaos. A particularly interesting case is where small stochastic perturbations might lead to quite different modes of behaviour (stochastic resonance). It has been suggested that the history of the global climate system exhibits features of stochastic resonance.

Coherence   A principle of probability theory that expresses that observations should be used in the best possible way to condition probabilities of a possible outcome. Can be applied most rigorously in ideal cases where input errors and model structural errors are negligible. Where such uncertainties are known to be significant it is not always obvious what is the best possible way. A less formal, but more generally useful, definition is that observations should not be used to condition probabilities of an outcome in a way that is clearly worse than some other conditioning process.

Conceptual Model   A hydrological model defined in the form of mathematical equations. A simplification of a *perceptual model.*

Conditioning   The process of refining a model structure, or a distribution of parameter values of a model structure as more data become available (see also *data assimilation* and *real-time forecasting*).

Confirmation   A process of evaluating model outputs to check that a model is still providing acceptable simulations. Now preferred to the terms *validation* or *verification.*

Copula   A method of transformation from a space of scaled unit axes to a complex multivariate distribution with dependencies.

Covariance Matrix   A way of expressing the statistical uncertainty for a set of multiple parameters or variables as a square matrix of coefficients. The diagonal elements in the matrix represent the variance of each individual member of the set; the off-diagonal elements the co-variation between pairs of members. The higher the degree of co-variance, the greater the interaction between pairs of parameters or variables. The covariance can be scaled to represent correlation between the members.

Crisp Set   A set for which the boundaries of the set are defined such that any potential member of the set will either be in the set or excluded.

Critical Rationalism   see *Rationalism.*

Data Assimilation   The process of using observational data to update model predictions (see Chapter 5, also *real-time forecasting* and *updating*).

Deduction   Inference from specific premises to some prediction. Theories based on physical assumptions without resort to empirical generalisations are examples of deductive reasoning. Examples common in mathematics and logic but rare in environmental science.

Deductive Reasoning The process of inference based on *deduction*.

Degree of Membership An expression of the strength with which a member of a fuzzy set is associated with that set. Normally takes the range [0–1], with non-zero values defining the range of support for the fuzzy set.

Deterministic Model A model that with a set of initial and boundary conditions has only one possible outcome or set of predictions.

Disaggregation/Downscaling The process of distributing variables calculated at large scales to estimate appropriate values at smaller scales. Required, for example, in modelling impacts of climate change available at the grid scale of a GCM at scales of local interest.

Distributed Model A model that predicts values of state variables varying in space (and normally time).

Effective Rainfall A part of the storm rainfall inputs to a catchment that is equivalent in volume to the "storm runoff" part of the hydrograph (but note that the storm runoff may not be all rainstorm water).

Empiricism Idea that knowledge must come from experience of the senses (classical form of empiricism from 17th century due to John Locke, 1632–1704). Recent version of constructive empiricism due to Bas van Fraasen in which empirical adequacy does not need to imply truth, especially for variables or entities that are not directly observable (some overlap with *instrumentalism*)

Entropy Used here as a measure of information due to Claude Shannon in 1948 using an analogy with thermodynamic entropy.

Epistemic Uncertainty The way in which the response of a system varies in ways that cannot be simply described by random stochastic variation. Often used in contrast to *aleatory uncertainty*. Also known as Knightian uncertainties (after Frank Knight (1885–1972) who himself referred to "true uncertainties" that could not be insured against, as opposed to risk that could be assessed probabilistically, see Knight, 1921).

Epistemology Study of the possibility and theory of knowledge. Evolutionary epistemology embodies the idea that knowledge will be revised and improved over time.

Equifinality 1 The concept that there may be many models of a system that are acceptably consistent with the observations available, derived from the General Systems theory of Ludvig von Bertalanffy (1968) and adopted in environmental modelling by Beven (1993, 2006).

Equifinality 2 The adaption of the von Bertalanffy concept to geomorphology by Culling (1957) that similar landforms might arise from different processes and histories.

Equifinality 3 The nonlinear dynamical systems version later expressed by Culling (1987, rejecting his earlier view). Culling distinguished between strict equifinality where a perturbed system will return to its original form after some transition time and weaker forms in which equifinality implies only persistence or stability of some property of the system in its trajectory in *state space* (as might be observed, for example, if there is some *attractor* in state space).

Extension Principle   The extension principle is used in fuzzy theory and allows the extrapolation of degrees of membership from members of a fuzzy set to functions of the values of those members. Thus, if the fuzzy set A is defined by the membership values of a discrete set of points in X, $\{x_1, x_2, x_3, \ldots x_n\}$ with membership values $\{\mu(x_1), \mu(x_2), \mu(x_3), \ldots \mu(x_n)\}$, then for any other fuzzy set $f(A)$ that is a function of A, membership will be defined by the membership values $\{\mu(f(x_1)), \mu(f(x_2)), \mu(f(x_3)), \ldots \mu(f(x_n))\}$.

Falsification   Answer to the problem of induction, primarily due to Karl Popper. The idea that science proceeds by setting up theories and then seeking evidence to falsify them. In a strong version, whereas no amount of evidence can completely confirm a theory, one false prediction might be sufficient to falsify it. This is the standard model of the "scientific method", which is now recognised to be largely an ideal that is rarely followed. To avoid falsification, it is more usual to neglect some evidence as "outliers" or modify a theory to take account of the new evidence, perhaps by changing auxiliary hypotheses or calibration parameters.

Formal likelihood   A quantitative measure of the acceptability of a particular model or parameter set in reproducing the system response being modelled based on a formal parametric function to represent the structure of the errors.

Foundationalism   A philosophical tradition based on the idea that there must be a set of basic beliefs from which all other beliefs can be derived. Both rationalism and empiricism are sometimes presented as forms of foundationalism: rationalism because it makes the claim that we must have some innate rational beliefs that cannot necessarily be tested by sense data; and empiricism because basic sense data provide justification for the most basic forms of belief.

Fuzzy Logic   A system of logical rules involving variables associated with a continuous fuzzy measure (normally in the range 0 to 1) rather than the binary measure (right/wrong, 0 or 1) of traditional logic. Rules are available for operations such as addition and multiplication of fuzzy measures and for variables grouped in fuzzy sets. Such rules can be used to reflect imperfect knowledge of how a variable will respond in different circumstances in terms of the possibilities of potential outcomes.

Fuzzy Measure   A degree of membership of a quantity to a fuzzy set (see *degree of membership, fuzzy logic*).

Fuzzy Set   A set of quantities, thought to have something in common, but for which membership of the set cannot be described precisely but only through a degree or grade of membership or fuzzy measure.

Gain   A multiplier applied to a transfer function to scale inputs to outputs in a linear systems analysis; may be made adaptive in *real-time forecasting*.

GCM   A global circulation model (or global climate model). Earlier GCMs modelled the atmosphere only but then had to impose "flux corrections" to take account of transfers of heat in the oceans. Recent models used coupled ocean–atmosphere components and do not require such flux corrections.

Genetic Algorithm   A method of optimisation based on treating parameter sets as "genes" that then "evolve" by melding, mutation and off-spring processes.

Global Optimum   A set of parameter values that gives the best fit possible to a set of observations

Hermeneutics   In its modern form (following Martin Heidegger, 1889–1976)

concerned with the interpretation of the being who interprets texts, theories and evidence. Such interpretations are considered to be always historically and socio-logically conditioned. Science is then no more (and no less) than an interpretation of the signs provided by the available texts and evidence. In this there is much in common with the "pragmaticism" of Charles Peirce (1839–1914). Main pro-ponents in science have been Richard Bernstein and Richard Rorty, who suggest that neither purely realist nor purely relativist approaches are appropriate but that it is possible to have a rational approach to scientific ideas within a developing historical framework within which choices and judgements of the participants are crucial.

Heteroscedastic Errors A time series of model residuals that exhibit a changing variance over a simulation period (see also *autocorrelated errors*).

Hypothesis A set of propositions about how a system works. Can be expressed either qualitatively or as a *theory* or model.

History Matching The calibration of a model by adjusting parameter values to reduce the differences between observations and predicted variables.

Idealism A philosophical position that holds that all experiences and understanding are to do with the mind. It is therefore not possible to have direct experience of an external reality, only a mental construct of what that reality might be.

Identifiability The ease with which particular parameter values in a model might be calibrated or conditioned by comparing model outputs to observed variables (see also *calibration* and *conditioning*).

Incoherence The use of observations to condition probabilities in a way that does not properly reflect the information content of the data (see *coherence*).

Incommensurate/Incommensurability Used here to refer to variables or param-eters with the same name that refer to different quantities because of a change in scale.

Independence Two variables are independent if a change in the value of one variable has no effect on the effect of the other.

Induction The inference from experimental evidence to general theory. The "prob-lem of induction" was originally identified by David Hume (1711–1776). This is that past evidence may not necessarily be a good guide to future experience (all swans are white . . .; groundwater flow is Darcian . . .) so that no theory can ever be verified by induction.

Informal Likelihood A quantitative, but subjectively chosen, measure of the accept-ability of a particular model or parameter set in reproducing the system response being modelled.

Initial Conditions The *auxiliary conditions* required at the start of a run of a model to define all initial model states.

Instrumentalism The idea that scientific theories are not true descriptions of reality but only useful constructs that allow experiences to be ordered. Allows that theor-ies do not have to be believed and that there may be different competing theories compatible with the evidence (although realists will counter that even if different theories are compatible with a body of evidence they may not all be equally well supported by that evidence).

Inverse Problem see *history matching*.

Kriging A method of spatial interpolation and variance estimation used in geostatistics. Originally developed for assessing yields in the mining industry by Danie G. Krige in South Africa. Assumes linearity and second-order stationarity and a constant spatial correlation structure as expressed in a *variogram*.

Lead Time   The time required for a forecast to be available so that it can be useful in making decisions about warnings etc.

Learning Set   A set of observed data used in the calibration of a Neural Net model.

Likelihood Measure   A quantitative measure of the acceptability of a particular model or parameter set in reproducing the system response being modelled.

Linearity   A model (or model component) is linear if the outputs are in direct proportion to the inputs.

Linear Store   A model component in which the output is directly proportional to the current storage value (equivalent to a bucket with a hole in the bottom!). The basic building block of the general linear transfer function model and the Nash cascade.

Linguistic Uncertainties   result from the fact that language, including the scientific vocabulary, is often underspecific, ambiguous, vague, context-dependent or exhibits theoretical indeterminacies. Linguistic uncertainties often overlap with *epistemic uncertainties*.

Local Optimum   A local peak in the parameter response surface where a set of parameter values gives a better fit to the observations than all parameter sets around it, but not as good a fit as the global optimum.

Logical Positivism   A representation of the progress of science as an increasingly good description of reality, including the reductionist argument that all science can be reduced to physics and that theories are verifiable. A major influence on modernist science in the 20th century, largely discredited due to Karl Popper.

Lumped Model   A model that treats the whole of a catchment as a single accounting unit and predicts only values of variables averaged over the catchment area.

Marginal Distribution   In a multivariate distribution, the marginal distribution obtained by integrating over all but the dimension associated with the particular variable for which the marginal distribution is required. It is the distribution of that variable conditioned on the distributions of all the other variables in the multivariate distribution function.

Mechanistic Model   A model based on physical principles or with a physical interpretation.

Model   A set of constructs, derived from explicit assumptions, about how a system responds to specified inputs. Quantitative models are normally expressed as sets of assumptions and mathematical equations and implemented as computer codes.

Model Space   A hyperspace defined by the ranges of feasible models and parameter values, with dimensions for each parameter within each model.

Monte Carlo Simulation   Simulation involving multiple runs of a model using different randomly chosen sets of parameter values.

Multiple Working Hypotheses   A scientific method, in the earth sciences commonly assigned to Chamberlin (1896) and Gilbert (1896), that is based on considering all possible explanations of a phenomenon and then subjecting each hypothesis to test. Given the limitations of data and measurement techniques it may not always be possible to reduce the multiple hypotheses to a single explanation.

Nomological System   A formally defined system of theories and concepts in science (as used, for example, by Cartright, 1999).

Non-identifiability   An expression of the problem of identifying parameter values in a model, given limited observational data (see also *equifinality*).

Nonlinear   A model is nonlinear if the outputs are not in direct proportion to the inputs but may vary with intensity or volume of the inputs or with antecedent conditions.

Nonparametric Method   A method of estimating distributions without making any assumptions about the mathematical form of the distribution.

Non-stationarity   A system in which the characteristics are expected to change over time; a model in which the parameters are expected to change over time.

Non-uniqueness   An expression of the problem of identifying parameter values in a model, given limited observational data (see also *equifinality*).

Normally Distributed   A variable is normally distributed if its distribution can be adequately fitted by the Normal or Gaussian distribution function that is symmetrical about the mean, bell-shaped and with infinite tails.

Objective Function   (Performance measure, goodness-of-fit) A measure of how well a simulation fits the available observations.

Ontology   The philosophical study of "being", of "what is there". There is a long history and extensive philosophical literature on ontology!

Open System   A system defined by uncontrolled boundary conditions that involve the exchanges of fluxes (of mass, energy, momentum etc.) with the rest of the world. The boundaries of open systems are often defined for convenience (for example, where a measurement is being made), rather because they can be physically well defined (e.g. a lake).

Optimisation   The process of finding a parameter set that gives the best fit of a model to the observations available. May be done manually or using an automatic calibration algorithm.

Optimisation of Expected Utility   Used in cost–benefit forms of decision making as the process of finding the best option amongst many in terms of utility.

Over-parameterisation   Problem induced by trying to calibrate the parameter values of a model that has too many parameters than can be supported by the information content of the calibration data.

Paradigm   A particular way of doing things. Introduced by the philosopher of science Thomas Kuhn in his explanation of how scientists work within a framework of unquestioned beliefs, concepts and theories that act to constrain the set of possible explanations. Development of science within a paradigm is "normal science". A change from one paradigm to a new set of theories is called a "paradigm shift". Quantities in competing paradigms may be *incommensurate*. Such changes may reflect sociological influences as well as purely scientific evidence (a classic example being the Copernican revolution).

Parameter   A constant that must be defined before running a model simulation.

Parameter Space   A space defined by the ranges of feasible model parameters, with one dimension for each parameter.

Pareto Front   The surface or manifold in the model space that joins all models in the *Pareto optimal set*.

Pareto Optimal Set   The set of models in a multi-objective evaluation that are not dominated by any other model on at least one evaluation measure, i.e. there is no model that performs better on that evaluation measure *and* on another evaluation measure (named after Vilfredo Pareto, 1848–1923, who originated the concept of Pareto efficiency in economics).

Parsimony The concept, sometimes known as Occam's razor, that a model should be no more complex than necessary to predict the observations sufficiently accurately to be useful.

Perceptual Model A qualitative description of the processes thought to be controlling the system response.

Performance Measure (Objective function, goodness-of-fit) A measure of how well a simulation fits the available observations (see also *objectvie function*).

Positivism The idea that ideas evolve over time to come closer to a true description of nature (originally due to Auguste Comte, 1798–1857) (see also *logical positivism*).

Possibility A non-statistical measure of the potential for an outcome or occurrence of an event as an alternative to probability theory. Used in fuzzy set theory, but also has a more general usage (see George Klir, 2006).

Posterior Distribution The statistical distribution of a model parameter or output variable after conditioning on the basis of observed data, for example, using a calculated likelihood in Bayes equation. In a Bayesian learning process, the posterior distribution after one conditioning step may become the *prior distribution* when new observational data are made available.

Prior Distribution The statistical distribution of a model parameter or output variable assumed or calculated on the basis of only knowledge about the characteristics of the system before data are collected.

Probability A statistical measure of the potential for an outcome or occurrence of an event. Many statisticians, most notably Dennis Lindley (2006), believe that probability is the *only* way of expressing uncertainty in a potential outcome. There are a variety of foundations for the estimation and interpretation of probabilities of which the main examples are the Frequentist and the Bayesian views. Frequentists hold that probabilities represent the likelihood of outcomes that would be found if it was possible to take a large number of samples over all potential outcomes. Bayesians recognise that this can only ever be an ideal and that prior (often subjective) estimates of probabilities might be useful as an input to estimating probabilities based on limited amounts of evidence.

Procedural Model A model represented as a computer program. May be an exact approximate solution of the equations defining the conceptual model of a system.

Random Sample A set of realisations of a model or variable generated by making choices from a feasible range of possibilities drawn by selecting pseudo-random numbers from specified distributions (see also *Monte Carlo Simulation*).

Rationalism The idea that things can be true without being directly experienced by the senses. Extended to suggest that certain truths can be known *a priori*. In its extreme form, Bertrand Russell (1872–1970) and Ludwig Wittgenstein (1889–1951) attempted to prove all of mathematics from simple logical axioms. The attempt was destroyed by Kurt Gödel's (1906–1978) incompleteness theorem of 1931.

Realisation One random sample taken from the set of all possible random samples in a Monte Carlo simulation.

Realism   The concept that things exist independently of our perceptions of those things, with extension to the idea that theories can be true representations of nature. Empirical realism, as proposed by Immanuel Kant (1724–1804), suggests that we can have knowledge of the nature of the world through empirical experience and perceptions. Since then there have been many variations on realism including: *transcedental realism* (that the existence and nature of things is independent of perception); critical realism (that scientific enquiry is theory-laden but that knowledge of a theory-independent world is possible); and entity realism (that non-observables should be interpreted as if they really exist). Underlying all these concepts is the idea that theories may be only approximately true but that over time they will become closer to meaningful descriptions of reality.

Real-time Forecasting   The use of a model to make predictions into the near future (over some *lead time*), taking account of data becoming available as the forecasting period progresses (see also *data assimilation* and *updating*). Includes numerical weather prediction and models used for flood forecasting (see Chapter 5).

Recursive Estimation   A form of model calibration and uncertainty estimation based on updating time step by time step as new observations become available. Commonly used in *data assimilation* and *real-time forecasting*.

Relativism   The idea that only man himself can judge the value of different beliefs, so that it is quite acceptable for different people to hold different beliefs. In science, relativist ideas have been entertainingly discussed by Paul Feyerabend (1924–1994), who was once part of the Vienna school of philosophy (with Karl Popper) but who later suggested that "anything goes".

Response Surface   The surface defined by the values of an objective function as it changes with changes in parameter values. May be thought of conceptually as a surface with "peaks" and "troughs" in the multidimensional space defined by the parameter dimensions, where the "peaks" represent good fits to the observations and the "troughs" represent poor fits to the observations (see also *parameter space*)

Risk   Uncertainty about responses of a real-world system that can be characterised in terms of probabilities. There is an ISO Standard on Risk Management Terminology (ISO, 2002) that uses the definition that risk is the combination of the probability of an event and its consequence, but the term is often used more generally.

Risk Analysis   The process of identifying sources of risk and assigning values of risk.

Risk Communication   The process of exchanging and translating information about risk between different stakeholder groups and decision makers.

Robustness   A decision is robust to future uncertainty if it leaves open the possibility of alternative strategies as the future unfolds.

Roughness Coefficient   A parameter of channel and overland flow models (most often in Manning's equation, but also Chezy equation or Darcy–Weisbach equation). Physically intended to represent the loss of energy due to friction at the boundaries of the flow. Effective values compensate for effects of irregular depths and cross-sections etc.

Semiotics   The study and interpretation of signs of all types (including words, graphics etc).

Sensitivity   The response of a model to a change in a parameter or input variable. Can either be assessed locally in the model space (when it is normally quantified as a gradient as a normalised rate of change of a model output to the rate of change of the parameter or input variable) or over a global sample of points in the model space (see Section 3.5).

Simulate Annealing   A method of optimisation based on an analogy with the organisation of molecules in cooling liquid metal. Initially, there is a random scatter of parameter sets but as the "temperature" cools, the optimisation becomes more and more structured. Useful when searching for a global optimum when there may be many local optima.

Stakeholder   An individual, group or community who might be affected by the outcome of a decision-making process.

State Space   The space of potential trajectories of a model (or, in a more limited sense, of a particular variable in a model).

Stochastic Model   A model that contains random elements as a way of expressing uncertainty in inputs, system characteristics or model response. Often constructed by proposing a certain (simple) structure for perturbations around some mean or control simulation. Introduces additional parameters within the stochastic structure to define the magnitude of the likely perturbations. Additive perturbations are often assumed for simplicity (but multiplicative perturbations can be transformed to an additive form by taking logs). If it is expected that the nature of the perturbations will change (e.g. with heteroscedastic variance) then more parameters will be required to represent such changes.

Support of a Fuzzy Set   The range of possible values that will have degree of membership greater than zero in a fuzzy set.

System   A part of the world that has been identified for study or for a decision. Occupies some physical space, some time period, and is separated from the rest of the world by the specification of certain boundary conditions. Environmental systems are usually *open systems*.

Theory   A set of constructs, derived from explicit assumptions, about the nature of the world. Formal theories are normally expressed as mathematical axioms and equations.

Transcendental Realism   see *realism*.

Transfer Function   A representation of the output from a system due to a unit input.

Trans-science   The idea, proposed by Alvin Weinberg in 1972, that some subjects, generally held to be within the realm of science, may not be open to rigorous scientific study. Arguably, this applies to very many environmental systems where boundary conditions cannot be controlled, experiments are necessarily place- and time-specific (see *uniqueness of place*), and details of processes are not fully known.

Type I Error   Accepting a false positive or, in the case of models, accepting a poor model that with better data might be rejected.

Type II Error   A false negative, i.e. rejecting a good model because poor data suggest that it does not give good simulations.

Type III Error   Using a model structure that does not properly account for important processes, perhaps because they are not yet recognised as being important.

Underdetermination Thesis   A theory (or model) is underdetermined if there is at least one other theory (or model) that is equally compatible with the available empirical evidence (observations with which a model can be compared).

Uniqueness of Place The concept that in applying environmental models to some particular location will involve the determination of particular boundary conditions and parameter values that are not easily estimated from past applications of the model elsewhere (see also *unknowability*).

Unknowabilty The concept that because of limitations of current measurement techniques there is much about environmental systems that cannot feasibly be known.

Updating The process of using data available now to condition forecasts of a variable into the future but changing values of parameters or state variables in the model. A form of *data assimilation*, commonly used in *real-time forecasting*. Includes different types of Kalman filter, and variational data assimilation (see Chapter 5).

Utility Function An expression of the benefits arising from implementing different levels of investment (costs) in formal decision making.

Validation I A process of evaluation of models to confirm that they are acceptable representations of a system. Philosophers of science have some problems with the concept of validation and verification (e.g. Oreskes et al., 1994) and it may be better to use "evaluation" or "confirmation" rather than validation or verification (which imply a degree of truth in the model).

Validation 2 Validation is sometimes used in the much more restrictive sense of validation of a computer code (model) to show that it does produce accurate solutions of the equations on which it is based. For complex models this can also be difficult to show in practice.

Variogram A spatial correlation function used in geostatistics, normally plotted as the variance at different distances between points in a random field. Normally increases with distance, up to the "range" where variance becomes constant. If variance continues to increase with distance, the field may be non-stationary (or fractal). Sometimes fitted by a functional form that includes a random effect at zero distance, the "nugget variance". Underlies the *Kriging* method of spatial interpolation in geostatistics. To estimate a variogram without a significant degree of uncertainty requires a large number of measured values, more if the measurements suggest that the correlation might vary with direction.

Verification see *validation*.

Verisimilitude A term used to express the concept of relative truthfulness of false theories. The term was used by Karl Popper (1963), but his notion of verisimilitude has since been criticised (e.g. by Tichy, 1974).

Bibliography

Adams, J. (1995), *Risk: The Policy Implications of Risk Compensation and Plural Rationalities*, Routledge: London.

Ajami, N. K., Duan, Q. & Sorooshian, S. (2007), An integrated hydrologic Bayesian multi-model combination framework: Confronting input, parameter, and model structural uncertainty in hydrologic prediction, *Water Resources Research*, 43, W01403, doi:10.1029/2005WR004745.

Allen, M., Stott, P., Mitchell, J., Schnur, R. & Delworth, T. (2000), Quantifying the uncertainty in forecasts of anthropogenic climate change, *Nature*, 407: 617–620.

Anderson, M. G. & Bates, P. D. (Eds.) (2001), *Model Validation: Perspectives in Hydrological Science*, Wiley: Chichester.

Anderson, M. P. & Woessner, W. W. (1992), The role of the post-audit in model validation, *Adv. In Water Resour.*, 15, 167–174.

Andersson, L. (2004), Riverine nutrient models in stakeholder dialogues. *Int. J. Water Resour. Dev.*, 20: 399–425.

Archer, G. E. B., Saltelli, A. & Sobol', I. M. (1997), Sensitivity measures, ANOVA-like techniques and the use of bootstrap, *J. Stat. Comput. Simul.*, 58: 99–120.

Aronica, G., Hankin, B. G., Beven, K. J. (1998), Uncertainty and equifinality in calibrating distributed roughness coefficients in a flood propagation model with limited data, *Adv. Water Resour.*, 22(4): 349–365.

Arulampalam, M. S., Maskell, S., Gordon, N. & Clapp, T. (2002), A tutorial on particle filters for online nonlinear/non-Gaussian Bayesian tracking, *IEEE Trans. Signal. Process.*, 50(2): 174–188.

Aven, T. (2003), *Foundations of Risk Analysis*, Wiley: Chichester.

Aven, T. (2007), On the ethical justification for the use of risk acceptance criteria, *Risk Analysis*, 27: 303–312.

Bair, E. S. (1994), Model (In)Validation: A view from the courtroom, *Ground Water*, 32(4): 530–531.

Babendreier, J. E. & Castleton, K. J. (2005), Investigating uncertainty and sensitivity in integrated, multimedia environmental models: tools for FRAMES-3MRA, *Environ. Modell. & Softw.*, 20: 1043–1055.

Bardossy, A. (2006), Copula-based geostatistical models for groundwater quality parameters, *Water Resources Research*, 42: W11416, doi:10.1029/2005WR004754.

Bardossy, A., Bogardi, I. & Duckstein, L. (1993), Fuzzy nonlinear regression analysis of dose-response curve, *Europ. J. Oper. Res.*, 66: 36–51.

Bardossy, A. & Duckstein, L. (1995), *Fuzzy-rule-based Modelling with Applications to Geophysical, Biological and Engineering Systems*, CRC Press: Boca Raton, FL.

Basili, M. (2006), A rational decision rule with extreme events, *Risk Analysis*, 26: 1721–1728.

Bates, B. C. & Campbell, E. P. (2001), A Markov chain Monte Carlo scheme for parameter estimation and inference in conceptual rainfall-runoff modelling, *Water Resources Research*, 37: 937–947.

Bayes, T. (1763), An essay towards solving a problem in the doctrine of chances. *Phil. Trans. R. Soc. Lond.*, 53: 370–418.

Beck, M. B. (1987), Water quality modelling: a review of the analysis of uncertainty, *Water Resources Research*, 23(8), 1393–1442.

Beck, M. B. & Halfon, E. (1991), Uncertainty, identifiability and the propagation of prediction errors: a case study of Lake Ontario, *J. Forecasting*, 10: 135–162.

Bedford, T. & Cooke, R. (2001), *Probabilistic Risk Analysis: Foundations and Methods*, Cambridge University Press: Cambridge.

Bell, D. E. (1982), Regret in decision making under uncertainty, *Operations Research*, 30: 961–981.

Ben-Haim, Y. (2005), Uncertainty, probability and information-gaps, *Reliability Engineering & System Safety*, 85: 249–266.

Ben-Haim, Y. (2006), *Info-Gap Decision Theory*, 2nd Edition, Academic Press: Amsterdam.

Benjamin, J. R. & Cornell, C. A. (1970), *Probability, Statistics and Decision for Civil Engineers*, McGraw-Hill: NY.

Bennet, A. F. & Thorburn, M. A. (1992), The generalized inverse of a nonlinear quasi-geostrophic ocean circulation model, *J. Phys. Oceanog.*, 22: 213–230.

Berger, J. (2006), The case for objective Bayesian analysis, *Bayesian Analysis*, 1: 385–402.

Bergin, M. S. & Milford, J. B. (2000), Application of Bayesian Monte Carlo analysis to a Lagrangian photochemical air quality model, *Atmos. Environ.*, 34: 781–792.

Bernado, J. M. & Smith, A. F. M. (1994), *Bayesian Theory*, Wiley: Chichester.

Beven, K. J. (1981), The effect of ordering on the geomorphic effectiveness of hydrologic events. Proceedings of the International Conference on Erosion and Sediment Transport in Pacific Rim Steeplands, Christchurch, New Zealand. *Int. Assoc. Scientific Hydrology Pub.*, No. 132: 510–526, IAHS Press: Wallingford, UK.

Beven, K. J. (1989), Changing ideas in hydrology: the case of physically based models. *J. Hydrol.*, 105: 157–172.

Beven, K. J. (1993), Prophecy, reality and uncertainty in distributed hydrological modelling, *Adv. Water Resour.*: 16, 41–51.

Beven, K. J. (1996), Equifinality and Uncertainty in Geomorphological Modelling, in B. L. Rhoads & C. E. Thorn (Eds.), *The Scientific Nature of Geomorphology*, Wiley: Chichester, 289–313.

Beven, K. J. (2000), Uniqueness of place and process representations in hydrological modelling, *Hydrology & Earth System Sciences*, 4: 203–213.

Beven, K. J. (2001a), *Rainfall-Runoff Modelling: the Primer*, Wiley, Chichester.

Beven, K. J. (2001b), How far can we go in distributed hydrological modelling? *Hydrology & Earth System Sciences*, 5: 1–12.

Beven, K. J. (2001c), On hypothesis testing in hydrology, *Hydrol. Process.*, 15: 1655–1657.

Beven, K. J. (2001d), On explanatory depth and predictive power, *Hydrol. Process.*, 15: 3069–3072.

Beven, K. J. (2002a), Towards a coherent philosophy for environmental modelling, *Proc. Roy. Soc. Lond.*, A458: 2465–2484.

Beven, K. J. (2002b), Towards an alternative blueprint for a physically-based digitally simulated hydrologic response modelling system, *Hydrol. Process.*, 16: 189–206.

Beven, K. J. (2005), On the concept of model structural error, *Water Science & Technology*, 52: 165–175.

Beven, K. J. (2006a), A manifesto for the equifinality thesis, *J. Hydrol.*, 320: 18–36

Beven, K. J. (2006b), The holy grail of scientific hydrology: $Q_t = H(SR)A$ as closure, *Hydrol. & Earth Syst. Sci.*, 10: 609–618.

Beven, K. J. (2006c), On undermining the science?, *Hydrol. Process. (HPToday)*, 20: 3141–3146.

Beven, K. J. (2007), Working towards integrated environmental models of everywhere: uncertainty, data, and modelling as a learning process. *Hydrology & Earth System Science*, 11: 460–467.

Beven, K. J. & Binley, A. M. (1992), The future of distributed models: model calibration and uncertainty prediction, *Hydrol. Process.*, 6: 279–298.

Beven, K. J., Freer, J., Hankin, B. & Schulz, K. (2000), The use of generalised likelihood measures for uncertainty estimation in high order models of environmental systems, in *Nonlinear and Nonstationary Signal Processing*, W. J. Fitzgerald, R. L. Smith, A. T. Walden & P. C. Young (Eds). Cambridge University Press, 115–151.

Beven, K. J. & Freer, J. (2001), Equifinality, data assimilation, and uncertainty estimation in mechanistic modelling of complex environmental systems, *J. Hydrol.*, 249: 11–29.

Beven, K. J. & Germann, P. F. (1982), Macropores and water flow in soils, *Water Resources Research*, 18: 1311–1325.

Beven, K. J., Smith, P. J. & Freer, J. E. (2008), So just why would a modeller choose to be incoherent?, *J. Hydrol.*, 354: 15–32.

Beven, K. J. & Young, P. C. (2003), Comment on "Bayesian Recursive Parameter Estimation for Hydrologic Models" by M. Thiemann, M. Trosset, H. Gupta & S. Sorooshian, *Water Resources Research*, 39(5): doi: 10.1029/2001WR001183.

Beven, K. J., Zhang, D. & Mermoud, A. (2006), On the value of local measurements on prediction of pesticide transport at the field scale. *Vadoze Zone Journal*, 5: 222–233.

Binley, A. M., Beven, K. J., Calver, A. & Watts, L. G. (1991), Changing responses in hydrology: assessing the uncertainty in physically-based model predictions, *Water Resources Research*, 27: 1253–1261.

Binley, A. M. & Beven, K. J. (2003), Vadose zone model uncertainty as conditioned on geophysical data, *Ground Water*, 41: 119–127.

Bhaskar, R., *A Realist Theory of Science*. Brighton: Harvester, 1980.

Bhaskar, R., *Reclaiming Reality*. London: Verso, 1989.

Blazkova, S., Beven, K. J. & Kulasova, A. (2002), On constraining TOPMODEL hydrograph simulations using partial saturated area information, *Hydrol. Process.*, 16: 441–458.

Blazkova, S. & Beven, K. J. (2002), Flood frequency estimation by continuous simulation for a catchment treated as ungauged (with uncertainty), *Water Resources Research*, 38: doi: 10.1029/2001/WR000500.

Boni, G., Entekhabi, D. & Castelli, F. (2001), Land data assimilation with satellite measurements for the estimation of surface energy balance components and surface control on evaporation, *Water Resources Research*, 37: 1713–1722.

Borgonovo, E. (2006), Measuring uncertainty importance: investigation and comparison of alternative approaches, *Risk Analysis*, 26: 1349–1361.

Borsuk, M. E., Stow, C. A. & Reckhow, K. H. (2003), Integrated approach to total maximum daily load development for the Neuse River estuary using a Bayesian probability network model (Neu-Bern), *J. Water Resour. Plann. Manag.*, 129: 271–282.

Borsuk, M. E., Stow, C. A. & Reckhow, K. H. (2004), A Bayesian network of eutrophication models for synthesis, prediction, and uncertainty analysis, *Ecol. Modell.*, 173: 219–239.

Borsuk, M. E., Reichert, P., Peter, A., Schager, E. & Burkhardt-Holm, P. (2006), Assessing the decline of brown trout (Salmo trutta) in Swiss rivers using a Bayesian probability network, *Ecol. Modell.*, 192: 224–244.

Box, G. E. P. & Cox, D. R. (1964), An analysis of transformations (with discussion), *J. Roy. Stat. Soc.*, B26: 211–252.

Box, G. E. P. & Jenkins, G. M. (1976), *Time Series Analysis: Forecasting and Control*, revised edition, Holden-Day: San Francisco.

Box, G. E. P. & Tiao, G. C. (1973), *Bayesian Inference in Statistical Analysis*, Addison-Wesley: Reading, MA.

Brashers, D. E. (2001), Communication and uncertainty management. *J. Communication*, 51(3): 477–497.

Brazier, R. E., Beven, K. J., Freer, J. E. & Rowan, J. S. (2000), Equifinality and uncertainty in physically-based soil erosion models: application of the GLUE methodology to WEPP, the Water Erosion Prediction Project – for sites in the UK and USA, *Earth Surf. Process. Landf.*, 25: 825–845.

Broad, D. R., Dandy, G. C. & Maier, H. R. (2005), Water distribution distribution system optimization using metamodels, *J. Water Resour. Planning Manag. ASCE*, 131: 172–180.

Bromley, J., Jackson, N. A., Clymer, O. J., Giacomello, A. M. & Jensen, F. V. (2005), The use of Hugin® to develop Bayesian networks as an aid to integrated water resource planning, *Environ. Modell. & Softw.*, 20: 231–242.

Brooks, S. P. & Gelman, A. (1998), General methods for monitoring convergence of iterative simulations, *J. Comp. Graph. Stat.*, 7: 434–455.

Brouwer, R., Akter, S., Brander, L. & Haque, E. (2007), Socioeconomic vulnerability and adaptation to environmental risk: a case study of climate change and flooding in Bangladesh, *Risk Analysis*, 27: 313–326.

Brown, J. D. & Heuvelink, G. (2007), The Data Uncertainty Engine (DUE): a software tool for assessing and simulating uncertain environmental variables, *Computer & Geosciences*, 33: 172–190.

Brusdal, K., Brankart, J., Halberstadt, G., Evenson, G., Brasseur, P., van Leeuwen, P. J., Domborwsky, E. & Verron, J. (2003), A demonstration of ensemble-based assimilation methods with a layered OGCM from the perspective of operational ocean forecasting systems, *J. Mar. Syst.*, 40–41: 253–289, doi:10.1016/S0924-7963(03)00021-6.

Buckley, D. (Ed.) (1992), *Engineering Safety*, McGraw-Hill: London.

Buizza, R., Miller, M. & Palmer, T. (1999), Stochastic representations of model uncertainties in the ECMWF Ensemble Prediction System, *Q. J. R. Meteorol. Soc.*, 125: 2887–2908.

Burgers, G., van Leeuwen, P. & Evensen, G. (1998), Analysis scheme in the ensemble Kalman filter, *Mon. Weath. Rev.*, 126: 1719–1724.

Burnham, K. P. & Anderson, D. R. (2002), *Model Selection and Multimodel Inference: a Practical Information-Theoretic Approach*, Springer-Verlag: NY.

Cameron, D. (2006), An application of the UKCIP02 climate change scenarios to flood estimation by continuous simulation for a gauged catchment in the northeast of Scotland, UK (with uncertainty). *J. Hydrol.*, 328: 212–226.

Cameron, D., Beven, K. & Naden, P. (2000), Flood frequency estimation under climate change (with uncertainty). *Hydrology & Earth System Sciences*, 4: 393–405.

Cappé, O., Guillin, A., Marin, J. M. & Robert, C. P. (2004), Population Monte Carlo, *J. Comp. Graph. Stats.*, 13: 907–929.

Carey, J. M., Beilin, R., Boxshall, A., Burgman, M. A. & Flander, L. (2007), Risk-based approaches to deal with uncertainty in a data-poor system: stakeholder involvement in hazard identification for marine national parks and marine sanctuaries in Victoria, Australia, *Risk Analysis*, 27: 271–281.

Carrera, J. & Neumann, S. P. (1986), Estimation of aquifer parameters under transient and steady-state conditions, *Water Resources Research*, 22: 199–242.

Cartwright, N. (1983), *How the Laws of Physics Lie*, Oxford University Press: Oxford.

Cartwright, N. (1999) *The Dappled World: a Study of the Boundaries of Science*. Cambridge University Press: Cambridge, UK.

Casella, B. (1992), Explaining the Gibbs sampler, *The American Statistician*, 46: 167–174.

Catelinois, O., Laurier, D., Verger, P., Rogel, A., Colonna, M., Ignasiak, M., Hémon, D. & Timarche, M. (2005), Uncertainty and sensitivity analysis in assessment of the thyroid cancer risk related to Chernobyl fallout in Eastern France, *Risk Analysis*, 25: 243–252.

Caya, A., Sun, J., Snyder, C. (2005), A comparison between the 4DVAR and the ensemble Kalman filter techniques for radar data assimilation, *Month. Weath. Rev.*, 133: 3081–3094.

Chamberlin, T. C. (1897), Studies for students, *J. Geol.* 5: 837–848, reprinted as: The method of multiple working hypotheses, *J. Geol.*, 39: 155–165, 1931.

Chalmers, A. (1989), Is Bhaskar's realism realistic? *Radical Phil.*, 49: 18–23.

Chen, C.-F., Ma, H.-W. & Reckhow, K. H. (2007), Assessment of water quality management with a systematic qualitative uncertainty analysis, *Science of the Total Environment*, 374: 13–25.

Choi, H. T. & Beven, K. J. (2007), Multi-period and multi-criteria model conditioning to reduce prediction uncertainty in distributed rainfall-runoff modelling within GLUE framework, *J. Hydrol.*, 332: 316–336

Christiaens, K., & Feyen, J. (2002), Constraining soil hydraulic parameter and output uncertainty of the distributed hydrological MIKE SHE model using the GLUE framework, *Hydrol. Process.*, 16: 373–391.

Christensen, S. & Cooley, R. L. (1999), Evaluation of prediction intervals for expressing uncertainties in groundwater flow model predictions, *Water Resources Research*, 35: 2627–2639

Clark, I. (1979), *Practical Geostatistics*, Elsevier: London.

Clark, J. S. (2006), Why environmental scientists are becoming Bayesians, *Ecol. Lett.*, 8: 2–14.

Clark, J. S., Ferraz, G., Oguge, N., Hays, H. & DiCostanzo, J. (2005), Hierarchical Bayes for structured, variable populations: from recapture data to life-history prediction, *Ecology*, 86: 2232–2244.

Clarke, R. T. (1994), *Statistical Modelling in Hydrology*, Wiley: Chichester.

Clarke, R. T. (2007), Hydrological prediction in a non-stationary world, *Hydro. Earth. Syst. Sci.*, 11(1): 408–414.

Collier, A. (1994), *Critical Realism*, Verso: London.

Collier, C. G. (2007), Flash flood forecasting: what are the limits of predictability? *Quart. J. R. Meteorol. Soc.*, 133: 3–23.

Congdon, P. (2007), *Bayesian Statistical Modelling*, Wiley: Chichester.

Cooke, R. M. (1991), *Experts in Uncertainty: Opinion and Subjective Probability in Science.* Oxford University Press: Oxford.

Courtier, P., Derber, J., Errico, R., Louis, J.-F. & Vukicevic, T. (1993), Important literature on the use of adjoint, variational methods and the Kalman filter in meteorology, *Tellus*, A45: 342–257.

Couso, I., Moral, S. & Walley, P. (2000), A survey of concepts of independence for imprecise probabilities, *Risk Decis. & Policy*, 5: 165–181.

Cox, E. (1994), *The Fuzzy Systems Handbook*, AP Professional: Boston, MA.

Cressie, N. A. (1993), *Statistics for Spatial Data* (revised edn.), Wiley: NY.

Croke, B. F. W., Ticehurst, J. L., Letcher, R. A., Norton, J. P., Newham, L. T. H. & Jakeman, A. J. (2007), Integrated assessment of water resources: Australian experiences, *Water Resour. Manag.*, 21: 351–373.

Crome, F. H. J., Thomas, M. R. & Moore, L. A. (1996), A novel Bayesian approach to assessing impacts of rain forest logging, *Ecol. Applic.*, 6: 1104–1123.

Crow, W. T. & Wood, E. F. (2003), The assimilation of remotely-sensed soil brightness temperature imagery into a land surface model using an Ensemble Kalman Filtering: a case study based on ESTAR measurements during SGP97, *Adv. Water Resour.*, 26(2): 137–149.

Cunnane, C. & Nash, J. E. (1971), Bayesian estimation of frequency of hydrological events, *IAHS Pubn No.*, 100: 47–55, IAHS Press: Wallingford.

Cushman, J. H. (1986), On measurement, scale and scaling, *Water Resources Research*, 22: 129–134.

Dagan, G. (1986), Statistical-theory of groundwater-flow and transport – pore to laboratory, laboratory to formation and formation to regional scale, *Water Resources Research*, 22 (suppl.): S120–S134.

Dakins, M. E., Toll, J. E., Small, M. J. & Brand, K. P. (1996), Risk-based environmental remediation: Bayesian Monte Carlo analysis and the expected value of sample information, *Risk Analysis*, 16: 67–79

Davies, P. A., Olague, N. E. & Goodrich, M. T. (1992), Application of a validation strategy to Darcy's experiment, *Adv. In Water Resour.*, 15: 175–180.

Dee, D. P. (2005), Bias and data assimilation *Q. J. R. Meterol. Soc.*, 131: 3323–3343.

Dee, D. P. & da Silva, D. M. (1998), Data assimilation in the presence of forecast bias, *Quart. J. R. Meteorol. Soc.*, 124: 269–296.

De Lannoy, G. J. M., Houser, P. R., Pauwels, V. R. N. & Verhoest, N. E. C. (2006), Assessment of model uncertainty for soil moisture through ensemble verification, *JGR-Atmos.*, 111(D10): D10101

De Michele, C. & Salvadori, G. (2003), A generalized Pareto intensity-duration model of storm rainfall exploiting 2-Copulas, *JGR-Atmos.*, 108(D2): D4067.

Demeritt, D. (2001), The construction of global warming and the politics of science. *Annals Assoc. Am. Geog.*, 91: 307–337.

Dilks, D. W., Canale, R. P. & Meier, P. G. (1992), Development of Bayesian Monte Carlo techniques for water quality model uncertainty, *Ecol. Modell.*, 62: 149–162.

Doherty, J. (2005), PEST: software for model-independent parameter estimation. Water Mark Numerical Computing, Australia. Available at http://www.sspa.com/pest.

Dorazio, R. M. & Johnson, F. A. (2003), Bayesian inference and decision theory – a framework for decision making in natural resource management, *Ecological Applications*, 13: 556–563.

Doucet, A., de Freitas, N. & Gordon, N. (2001), Sequential Monte Carlo methods in practice, in M. Jordan (Ed.), *Statistics for Engineering and Information Science*, Springer: NY.

Draper, D. (1995), Assessment and propagation of model uncertainty, *J. Roy. Stat. Soc.*, B37: 45–98.

Draper, D. (2006), Coherence and calibration: comments on subjectivity and objectivity in Bayesian analysis (Comment on articles by Berger & Goldstein), *Bayesian Analysis*, 1: 423–428.

Draper, N. R. & Smith, M. (1998), *Applied Regression Analysis*, Wiley: Chichester.

Duan, Q. S., Soorooshian, S. & Gupta, H. J. (1992), Effective and efficient global optimisation for conceptual rainfall-runoff models, *Water Resources Research*, 28: 1015–1031.

Duan, Q., Gupta, V. K. & Sorooshian, S. (1993), A shuffled complex evolution approach for effective and efficient global minimization, *J. Optim. Theory & Applic.*, 76: 501–521.

Duan, Q., Sorooshian, S. & Gupta, V. K. (1994), Optimal use of the SCE-UA global optimization method for calibrating watershed models, *J. Hydrol.*, 158: 265–284.

Duan, Q., Ajami, N. K., Gao, X. & Sorooshian, S. (2007), Multi-model ensemble hydrologic prediction using Bayesian model averaging, *Adv. Water Resour.*, 30: 1371–1386.

Dubois, D. (1990), Rough fuzzy sets and fuzzy rough sets. *Int. J. Gen. Syst.*, 17: 191–209.

Dubois, D. & Prade, H. (1980), *Fuzzy Sets and Systems: Theory and Applications*, Academic Press: San Diego, CA.

Dunne, S. & Entekhabi, D. (2005), An ensemble-based reanalysis approach to land data assimilation, *Water Resources Research*, 41: W02013, doi: 10.1029/2004WR003449.

Dunne, S. & Entekhabi, D. (2006), Land surface state and flux estimation using the ensemble Kalman smoother during the Southern Great Plains 1997 field experiment, *Water Resources Research*, 42: W01407, doi: 10.1029/2005WR004334.

Ely, A. (2004). *Handling Uncertainty in Scientific Advice*. Parliamentary Office of Science and Technology: London, 4 pp.

Earman, J. (1992), *Bayes or Bust: a Critical Examination of Bayesian Confirmation Theory*, MIT Press: Cambridge, MA.

Efron, B. & Tibshiriani, R. (1993), *An introduction to the bootstrap*, Chapman-Hall: New York.

Egbert, G. D., Bennett, A. F. & Foreman, M. G. G. (1994), TOPEX/POSEIDON tides estimated using a global inverse model, JGR-Oceans, 99(C12): 24821–24852.

Ellison, A. (1996), An introduction to Bayesian inference for ecological research and environmental decision-making, *Ecol. Applic.*, 6: 1036–1046.

Elshorbagy, A. (2006), Multicriterion decision analysis approach to assess the utility of watershed modelling for management decisions, *Water Resources Research*, 42, W09407, doi:10.1029/2005WR004264.

Engeland, K., Xu, C.-Y. & Gottschalk, L. (2005), Assessing uncertainties in a conceptual water balance model using Bayesian methodology, *Hydrol. Sci. J.*, 50(1): 45–63.

Ersdal, G. & Aven, T. (2007), Risk-informed decision-making and its ethical basis, *Reliability Engineering & System Safety*, 93: 197–205 RESS515.

EU. (2000), Water Framework Directive No. 2000/60/EC, European Community: Brussels.

Evensen, G. (1994), Sequential data assimilation with a nonlinear quasi-geostrophic model using Monte Carlo methods to forecast error statistics, *J. Geophys. Res.*, 99(C5): 10243–10162.

Evensen, G. (2004), Sampling strategies and square root analysis schemes for the EnKF, *Ocean Dynamics*, 54, 539–560, doi: 10.1007/s10236–004–0099–2.

Evensen, G. (2006), *Data Assimilation: The Ensemble Kalman Filter*, Springer: Berlin.

Evensen, G. & van Leeuwen, P. V. (2000), An ensemble Kalman smoother for nonlinear dynamics, *Mon. Weath. Rev.*, 128(6): 1852–1867, doi: 10.1175/1520–0493(2000)128 <1852:AEKSFN>2.0.CO;2.

Failing, L., Horn, G. & Higgins, P. (2004), Using expert judgement and stakeholder values to evaluate management options, *Ecology & Society*, 9:art 13 (www.ecologyandsociety.org/vol9/iss1/art13).

Faulkner, H., Parker, D., Green, C., & Beven, K. J. (2007), Developing a translational discourse to communicate uncertainty in flood risk between science and the practitioner, *Ambio*: 36(8): 692–703.

Favre, E. M., El Adlouni, S., Perrault, N., Thiémonge, N. & Bobée, B. (2004), Multivariate hydrological frequency analysis using copulas, *Water Resources Research*, 40: W01101, doi. 10.1029/2003WR00256.

Ferson, S. (2002), *RAMAS Risk Calc 4.0 Software: Risk Assessment with Uncertain Numbers*, Lewis Publishers: Boca Raton, FL.

Ferson, S. & Ginzburg, L. R. (1996), Different methods are needed to propagate ignorance and variability, *Reliab. Engin. & System Safety*, 54: 133–144.

Ferson, S. & Hajagos, J. G. (2004), Arithmetic with uncertain numbers: rigorous and (often) best possible answers, *Reliab. Engin. & System Safety*, 85: 135–152.

Feyen, L., Beven, K. J., De Smedt, F. & Freer, J. E. (2001), Stochastic capture zones delineated within the Generalised Likelihood Uncertainty Estimation methodology: conditioning on head observations, *Water Resources Research*, 37: 625–638.

Feyen, L., Ribeiro, P. J., Gomez-Hernandez, J. J., Beven, K. J. & De Smedt, F. (2003), Bayesian methodology for stochastic capture zone delineation incorporating transmissivity measurements and hydraulic head observations, *J. Hydrol.*, 271: 156–170.

Feyen, L., Vrugt, J. A., O., Nuallain, B., van der Knijff, J. & De Roo, A. (2007), Parameter optimisation and uncertainty assessment for large-scale streamflow simulation with the LISFLOOD model, *J. Hydrol.*, 332: 276–289.

Feyerabend, P. (1978), *Against Method*, Verso: London

Fienen, M. N., Luo, J. & Kitanidis, P. K. (2006), A Bayesian geostatistical transfer function approach to tracer test analysis, *Water Resources Research*, 42, W07426, doi:10.1029/2005WR004576.

Fletcher, S. J. & Zupanski, M. (2006), A data assimilation method for log-normally distributed observation errors, *Q. J. R. Meteorol. Soc.*, 132: 2505–2519.

Fortin, V., Bernier, J., & Bobe, B. (1997), Simulation, Bayes, and bootstrap in statistical hydrology, *Water Resources Research*, 33: 439–448.

Fox, C. R. & Irwin, J. R. (1998), The role of context in the communication of uncertain beliefs. *Basic & Applied Social Psychology*, 20(1): 57–70.

Frame, D. J., Faull, N. E., Joshi, M. M. & Allen, M. R. (2007), Probabilistic climate forecasts and inductive problems, *Phil. Trans. R. Soc.*, A365: 1971–1992.

Franks, S. W., Gineste, P., Beven, K. J. & Merot, P. (1998), On constraining the predictions of a distributed model: the incorporation of fuzzy estimates of saturated areas into the calibration process, *Water Resources Research*, 34, 787–797.

Franks, S. W. & Beven, K. J. (1999), Conditioning a multiple patch SVAT model using uncertain time-space estimates of latent heat fluxes as inferred from remotely-sensed data, *Water Resources Research*, 35(9): 2751–2761.

Freer, J. E., Beven, K. J. & Ambroise, B. (1996), Bayesian estimation of uncertainty in runoff prediction and the value of data: an application of the GLUE approach, *Water Resources Research*, 32(7): 2161–2173.

Freer, J. E., Beven, K. J., & Peters, N. E. (2003), Multivariate seasonal period model rejection within the generalised likelihood uncertainty estimation procedure, in *Calibration of Watershed Models*, edited by Q. Duan, H. Gupta, S. Sorooshian, A. N. Rousseau & R. Turcotte, AGU Books: Washington, 69–87.

Freer, J. E., McMillan, H., McDonnell, J. J. & Beven, K. J. (2004), Constraining Dynamic TOPMODEL responses for imprecise water table information using fuzzy rule-based performance measures, *J. Hydrol.*, 291: 254–277

Freeze, R. A., James, B., Massmann, J., Sperling, T. & Smith, L. (1992), Hydrogeological decision analysis: 4. The concept of data worth and its use in the development of site investigation strategies, *Ground Water*, 30: 574–588.

Freeman, M. F. & Tukey, J. W. (1950), Transformations related to the angular and the square root, *Ann. Math. Statist.*, 21: 607–611.

Freissinet, C., Vauclin, M. & Erlich, M. (1999), Comparison of first-order analysis and fuzzy set approach for the evaluation of imprecision in a pesticide groundwater pollution screening model, *J. Contam. Hydrol.*, 37: 21–43.

French. (1986), *Decision Theory: an Introduction to the Mathematics of Rationality*, Ellis Horwood: Chichester.

Funtowicz, S. O. & Ravetz, J. R. (1990), *Uncertainty and Quality in Science for Policy*, Kluwer Academic: Dordrecht.

Funtowicz, S. O. & Ravetz, J. R. (1999), Post-normal science: an insight now maturing, *Futures*, 25: 735–755.

Furmston, M. P. (1992), Reliability and the Law, in Buckley, D. (Ed.) (1992), *Engineering Safety*, McGraw-Hill: London, 385–401.

Gallagher, M. & Doherty, J. (2007), Parameter estimation and uncertainty analysis for a watershed model, *Environ. Modell. & Software*, 22: 1000–1020.

Gamerman, D. (1997), *Markov Chain Monte Carlo: Stochastic Simulation for Bayesian Inference*, Chapman & Hall/CRC: Boca Raton, FL.

Gan, T. Y. & Biftu, G. F. (1996), Automatic calibration of conceptual rainfall-runoff models: Optimization algorithms, catchment conditions, and model structure, *Water Resources Research*, 32: 3513–3524.

Gardner, R. H. & O'Neill, R. V. (1983), Parameter uncertainty and model predictions: a review of Monte Carlo results, in M. B. Beck & G. van Straten (Eds.), *Uncertainty and Forecasting of Water Quality*, Springer: Berlin, 245–257.

Geman, S. & Geman, D. (1984), Stochastic relaxation, Gibbs distributions, and the Bayesian restoration of images, *IEEE Trans. Pattern Anal. Mach. Intell.*, 6: 721–741.

Ghil, M., & Manalotte-Rizzoli, P. (1991), Data assimilation in meteorology and oceanography, *Adv. Geophys.*, 33: 141–266.

Gilbert, G. K. (1886), The inculcation of scientific method by example, *Am. J. Sci.*, 31: 284–299.

Gilbert, G. K. (1896), The origin of hypotheses, *Science*, N.S. 3: 1–13.

Goldstein, M. (2006), Subjective Bayesian analysis: principles and practice, *Bayesian Analysis*, 1: 403–420.

Gordon, N. J., Salmond, D. J. & Smith, A. F. M. (1993), Novel approach to nonlinear/non-Gaussian Bayesian state estimation, *Proc. Inst. Electr. Eng.*, 140: 107–113.

Green, P. J. (1995), Reversible jump Markov chain Monte Carlo computation and Bayesian model determination, *Biometrika*, 82: 711–732.

Gregory, R., Ohlson, D. & Arval, J. (2006), Deconstructing adaptive management: criteria for applications to environmental management, *Ecol. Applic.*, 16: 2411–2425

Gronnevik, R. & Evensen, G. (2001), Application of ensemble-based techniques in fish stock assessment, *Sarsia*, 86: 517–526.

Groves, D. G. & Lempert, R. J. (2007), A new analytic method for finding policy-relevant scenarios, *Global Environ. Change*, 17: 73–85.

Guan, B. T., Gertner, G. Z. & Parysov, P. (1997), A framework for uncertainty assessment of mechanistic forest growth models: A neural network example. *Ecol. Modell.*, 98: 47–58.

Gupta, V. K. & Sorooshian, S. (1985), The relationship between data and the precision of parameter estimates of hydrological models, *J. Hydrol.*, 81: 57–77.

Gupta, V. K. & Sorooshian, S. (1986), The influence of data length, information content, and noise characteristics on model calibration, invited chapter in *Multivariate Analysis of Hydrologic Processes*, edited by H. W. Shen et al., Colorado State University Press: Fort Collins, Colorado, 434–449.

Gupta, H. V., Sorooshian, S. & Yapo, P. O. (1998), Towards improved calibration of hydrologic models: multiple and incommensurable measures of information, *Water Resources Research*, 34: 751–763.

Gupta, H. V., Bastidas, L., Sorooshian, S., Shuttleworth, W. J. & Yang, Z. L. (1999), Parameter estimation of a land surface scheme using multicriteria methods, *JGR Atmospheres*, 104(D16): 19491–19503.

Guven, B. & Howard, A. (2007), Identifying the critical parameters of a cyanobacterial growth and movement model by using generalised sensitivity analysis, *Ecol. Modell.*, doi:10.1016/j.ecolmodel.2007.03.024.

Guymon, G. L., Harr, M. E., Berg, R. L. & Hromadka, T. V. (1981), A probabilistic-deterministic analysis of one-dimensional ice segregation in a freezing soil column, *Cold Reg. Sci. Technol.*, 5: 127–140.

Haack, S. (2003), *Defending Science—Within Reason: Between Scientism and Cynicism*. Prometheus Books: Amherst, MA.

Haag, D. & Kaupenjohann, M. (2001), Parameters, prediction, post-normal science and the precautionary principle – a road map for modelling for decision-making, *Ecol. Modell.*, 144: 45–60.

Hahn, G. J. & Meeker, W. Q. (1991), *Statistical Intervals*, Wiley: New York.

Hall, J. W. (2003), Handling uncertainty in the hydroinformatic process, *J. Hydroinformatics*, 5: 215–232.

Hall, J. W. (2006), Uncertainty-based sensitivity indices for imprecise probabilities, *Reliability Engineering & Systems Safety*, 91: 1443–1451.

Hall, J. W. (2007), Probabilistic climate scenarios may misrepresent uncertainty and lead to bad adaptation decisions, *Hydrol. Process.*, 21: 1127–1129.

Hall, J. W., Tarantola, S., Bates, P. D. & Horritt, M. S. (2005), Distributed sensitivity analysis of flood inundation model calibration, *J. Hydraul. Eng. ASCE*, 131: 117–126, doi:10.1061/(ASCE)0733-9429(2005)131:2(117).

Hall, J., Fu, G. & Lawry, J. (2006), Imprecise probabilities of climate change: aggregation of fuzzy scenarios and model uncertainties, *Climate Change*, 81: 265–281 doi: 10.10007/s10584-006-9175-6.

Harremöes, P., Gee, D., MacGarvin, M., Stirling, A., Keys, J., Wynne, B. & Guedes Vaz, S. (Eds.) (2001), *Late Lessons from Early Warnings: the Precautionary Principle* 1896–2000. Office for Official Publications of the European Communities: Luxembourg.

Hartley, R. V. L. (1928), Transmission of information, *Bell System Technical J.*, 7: 535–563.

Hastings, W. (1970), Monte Carlo sampling methods using Markov Chains and their applications, *Biometrika*, 57: 97–106.

Haszeldine, R. S. & McKeown, C. (1995), A model approach to radioactive waste disposal at Sellafield, *Terra Nova*, 7: 87–95.

Haynes-Young, R. & Petch, G. (1986), *Physical Geography: Its Nature and Methods*, Harper and Row: London.

Henriksen, H. J., Troldborg, L., Nyegaard, P., Sonnenborg, T. O., Refsgaard, J. C. & Madsen, B. (2003), Methodology for construction, calibration and validation of a national hydrological model for Denmark, *J. Hydrol.*, 280: 52–71.

Herskowitz, P. J. (1991), A theoretical framework for simulation validation: Popper's falsificationism. *Int. J. Modell. & Simul.*, 11: 56–58.

Hill, M. C., Banta, E. R., Harbaugh, A. W. & Anderan, E. R. (2000), MODFLOW-2000, the US Geological Survey modular groundwater model – User guide to the observations, sensitivity, and parameter estimation processes, *US Geological Survey Open-File Report 00-184*.

Hill, M. C. & Tiedeman, C. R. (2007), *Effective Groundwater Model Calibration*, Wiley: Hoboken, NJ.

Hills, R. C. and Reynolds, S. G. (1969), Illustrations of soil moisture variability in selected areas and plots of different sizes. *J. Hydrol.*, 8: 27–47.

Hine, D. J. (2007), *Robust Flood-Risk Management Decisions*, unpublished PhD thesis, Newcastle University, UK.

Hine, D. J. and Hall, J. W. (2005), Convex analysis of flood inundation model uncertainties and info-gap flood management decisions, *Proceedings ISSSH Stochastic Hydraulics 2005*, Nijmegen, The Netherlands.

Hipel, K. W. (1995), *Stochastic and Statistical Methods in Hydrology*, Springer: Berlin.

Hofmann, J. R. & Hofmann, P. A. (1992), Darcy's law and structural explanation in hydrology, *PSA 1992*, 1: 23–35, Philosophy of Science Association.

Hojati, M., Bector, C. R. & Smimou, K. (2005), A simple method for computation of fuzzy linear regression, *Europ. J. Oper. Res.*, 166: 172–184.

Holling, C. S. (Ed.) (1978), *Adaptive Environmental Assessment and Management*, Wiley: NY.

Hooper, R. P., Stone, A., Christophersen, N., de Grosbois, E. & Seip, H. (1986), Assessing the Birkenes model of stream acidification using a multi-signal calibration methodology, *Water Resources Research*, 22: 1444–145.

Hornberger, G. M. & Spear, R. C. (1981), An approach to the preliminary analysis of environmental systems, *J. Environmental Management*, 12: 7–18.

Hornberger, G. M., Beven, K. J., Cosby, B. J. & Sappington, D. E. (1985), Shenandoah watershed study: calibration of a topography-based, variable contributing area hydrological model to a small forested catchment, *Water Resources Research*, 21: 1841–1850.

Horne, J. (2006), *Unnatural Disaster: Hurricane Katrina and the Drowning of New Orleans*, Random House: NY.

Horritt, M. S. & Bates, P. D. (2001), Predicting flood plain inundation: raster-based modelling versus the finite element approach, *Hydrol. Process.*, 15: 825–842.

Houtekamer, P. L. & Mitchell, H. L. (2001), A sequential ensemble Kalman filter for atmospheric data assimilation, *Month. Weath. Rev.*, 129: 123–137.

Houtekamer, P. L., Mitchell, H. L., Pellerin, G., Buehner, M., Charron, M., Spacek, L. & Hansen, M. (2005), Atmospheric data assimilation with an ensemble Kalman filter: results with real observations, *Month. Weath. Rev.*, 133: 604–620

Houtekamer, P. L., Lefabre, L., Derome, J., Ritchie, H. & Mitchell, H. (1996), A system simulation approach to ensemble prediction, *Month. Weath. Rev.*, 124: 1225–1242.

Howson, C. & Urbach, P. (1993), *Scientific Reasoning: the Bayesian Approach*, 2nd edition, Open Court: Chicago.

Ide, K., Courtier, P., Ghil, M. & Lorenc, A. C. (1997), Unified notation for data assimilation: operational, sequential and variational, *J. Meteorol. Soc. Japan*, 75 (1B): 181–189.

Iman, R. L. & Conover, W. J. (1982), A distribution-free approach to inducing rank correlation among input variables, *Communications in Statistics*, Ser. B, 311–334.

Inter-governmental Panel on Climate Change (IPCC). (2007), *Climate Change 2007: The Physical Science Basis*. Summary for policy makers, WMO: Geneva.

Iooss, B., van Dorpe, F. & Devictor, N. (2006), Response surfaces and sensitivity analyses for an environmental model of dose calculations, *Reliability Surfaces & System Safety*, 91: 1241–1251.

Iorgulescu, I., Beven, K. J. & Musy, A. (2005), Data-based modelling of runoff and chemical tracer concentrations in the Haute-Mentue (Switzerland) Research Catchment, *Hydrol. Process.*, 19: 2257–2574.

Iorgulescu, I., K. J. Beven, & A. Musy. (2007), Flow, mixing, and displacement in using a data-based hydrochemical model to predict conservative tracer data, *Water Resources Research*, 43: W03401, doi:10.1029/2005WR004019.

ISO. (2002), *Risk Management Vocabulary*, International Organisation for Standardization, ISO/IEC Guide 73.

Jackson, B. M., Wheater, H. S., McIntyre, N. & Whitehead, P. (2004), Application of Markov Chain Monte Carlo calibration and uncertainty framework to a process-based integrated nitrogen model (INCA). *Proc. Sensitivity Analysis of Model Outputs*, SAMO-04-37.

Jacquin, A. P. & Shamseldin, A. Y. (2007), Development of a possibilistic method for the evaluation of predictive uncertainty in rainfall-runoff modelling, *Water Resources Research*, 43: W04425, doi:10.1029/2006WR005072.

Janssen., P. H. M., Petersen, A. C., van der Sluijs, J. P., Risbey, J. S. & Ravetz, J. R. (2005), A guidance for assessing and communicating uncertainties, *Water Sci. & Tech.*, 52: 125–131.

Jensen, F. V. (2001), *Bayesian Networks and Decision Graphs*. Springer-Verlag: NY.

Josephson, D. H. (1994), The Great Midwest Flood of 1993, *Natural Disaster Survey Report*, Department of Commerce, NOAA, National Weather Service, Silver Spring, Maryland.

Kaheil, Y. H., Gill, M. K., McKee, M. & Bastidas, L. (2006), A new Bayesian recursive technique for parameter estimation, *Water Resources Research*, 42: W08423.

Kalman, R. (1960), New approach to linear filtering and prediction problems, *J. Basic Eng.*, 82D: 35–45.

Kalnay, E. (2003), *Atmospheric Modelling, Data Assimilation and Predictability*, Cambridge University Press: Cambridge.

Kaplan, S. & Garrick, B. J. (1981), On the quantitative definition of risk, *Risk Analysis*, 1: 11–27.

Kavetski, D., Kuczera, G. & Franks, S. W. (2005), Bayesian analysis of input uncertainty in hydrological modeling: 2. Application, *Water Resources Research*, 42: W03408, doi: 10.1029/2005WR004376.

Keeny, R. L. & Raiffa, H. (1993), *Decisions with Multiple Objectives: Preferences and Value Tradeoffs* (2nd edition), John Wiley: NY.

Keesman, K. & van Straten, G. (1990), Set membership approach to identification and prediction of lake eutrophication, *Water Resources Research*, 26: 2643–2652.

Kelly, K. S. & Krzysztofowicz, R. (1997), A bivariate meta-Gaussian density for use in hydrology, *Stochastic Hydrology and Hydraulics*, 11: 17–31.

Kennedy, M. C. & O'Hagan, A. (2001), Bayesian calibration of mathematical models, *J. Roy. Statist. Soc.*, D63(3): 425–450.

Khu, S.-T. & Madsen, H. (2005), Multi-objective calibration with Pareto preference ordering. An application to rainfall-runoff model calibration, *Water Resources Research*, 41, doi: 10.1029/2004WR003041.

Kirchner, J. W. (2006), Getting the right answers for the right reasons: linking measurements, analyses and models to advance the science of hydrology, *Water Resources Research*, 42, W03S04, doi. 10.1029/2005WR004362.

Kirchner, J. W., Feng, X. & Neal, C. (2001), Catchment-scale advection and dispersion as a mechanism for fractal scaling in stream tracer concentrations, *J. Hydrol.*, 254: 82–101.

Kirwan, B. (1994), *A Guide to Practical Human Reliability Assessment*, Taylor & Francis: London.

Kitanidis, P. K. (1997), *Introduction to Geostatistics: Applications in Hydrogeology*, Cambridge University Press: Cambridge.

Kleindorfer, G. B., O'Neill, L. & Ganeshan, R. (1998), Validation in simulation: various positions in the philosophy of science, *Management Science*, 44: 1087–1099.

Klemeš, V. (1986), Operational testing of hydrologic simulation models, *Hydrol. Sci. J.*, 31: 13–24.

Klepper, O. & Hendrix, E. M. T. (1994), A comparison of algorithms for global characterisation of confidence regions for nonlinear models, *Environmental Toxicology & Chemistry*, 13(12): 1887–1899.

Klir, G. (2006), *Uncertainty and Information*, Wiley: Chichester.

Klir, G. & Folger, T. (1988), *Fuzzy Sets, Uncertainty and Information*, Prentice Hall: Englewood Cliffs, NJ.

Knight, F. H. (1921), *Risk, Uncertainty and Profit*, Houghton-Mifflin Co. (reprinted University of Chicago Press, 1971).

Konikow, L. F. & Bredehoeft, J. D. (1992), Groundwater models cannot be validated?, *Adv. Water Resour.*, 15, 75–83.

Koren, V., Reed, S., Smith, M., Zhang, Z. & Seo, D. J., Hydrology laboratory research modelling system (HL-RMS) of the US National Weather Service, *J. Hydrol.*, 291: 297–318.

Krayer von Krauss, M., van Asselt, M. B. A., Henze, M., Ravetz, J. & Beck, M. B. (2005), Uncertainty and precaution in environmental management, *Water Sci. & Technol.*, 52: 1–9.

Krzysztofovitz, R. (1999), Bayesian theory of probabilistic forecasting via deterministic hydrologic model, *Water Resources Research*, 35: 2739–2750.

Krzysztofowicz, R. (2002a), Bayesian system for probabilistic river stage forecasting, *J. Hydrol.*, 268: 16–40.

Krzysztofowicz, R. (2002b), Probabilistic floods forecast: bounds and approximations, *J. Hydrol.*, 268: 41–55.

Kuczera, G. & Parent, E. (1998), Monte Carlo assessment of parameter uncertainty in conceptual catchment models: the Metropolis algorithm, *J. Hydrol.*, 211: 69–85.

Kuczera, G., Kavetski, D., Franks, S. & Thyer, M. (2006), Towards a Bayesian total error analysis of conceptual rainfall-runoff models: Characterising model error using storm-dependent parameters, *J. Hydrol.*, 331: 161–177.

Kundzewicz, Z. W. & Robson, A. J. (2000), Change detection in hydrological records – a review of the methodology, *Hydrol. Sci. J.*, 49(1): 7–19.

Kuparinen, A., Snäll, T., Vänskä, S. & O'Hara, R. (2007), The role of model selection in describing stochastic ecological processes, *Oikos*, 116: 966–974.

Kurowicka, D. & Cooke, R. (2006), *Uncertainty Analysis with High-Dimensional Dependence Modelling*. Wiley: Chichester.

Lamb, R., Beven, K. J. & Myrabø, S. (1998), Use of spatially distributed water table observations to constrain uncertainty in a rainfall-runoff model., *Adv. Water Resour.*, 22: 305–317.

Landau, S., Mitchell, R. A. C., Barnett, V., Colls, J. J., Craigon, J., Moore, K. L. & Payne, R. W. (1998), Testing winter wheat simulation models predictions against observed UK grain yields, *Agric. Forest Meteorol.*, 89: 85–99.

Laplace, P. S. (1774), Mémoire sur la probabilité des causes par les événements. *Mémoires de l'Académie de Science de Paris*, 6: 621–656.

Larssen, T., Høgåsen, T. & Cosby, B. J. (2007), Impact of time series data on calibration and prediction uncertainty for a deterministic hydrogeochemical model, *Ecol. Modell.*, doi: 10.1016/j.ecolmodel.2007.03.016.

Lawes, J. B., Gilbert, J. H. & Warrington, R. (1882), *On the Amount and Composition of the Rain and Drainage Water Collected at Rothamsted*. Williams, Clowes & Sons Ltd: London.

Leavesley, G. H., Markstrom, S. L., Restrepo, P. J. & Viger, R. J. (2002), A modular approach to addressing model design, scale, and parameter estimation issues in distributed hydrological modelling, *Hydrol. Process.*, 16: 173–187.

Lees, M., Young, P. C., Ferguson, S., Beven, K. J. & Burns, J. (1994), An adaptive flood warning scheme for the River Nith at Dumfries, in W. R. White & J. Watts (Eds.), *River Flood Hydraulics*, 65–75, Wiley, Chichester.

Lemos, M. C., Finan, T. J., Fox, R. W., Nelson, D. R. & Tucker, J. (2002), The use of seasonal climate forecasting in policymaking: lessons from Northeast Brazil. *Climatic Change*, 55(4): 479–507.

Lindley, D. V. (2006), *Understanding Uncertainty*, Wiley: Chichester.

Link, W. A. & Barker, R. J. (2006), Model weights and the foundations of multimodel inference, *Ecology*, 87: 2626–2635.

Linkov, I. & Burmistrov, D. (2003), Model uncertainty and choices made by modelers: lessons learned from the International Atomic Energy Agency model intercomparisons. *Risk Analysis*, 23: 1335–1346.

Liu, J. S. & Chen, R. (1998), Sequential Monte Carlo methods for dynamic systems, *J. Am. Stat. Assoc.*, 93: 1032–1044.

Lo, E. (2005), Gaussian error propagation applied to ecological data: post-ice-storm-down woody biomass, *Ecol. Monog.*, 75: 451–466.

Lohmann, D. & 28 others. (1998), The Project for Intercomparison of Land-surface Parameterization Schemes (PILPS), phase 2c Red–Arkansas River basin experiment. 3. Spatial and temporal analysis of water fluxes, *Global & Planetary Change*, 19: 161–179.

Lorenc, A. C. (1986), Analysis methods for numerical weather prediction, *Q. J. R. Meteorol. Soc.*, 112: 1177–1194.

Lorenc, A. C. (1997), Development of an operational variational assimilation scheme, *J. Meteorol. Soc. Japan*, 75(1B): 339–346.

Lorenc, A. C. (2003a), Modelling of error covariances by 4D-Var data assimilation, *Q. J. R. Meteorol. Soc.*, 129: 3167–3182.

Lorenc, A. C. (2003b), The potential of the ensemble Kalman filter for NWP – a comparison with 4d-Var, *Q. J. R. Meteorol. Soc.*, 129: 3183–3203.

Lorenc, A. C., Ballard, S. P., Bell, R. S., Ingleby, N. B., Andrews, P. L. F., Barker, D. M., Bray, J. R., Clayton, A. M., Dalby, T., Li, D., Payne, T. J. & Saunders, F. W. (2000), The Met Office global 3-dimensional variational data assimilation scheme, *Quart. J. R. Meteorol. Soc.*, 126: 2991–3012.

Lorenz, E. (1963), The predictability of a flow which possesses many scales of motion, *Tellus*, 21: 289–307.

Lorenz, E. (1993), *The Essence of Chaos*, University of Washington Press: Seattle.

Luseno, W. K., McPeak, J. G., Barrett, C. B., Little, P. D. & Gebru, G. (2003), Assessing the value of climate forecast information for pastoralists: evidence from southern Ethiopia and northern Kenya. *World Development*, 31: 1477–1494.

Madsen, H. (2003), Parameter estimation in distributed hydrological catchment modelling using automatic calibration with multiple objectives, *Adv. Water Resour.*, 26: 205–216.

Madsen, H. & Cañizares, R. (1999), Comparison of extended and ensemble Kalman filter for data assimilation in coastal area modelling, *Int. J. Numer. Methods Fluids*, 31: 961–981.

Madsen, H. & Khu, S.-T. (2006), On the use of Pareto optimisation for multi-criteria calibration of hydrological models, in *Calibration and Reliability in Groundwater Modelling: from Uncertainty to Decision Making*, M. F. P. Bierkens, J. C. Gehrels and K. Kovar (Eds.), *IAHS Pub. No. 304*: 93–99, IAHS Press: Wallingford.

Makowski, D., Wallach D., Tremblay M. (2002), Using a Bayesian approach to parameter estimation; comparison of the GLUE and MCMC methods, *Agronomie*, 22 (2): 191–203.

Mansilla, V. B., Dillon, D., & Middlebrooks, S. (2006), *Building Bridges Across Disciplines: Organisational and Individual Qualities of Exemplary Interdisciplinary Work. "Goodwork Project"*, Harvard Graduate School of Education: Cambridge, MA.

Mantovan, P. & Todini, E. (2006), Hydrological forecasting uncertainty assessment: incoherence of the GLUE methodology, *J. Hydrol.*, 330: 368–381.

Margulis, S. A., McLaughlin, D., Entekhabi, D. & Dunne, S. (2002), Land data assimilation and estimation of soil moisture using measurements from the Southern Great Plains 1997 field experiment, *Water Resources Research*, 38(12): W01299, doi:10.1029/2001WR001114.

Marin, C., Medina M. & Butcher, J. (1989), Monte Carlo analysis and Bayesian decision theory for assessing the effects of waste sites on groundwater, *J. Contam. Hydrol.*, 5: 1–13.

Markus, M., Tsai, C. W.-S. & Demissie, M. (2003), Uncertainty of weekly nitrate-nitrogen forecasts using Artificial Neural Networks, *J. Environ. Eng, ASCE*, 129: 267–274

Marshall, L., Nott, D. & Sharma, A. (2004), A comparative study of Markov chain Monte Carlo methods for conceptual rainfall-runoff modelling, *Water Resources Research*, 40: W02501.

Martin, T. G., Kuhnert, P. M., Mengersen, K. & Possingham, H. P. (2005), The power of expert opinion in ecological models using Bayesian methods: impact of grazing on birds. *Ecol. Applic.*, 15: 266–280.

Matsumoto, M. & Nishimura, T. (1998), Mersenne twister: a 623-dimensionally equidistributed uniform pseudorandom number generator, *ACM Trans. Model. Comput. Simul.*, 8: 3.

Mayes, W. M., Wlash, C. L., Batchurst, J. C., Kilsby, C. G., Quinn, P. F., Wilkinson, M. E., Daugherty, A. J. and O'Connell, P. E. (2006), Monitoring a flood event in a densely instrumented catchment, the Upper Eden, Cumbria, UK, *Water & Environ. J.*, 20: 217–226.

MacGarigal, K., Stafford, S. G., Cushman, S. (2000), *Multivariate Statistics for Wildlife and Ecology Research*, Springer: Berlin.

McCarthy, S., Tunstall, S., Parker D., Faulkner, H. & Howe, J. (2007), Risk communication in emergency response to a simulated extreme flood, *Environmental Hazards*, 7(3): 179–192.

McIntyre, N., Jackson B., Wade A. J., Butterfield D., & Wheater H. S. (2005), Sensitivity analysis of a catchment-scale nitrogen model, *J. Hydrol.*, 315: 71–92.

Metropolis, N., Rosenblueth, A. W., Rosenblueth, M. N., Teller, A. H. & Teller, E. (1953), Equations of state calculations by fast computing machines, *J. Chem. Phys.*, 21: 1087–1091.

Min, S.-K., Simonis, D. & Hense, A. (2007), Probabilistic climate change predictions applying Bayesian model averaging, *Phil. Trans. R. Soc.*, A365: 2103–2116.

Montanari, A. (2005), Large sample behaviors of the generalized likelihood uncertainty estimation (GLUE) in assessing the uncertainty of rainfall-runoff simulations, *Water Resources Research*, 41, W08406, doi:10.1029/2004WR003826.

Montanari, A. & Brath, A. (2004), A stochastic approach for assessing the uncertainty of rainfall-runoff simulations, *Water Resources Research*, 40, W01106, doi:10.1029/2003WR002540.

Marin, C. M., Guvanasen V. & Saleem Z. A. (2003), The 3MRA risk assessment framework – a flexible approach for performing multimedia, multipathway and multireceptor risk assessments under uncertainty, *Human & Econ. Risk Assess*, 9: 1655–1678.

Moore, C. & Doherty, J. (2006), The cost of uniqueness in groundwater model calibration, *Adv. Water Resour.*, 29: 605–623.

Moore, R. E. (1979), *Methods and Applications of Interval Analysis*, SIAM: Philadelphia.

Moradkhani, H., Sorooshian, S., Gupta, H. V. & Hauser, P. R. (2005a), Dual state-parameter estimation of hydrological models using ensemble Kalman filter, *Adv. Water Resour.*, 28: 135–147.

Moradkhani, H., Hsu, K., Gupta, H. V. & Sorooshian, S. (2005b), Uncertainty assessement of hydrologic model states and parameters: sequential data assimilation using particle filter, *Water Resources Research*, 41: W05012, doi: 10.1029/2004WR003604.

Morgan, M. & Henrion, M. (1990), *Uncertainty: a Guide to Dealing with Uncertainty in Quantitative Risk and Policy Analysis*, Cambridge University Press: NY.

Morgan, R. P. (1994), A predictive model for the assessment of soil erosion risk. *J. Agric. Engin. Res.*, 30: 245–253.

Morton, A. (1993), Mathematical models: questions of trustworthiness, *Brit. J. Phil. Sci.*, 44: 659–674.

Mysiak, J. & Sigel, K. (2005), Sources of uncertainty in economic analysis of the Water Framework Directive, *Water Sci. & Techn.*, 52: 161–166.

Nash, J. E. & Sutcliffe, J. V. (1970), River flow forecasting through conceptual models 1. A discussion of principles, *J. Hydrol.*, 10: 282–290.

Nelson, R. (1999), *An Introduction to Copulas*, Springer: New York.

Neumaier, A. (1990), *Interval Methods for Systems of Equations*, Cambridge University Press: Cambridge.

Neuman, S. P. (1990), Universal scaling of hydraulic conductivities and dispersivities in geologic media, *Water Resources Research*, 26: 1749–1758.

Neuman, S. P. (2003), A comprehensive strategy of hydrogeologic modelling an uncertainty analysis for nuclear facilities and sites. NUREG/CR-6805. U.S. Nuclear Regulatory Commission, Washington, DC 20555 (pdf available at http://www.nrc.gov/reading-rm/doc-collections/nuregs/contract/cr6805/).

New, M., Lopez, A., Dessai, D. & Hense, A. (2007), Challenges in using probabilistic climate change information for impact assessments: an example from the water sector, *Proc. Roy. Soc.*, A365: 2117–2131.

Newig, J., Pahl-Wostl, C. & Sigle, K. (2005), The role of public participation in managing uncertainty in the implementation of the Water Framework Directive, *European Environment*, 15: 333–343.

Newson, M. D. (1980), The geomorphological effectiveness of floods – a contribution stimulated by two recent floods in mid-Wales, *Earth Surf. Process.*, 5: 1–16.

Nielsen, D. R., Biggar, J. W. and Erh, K. T. (1973), Spatial variability of field-measured soil-water properties, *Hilgardia*, 42: 215–259.

Nijssen, B. & 25 others. (2003), Simulation of high latitude hydrological processes in the Torne-Kalix basin: PILPS, phase 2(e) 2. Comparison of model results with observations, *Global & Planetary Change*, 38: 31–53.

Norton, J. P. (1986), *An Introduction to Identification*, Academic Press: London.

Novotny, H., Scott, P. & Gibbons, M., *Re-thinking Science. Knowledge and the Public in an Age of Uncertainty*, Polity Press: Cambridge.

Oakley, J. & O'Hagan, A. (2004), Probabilistic sensitivity analysis of complex models: a Bayesian approach, *J. Roy. Statist. Soc.*, B66: 751–769.

O'Hagan, A. (2006a), Bayesian analysis of computer code outputs: a tutorial, *Reliability Engin. & System Safety*, 91: 1290–1300.

O'Hagan, A. (2006b), Science, subjectivity and software (Comment on articles by Berger & Goldstein), *Bayesian Analysis*, 1: 445–450.

O'Hagan, A. & Oakley, A. E. (2004), Probability is perfect but we can't elicit it perfectly, *Reliability Engineering & System Safety*, 85: 239–248.

O'Hagan, A., Buck, C. E., Daneshkhah, A., Eiser, J. R., Garthwaite, P. H., Jenkinson, D. J., Oakley, J. E. & Rakow, T. (2006), *Uncertain Judgements: Eliciting Expert's Probabilities*, Wiley: Chichester.

Olsson, J. A. & Berg, K. (2005), Local stakeholders' acceptance of model-generated data used as a communication tool in water management: the Rönneå study, *Ambio*, 34: 507–512.

Olsson, J. A. & Andersson, L. (2007), Possibilities and problems with the use of models as a communication tool in water resources management, *Water Resour. Manag.*, 21: 97–110.

Omlin, M. & Reichert P. (1999), A comparison of techniques for the estimation of model prediction uncertainty, *Ecol. Modell.*, 115: 45–59.

Oreskes, N., Shrader-Frechette, K. & Belitz, K. (1994), Verification, validation and confirmation of numerical models in the earth sciences, *Science*, 263, 641–646.

Osidele, O. O., Zeng, W. & Beck, M. B. (2006), A random search methodology for examining parametric uncertainty in water quality models, *Water Sci. & Techn.*, 53(1): 33–40.

Page, T., Beven, K. J., Freer, J. & Jenkins, A. (2003), Investigating the uncertainty in predicting responses to atmospheric deposition using the Model of Acidification of Groundwater in Catchments (MAGIC) within a Generalised Likelihood Uncertainty Estimation (GLUE) framework, *Water, Air, Soil Pollution*, 142: 71–94.

Page, T., Whyatt, D., Beven, K. J. & Metcalfe, S. E. (2004), Uncertainty in modelled estimates of acid deposition across Wales: a GLUE approach. *Atmos. Environ.*, 38: 2079–2090.

Page, T., Beven, K. J. & Freer, J. (2007), Modelling the chloride signal at the Plynlimon Catchments, Wales using a modified dynamic TOPMODEL. *Hydrol. Process.*, 21, 292–307.

Pahl-Wostl, C. (2002), Towards sustainability in the water sector: the importance of human actors and processes of social learning. *Aquatic Sciences*, 64: 394–411.

Pahl-Wostl, C. (2007), Transitions towards adaptive management of water facing climate and global change, *Water Resour. Manag.*, 21: 49–62.

Papoulis, A. (1965), *Probability, Random Variables and Stochastic Processes*, McGraw-Hill: NY.

Pappenberger, F. & Beven, K. J. (2006), Ignorance is bliss: 7 reasons not to use uncertainty analysis, *Water Resources Research*, 42: W05302, doi:10.1029/2005WR004820, 2006.

Pappenberger, F., Harvey, H., Beven K., Hall, J. & Meadowcroft, I. (2006a), Decision tree for choosing an uncertainty analysis methodology: a wiki experiment http://www.floodrisknet.org.uk/methods http://www.floodrisk.net, *Hydrol. Process.*, 20: 3793–3798.

Pappenberger, F., Iorgulescu, I. & Beven, K. J. (2006b), Sensitivity analysis based on regional splits (SARS) and regression trees, *Environmental Modelling & Software*, 21: 976–990.

Pappenberger, F., Frodsham, K., Beven, K. J., Romanovicz, R. & Matgen, P. (2006c), Fuzzy set approach to calibrating distributed flood inundation models using remote sensing observations. *Hydrology & Earth System Sciences*, 10: 1–14.

Pappenberger, F., Beven, K. J., Frodsham, K., Romanovicz, R. & Matgen, P. (2006d), Grasping the unavoidable subjectivity in calibration of flood inundation models: a vulnerability weighted approach. *J. Hydrol.*, 333: 275–287.

Park, S. K. & Miller K. W. (1988), Random number generators: good ones are hard to find, *Comm. ACM*, 31: 1192–1201.

Parkin, G., O'Donnell, G., Ewen, J., Bathurst, J. C., O'Connell, P. E. & Lavabre, J. (1996), Validation of catchment models for predicting land-use and climate change impacts. 2. Case study for a Mediterranean catchment, *J. Hydrol.*, 175, 595–613.

Patt, A. & Dessai, S. (2005), Communicating uncertainty: lessons learned and suggestions for climate change assessment. *Comptes Rendus Géoscience*, 337(4): 425–441.

Patwaedhan, A. & Small, M. J. (1992), Bayesian methods for model uncertainty analysis with application to future sea level rise, *Risk Analysis*, 12: 513–523.

Pawlak, Z. (1991), *Rough Sets: Theoretical Aspects of Reasoning About Data*. Kluwer: Dordrecht.

Pearl, J. (1988), *Probabilistic Reasoning in Intelligent Systems*, Morgan-Kaufman: San Mateo, CA.

Peterson, G. D., Carpenter, S. R. & Brock, W. A. (2003), Uncertainty and the management of multistate ecosystems: an apparently rational route to collapse, *Ecology*, 84: 1403–1411.

Philip, J. R. (1980), Field heterogeneity: some basic issues. *Water Resources Research*, 16: 443–448.

Piani, C., Frame, D. J., Stainforth, D. A. & Allen, M. R. (2005), Constraints on climate change from a multi-thousand member ensemble of simulations, *Geophysical Review Letters*, 32: L23825.

Pielke, R. A., Jr. (1999), Who decides? Forecasts and responsibilities in the (1997) Red River Floods, *Appl. Behav. Sci. Rev.*, 7: 83–101.

Pielke, R. A., Jr & Conant, R. T. (2003), Best practices in prediction for decision-making: lessons from the atmospheric and earth sciences, *Ecology*, 84(6): 1351–1358.

Pielke, R. A., Sr. (2002), Overlooked issues in the US National Climate and IPCC Assessments, *Climate Change*, 52: 1–11.

Piñol, J., Beven, K. J. & Freer, J. (1997), Modelling the hydrological response of Mediterranean catchments, Prades, Catalonia – the use of distributed models as aids to hypothesis formulation, *Hydrol. Process.*, 11(9): 1287–1306.

Piñol, J., Beven, K. J. & Viegas, D. X. (2004), Modelling the effect of fire-exclusion and prescribed fire on wildfire size in Mediterranean ecosystems. *Ecol. Model.*, 183, 397–409.

Piñol, J., Castellnou, M. & Beven, K. J. (2007), Conditioning uncertainty in ecological models: assessing the impact of fire management strategies, *Ecol. Model.*, 207: 34–44, doi:10.1016/j.ecolmodel.2007.03.020

Poeter, E. P., Hill, M. C., Banta, E. R. & Mehl, S. W. (2005), UCODE_2005 and three post-processors – computer codes for universal sensitivity analysis, inverse modelling and uncertainty evaluation. *US Geological Survey Techniques & Methods Report TM 6-A11*, Washington, DC.

Poole, D. & Raftery, A. E. (2000), Inference for deterministic simulation models: the Bayesian melding approach, *J. Am. Statist. Assoc.*, 95: 1244–1255.

Popper, K. R. (1963), *Conjectures and Refutations: The Growth of Scientific Knowledge*, (Republished in paperback by Routledge Classics, 1991).

Popper, K. R. (1979), *Objective Knowledge*, 2nd Edition, Oxford University Press: Oxford.

Press, F. (1968), Earth models obtained by Monte Carlo inversion, *J. Geophys. Res.*, 73: 5223–5234.

Press, S. J. & Tanur, J. M. (2001), *The Subjectivity of Scientists and the Bayesian Approach*, Wiley: Chichester.

Press, W. H., Teukolsky S. A., Vetterling W. T. & Flannery B. P. (2007), *Numerical Recipes: the Art of Scientific Computing*, 3rd Edition, Cambridge University Press: Cambridge.

Prigogine, I. (1997) *The End of Certainty*. The Free Press: New York.

Qian, S. S., Stow, C. A. & Borsuk, M. E. (2003), On Monte Carlo methods for Bayesian inference, *Ecol. Modell.*, 159: 269–277

Rabier, F. (2005), Overview of global data assimilation developments in numerical weather-prediction centres, *Q. J. R. Meteorol. Soc.*, 131: 3215–3233.

Raftery, A. E., Gneiting, T., Balabdaoui, F. & Polakowski, M. (2005), Using Bayesian model averaging to calibrate forecast ensembles, *Month. Weath. Rev.*, 133: 1155–1174.

Ratto, M., Tarantola, S. & Saltelli, A. (2001), Sensitivity analysis in model calibration: GSA-GLUE approach, *Comp. Phys. Comms.*, 136: 212–224.

Ratto, M., Tarantola, S., Saltelli, A. & Young, P. C. (2005), Accelerated estimation of sensitivity indices using Sate-Dependent Parameter Models, in K. M. Hanson & F. M. Hemez (Eds.), *Sensitivity Analysis of Model Output*, Los Alamos National Laboratory, NM: 61–70.

Rawlins, F., Ballard, S. P., Bovis, K. J., Clayton, A. M., Li, D., Inverarity, G. W., Lorenc, A. C. & Payne, T. J. (2007), The Met Office global four-dimensional variational data assimilation scheme, *Q. J. R. Meteorol. Sco.*, 133: 347–362.

Reckhow, K. H. (1999), Water quality prediction and probability network models, *Can. J. Fish Aquat. Sci.*, 56: 1150–1158.

Refsgaard, J.-C. (1997), Parameterisation, calibration and validation of distributed hydrological models, *J. Hydrol.*, 198: 69–97.

Refsgaard, J.-C., Nilsson, B., Brown, J., Klauer, B., Moore, R. V., Bech, T., Vurro, M., Blind, M., Castilla, G., Tsanis, I. & Biza, P. (2005), Harmonised techniques and representative river basin data for assessment and use of uncertainty information in integrated water management, *Environmental Science & Policy*, 8: 267–277.

Refsgaard, J.-C., van der Sluijs, J. P., Brown, J. & van de Keur, P. (2006), A framework for dealing with uncertainty due to model structural error, *Adv. Water Resour.*, 29: 1586–1597.

Regan, H. M., Colyvan, M. & Burgman, M. A. (2002), A taxonomy and treatment of uncertainty for ecology and conservation biology, *Ecol. Applic.*, 12: 618–628.

Regan, H. M., Ben-Haim, Y., Langford, B., Wilson, W. G., Lundberg, P., Andelman, S. J. & Burgman, M. A. (2005), Robust decision-making under severe uncertainty for conservation management, *Ecological Applications*, 15: 1471–1477.

Reichert, P. & Omlin M. (1997), On the usefulness of overparameterised ecological models, *Ecol. Modell.*, 95: 289–299.

Reichert, P. & Borsuk, M. E. (2005), Does high forecast uncertainty preclude effective decision support? *Environ. Modell. Softw.*, 20: 991–1101.

Reichle, R., Entekhabi, D. & McLaughlin, D. B. (2001), Downscaling of radiobrightness measurements for soil moisture estimation: a four-dimensional variational data assimilation approach, *Water Resources Research*, 37: 2353–2364.

Reichle, R., McLaughlin, D. B. & Entekhabi, D. (2002), Hydrologic data assimilation with the ensemble Kalman filter, *Month. Weath. Rev.*, 130: 103–114.

Reinert, J. M. & Apostolakis, G. E. (2006), Including model uncertainty in risk-informed decision making, *Annals of Nuclear Energy*, 33: 354–369.

Renard, B. & Lang, M. (2007), Use of a Gaussian copula for multivariate extreme value analysis: some case studies in hydrology, *Adv. Water Resour.*, 30: 897–912.

Richards, L. A. (1931), Capillary conduction of liquids through porous mediums, *Physics*, 1: 318–333

Richardson, L. F. (1922), *Weather Prediction by Numerical Process*, Cambridge University Press: Cambridge.

Robert, C. P. & Casella, G. (2004), *Monte Carlo Statistical Methods* (2nd edition), Springer: Berlin

Roberts, G. O., & Smith, A. F. M. (1994), Simple conditions for the convergence of the Gibbs sampler and Metropolis–Hastings algorithms, *Stoch. Process. Applic.*, 49: 207–216.

Roberts, G. O., & Rosenthal, J. S. (2001), Optimal scaling for various Metropolis–Hastings algorithms, *Stat. Sci.*, 16: 351–367.

Robson, A. (2002), Evidence for trends in UK flooding. *Philosophical Transactions of the Royal Society of London*. A360: 1327–1343, doi: 10.1098/rsta.2002.1003.

Romanowicz, R., Beven, K. J. & Tawn, J. (1994), Evaluation of predictive uncertainty in non-linear hydrological models using a Bayesian approach, in V. Barnett & K. F. Turkman (Eds.) *Statistics for the Environment II . Water Related Issues*, Wiley: Chichester, 297–317.

Romanowicz, R., Beven, K. J. & Tawn, J. (1996), Bayesian calibration of flood inundation models, in M. G. Anderson, D. E. Walling & P. D. Bates (Eds.), *Floodplain Processes*, 333–360.

Romanowicz, R. & Beven, K. J. (1998), Dynamic real-time prediction of flood inundation probabilities, *Hydrol. Sci. J.*, 43: 181–196.

Romanowicz, R. & Beven, K. J. (2003), Bayesian estimation of flood inundation probabilities as conditioned on event inundation maps, *Water Resources Research*, 39:W01073 doi:10.1029/2001WR001056.

Romanowicz, R., Young, P. C. & Beven, K. J. (2006), Data assimilation and adaptive forecasting of water levels in the River Severn catchment, UK, *Water Resources Research*, 42: W06407, doi:10.1029/2005WR004373

Rosenblueth, E. (1975), Point estimates for probability moments, *Proc.Natl. Acad. Sci. USA*, 72: 3812–3814.

Ross, T. J. (1995), *Fuzzy Logic with Engineering Applications*, McGraw-Hill: New York.

Rougier, J. (2007), Probabilistic inference for future climate using an ensemble of climate model evaluations, *Climate Change*, 81: 247–264.

Rougier, J. & Sexton D. M. H. (2007), Inference in ensemble experiments, *Phil. Trans. R. Soc.*, A365: 2133–2143.

Roulston, M. S. & Smith, L. A. (2004), The boy who cried wolf revisited: the impact of false alarm intolerance on cost-loss scenarios, *Weather and Forecasting*, 19: 391–397.

Rutherford, J. (1994), *River Mixing*, Wiley: Chichester.

Rykiel, E. J. (1996), Testing ecological models: the meaning of validation, *Ecological Modelling*, 90: 229–244.

Salas, J. D., Markus, M. & Tokar, A. S. (2000), Streamflow forecasting based on artificial neural networks. *Artificial Neural Networks in Hydrology*, R. S. Govindaraju & A. Ramachandra Rao (Eds.), Kluwer, Dordrecht, The Netherlands, 23–51.

Saltelli, A., Tarantola, S., Campolongo, F. & Ratto, M. (2004), *Sensitivity Analysis in Practice: a Guide to Assessing Scientific Models*, Wiley: Chichester.

Saltelli, A., Ratto, M., Tarantola, S., Campolongo, F. (2006), Sensitivity analysis practices: strategies for model-based inference, *Reliability Engineering System Safety*, 91: 1109–1125.

Sander, P., Bergbäck, B. & Öberg, T. (2006), Uncertain numbers and uncertainty in the selection of input distributions – consequences for a probabilistic risk assessment of contaminated land, *Risk Analysis*, 26: 1363–1375.

Savage, L. J. (1951), The theory of statistical decision, *J. Am. Stat. Assoc.*, 46: 55–67.

Savage, L. J. (1972), *The Foundations of Statistics*, Dover: NY.

Sayers, P. B., Hall, J. W. & Meadowcroft, I. C. (2002), Towards risk-based flood hazard management in the UK, *Proc. ICE, Civil Engineering*, 150: 36–42.

Schecher, W. D. & Driscoll, C. T. (1988), An evaluation of the equilibrium calculations within acidification models: the effect of uncertainty in measured chemical components, *Water Resources Research*, 24: 533–540.

Schulz, K. & Huwe, B. (1997), Water flow modelling in the unsaturated zone with imprecise parameters using a fuzzy approach, *J. Hydrol.*, 201: 211–229.

Schulz, K., Huwe, B. & Peiffer, S. (1999), Parameter uncertainty in chemical equilibirium calculations using fuzzy set theory, *J. Hydrol.*, 217: 119–134.

Schulz, K., Beven, K. & Huwe, B. (1999), Equifinality and the problem of robust calibration in nitrogen budget simulations, *Soil Sci. Soc. Amer. J.*, 63(6): 1934–1941.

Schulz, K., & Beven, K. J. (2003), Towards simplified robust model structures in land surface – atmosphere flux predictions, *Hydrol. Process.*, 17, 2259–2277.

Schumm, S. A. (1998), *To Interpret the Earth: Ten Ways to be Wrong*, Cambridge University Press: Cambridge.

Sen, M. & Stoffa, P. L. (1995), *Global Optimisation Methods in Geophysical Inversion*, Elsevier: Amsterdam.

Seo, D. J., Koren, V. & Cajina. (2003), Real-time variational assimilation of hydrologic and hydrometeorological data into operational hydrological forecasting, *J. Hydrometerol.*, 4: 627–641.

Schärer, M., Page, T., Beven, K. J. (2006), A fuzzy decision tree to predict phosphorus export at the catchment scale. *J.Hydrol.*, 331: 484–494.

Shackle, G. L. S. (1949), *Expectation in Economics*, Cambridge University Press: Cambridge.

Shackle, G. L. S. (1955), *Uncertainty in Economics and Other Reflections*, Cambridge University Press: Cambridge.

Shafer, G. (1976), *A Mathematical Theory of Evidence*, Princeton University Press: Princeton, NJ.

Shrader-Frechette, K. (1989), Idealized laws, antirealism and applied science: a case in hydrogeology, *Synthese*, 81, 329–352.

Shrestha, R. R., Bardossy, A. & Nestmann, F. (2007), Analysis and propagation of uncertainties due to stage discharge relationship: a fuzzy set approach, *Hydrol. Sci. J.*, in press.

Siler, W. & Buckley, J. J. (2005), *Fuzzy Expert Systems and Fuzzy Reasoning*, Wiley: Chichester.

Skaugen, T., Langsholt E. G., Hisdla, H., Langsrud, Ø., Follestad, T. & Høst, G. (2005), Uncertainty in flood forecasting, in Balbanis, P., Lambroso, D. & Samuels, P., Innovation, Advances and Implementation of Flood Forecasting Technology, Proceedings of the ACTIF meeting, Tromsø on CD (ISBN 978–1–898485–12–4), HRWallingford: Wallingford, UK.

Smemoe, C. M., Nelson, E. J., Zundel, A. K., & Miller, A. W. (2007), Demonstrating floodplain uncertainty using flood probability maps, *J. Am. Water Resour. Assoc.*, 43: 359–371.

Smith, L. A. (2000), Disentangling Uncertainty and Error: on the predictability of nonlinear systems, in A Mees (Ed.), *Nonlinear Dynamics & Statistics*, Springer: Berlin, 31–64.

Smith, L. A. (2007), *A Very Brief Introduction to Chaos*, Oxford University Press: Oxford.

Smith, P. J., Beven, K. J., Tawn, J., Blazkova, S. & Merta, L. (2006), Discharge-dependent pollutant dispersion in rivers: estimation of ADZ parameters with surrogate data. *Water Resources Research*, 42: W04412, doi:10.1029/2005WR004008.

Smith, P. J., Beven, K. J. & Tawn, J. (2007), The detection of structural inadequacy in process-based hydrological models: a particle filtering approach, *Water Resources Research*, 44(1), W01410 doi: 10.1029/2006WR005205.

Smithson, M. (1989), *Ignorance and Uncertainty*, Springer-Verlag: New York.

Sobol', I. M. (2001), Global sensitivity indices for nonlinear mathematical models and their Monte Carlo estimates. *Mathematics and Computers in Simulation*, 55: 271–280.

Sohn M. D., Small M. J., & Pantazidou M. (2000), Reducing uncertainty in site characterization using Bayes Monte Carlo methods, *J. Environ. Eng. ASCE*, 126: 893–902

Sokal, A. & Bricmont, D. (1997), *Impostures Intellectuelles*, Editions Odile Jacob, Paris, also published with a discussion of the reaction generated by the original French edition as *Intellectual Impostures*, Profile Books, London, 1998.

Sørensen, J. V. T., Madsen, H. & Madsen, H. (2004), Data assimilation in hydrodynamic modeling: on the treatment of nonlinearity and bias. *Stoch. Envir. Res. Risk Ass.*, 18: 228–244.

Sorooshian, S., Duan, Q. & Gupta, V. K. (1993), Calibration of rainfall-runoff models: application of global optimization to the Sacramento soil moisture accounting model, *Water Resources Research*, 29: 1185–1194.

Spear, R. C. (1997), Large simulation models: calibration, uniqueness and goodness of fit, *Environ. Modell. Software*, 12: 219–228.

Spear, R. C. & Hornberger, G. M. (1980), Eutrophication in Peel Inlet. II. Identification of critical uncertainties via generalized sensitivity analysis, *Water Resources Research*, 14: 43–49.

Spear, R. C. Grieb, T. M. & Shang, N. (1994), Parameter uncertainty and interaction in complex environmental models, *Water Resources Research*, 30: 3159–3170.

Stainforth, D. A., Allen M. R., Tredger E. R. & Smith L. A. (2007a), Confidence, uncertainty and decision-support relevance in climate predictions, *Phil. Trans. R. Soc.*, A365: 2145–2161.

Stainforth, D. A., Downing T. E., Washington R., Lopez A. & New M. (2007b), Issues in the interpretation of climate model ensembles to inform decisions, *Phil. Trans. R. Soc.*, A365: 2163–2177.

Stephenson, G. R. & Freeze, R. A. (1974), Mathematical simulation of subsurface flow contributions to snowmelt runoff, Reynolds Creek, Idaho, *Water Resources Research*, 10(2): 284–298.

Stern, N. (2007), *The Economics of Climate Change*, Cambridge University Press: Cambridge.

Sugeno, M. (1977), Fuzzy measures and fuzzy integrals: a survey, in M. M. Gupta, G. N. Saridis & B. R. Gaines (Eds.), *Fuzzy Automata and Decision Processes*, North-Holland: Amsterdam, 89–102.

Sykes, J. F., Harvey, D. J. M. & Wilger, C. (1996), The risk associated with aquifer remediation, *IAHS Publication*, 237: 533–542, IAHS Press: Wallingford, UK.

Talnay, E. (2003), *Atmospheric Modelling: Data Assimilation and Predictability*, Cambridge University Press: Cambridge.

Tang, Y., Reed, P., van Werkhoven, K. & Wagener, T. (2007a), Advancing the identification and evaluation of distributed rainfall-runoff models using global sensitivity analysis, *Water Resources Research*, 43, W06415, doi:10.1029/2006WR005813.

Tang, Y., Reed, P., Wagener, T. & van Werkhoven, K. (2007b), Comparing sensitivity analysis methods to advance lumped watershed model identification and evaluation, *Hydro. Earth. Sys. Sci.*, 11: 793–817.

Tarantola, A. (2005), *Inverse problem theory and model parameter estimation*, SIAM: Philadelphia, PA.

Tarantola, A. (2006), Popper, Bayes and the inverse problem, *Nature Physics*, 2: 492–494.

Taylor, C. J., Pedregal, D. J., Young, P. C. & Tych, W. (2007), Environmental time series analysis and forecasting with the Captain toolbox, *Environ. Modell. & Softw.*, 22: 797–814.

Tebaldi, C., Smith R. L., Nychka D. & Mearns L. O. (2005), Quantifying uncertainty in projections of regional climate change: a Bayesian approach to the analysis of multimodel ensembles, *J. Clim.*, 18: 1524–1540.

Tebaldi, C. & Knutti, R. (2007), The use of the multi-model ensemble in probabilisitic climate projections, *Phil. Trans. R. Soc.*, A365: 2053–2075.

Thiemann, M., Trosset, M., Gupta, H. & Sorooshian, S. (2001), Bayesian recursive parameter estimation for hydrologic models, *Water Resources Research*, 37: 521–2535.

Tiedeman, C. R., Ely, D. M., Hill, M. C. & O'Brian, G. M. (2004), A method for evaluating the importance of system state observations to model predictions, with application to the Death Valley regional groundwater flow system, *Water Resources Research*, 40: W12411, doi:10.1029/2004WR003313.

Tichý, P. (1974), On Popper's definitions of verisimilitude. *The British Journal for the Philosophy of Science*, 25: 155–160.

Todini, E. (1999), Using phase-state modelling for inferring forecasting uncertainty in nonlinear stochastic decision schemes, *J. Hydroinformatics*, 1(2): 75–82.

Todini, E. (2004), Role and treatment of uncertainty in real-time flood forecasting, *Hydrological Process.*, 18: 2743–2746.

Toivonen, H. T. T., Mannila, H., Korhola, A. & Olander, H. (2001), Applying Bayesian statistics to organism-based environmental reconstruction, *Ecol. Applic.*, 11: 618–630.

Toride, N., Leij, F. J. & van Genuchten, MTh. 1995, The CXTFIT code for estimating transport parameters from laboratory or field tracer experiments – Version 2.0. USSal. Lab., Res. Rep. No. 137, 121pp.

Toth, Z. & Kalnay, E. (1993), Ensemble forecasting at NMC: the generation of perturbations, *Bull. Amer. Meteor. Soc.*, 74: 23127–2330.

Trémolet, Y. (2006), Accounting for an imperfect model in 4D-Var, *Q. J. R. Meteorol. Soc.*, 132: 2483–2504.

Tyszka, T. & Zaleskiewicz, T. (2006), When does information about probability count in choices under risk, *Risk Analysis*, 26: 1623–1636.

US Nuclear Regulatory Commission. (2004), An approach for determining the technical adequacy of probabilisitic risk assessment results for risk-informed activities. Regulatory Guide 1.200, Washington: DC.

Uusitalo, L. (2007), Advantages and challenges of Bayesian networks in environmental modelling, *Ecol. Modell.*, 203, 312–318.

Van Asselt, M. B. A., & Rotmans, J. (2002), Uncertainty in integrated assessment modelling, from Positivism to Pluralism, *Climate Change*, 54: 75–105.

Van der Brugge, R., & Rotmans, J. (2007), Towards transition management of European water resources, *Water Resour. Manag.*, 21: 249–267.

Van der Giessen, A. (Ed.). (2005), Naar en gezamenlijk national hydrologiisch modelinstrunnetarium. Report 500026002/2005, MNP: Bilhoven: NL.

Van der Sluijs, J. P. (2002), A way out of the credibility crisis of models used in integrated environmental assessment, *Futures*, 34: 133–146.

Van der Sluijs, J. P., Risbey, J. S., Kloprogge, P., Ravetz, J. R., Guntowicz, S. O., Corral Quintana, S., Guimaraes Pereira, A., De Marchi, B., Petersen, A. C., Janssen, P. H. M., Hoppe, R. & Huijs, S. W. F. (2003), RIVA/MNP Guidance for Uncertainty Assessment and Communication: Detailed Guidance, Copernicus Institute for Sustainable Development, Utrecht University & RIVM/MNP: The Netherlands, available at www.nusap.net/sections. php?op=viewarticle&artid=17.

Van der Sluijs, J. P., Craye, M., Funtowicz, S., Kloprogge, P., Ravetz, J. R. & Risbey, J. (2005a), Experiences with the NUSAP system for multidimensional uncertainty assessment. *Water Science & Technology*, 52(6): 133–144.

Van der Sluijs, J. P., Craye, M., Funtowicz, S., Kloprogge, P., Ravetz, J. R. & Risbey, J. (2005b), Combining quantitative and qualitative measures of uncertainty in model-based environmental assessment: the NUSAP system, *Risk Analysis*, 25: 481–492.

Van Fraassen, B. C. (1980) *The Scientific Image*. Oxford: Clarendon.

Van Heerden, I. & Bryan, M. (2006), *The Storm: What Went Wrong and Why During Hurricane Katrina*, Viking: NY.

Van Leeuwen, P. V. & Evensen, G. (1998), Data assimilation and inverse methods in terms of a probabilisitic formulation, *Mon. Weath. Rev.*, 124: 2898–2913, doi: 10.1175/1520–0493(1996)124<2898:DAAIMI>2.0.CO;2.

Van Oijen, M., Rougier, J. & Smith, R. (2005), Bayesian calibration of process-based forest models: bridging the gap between models and data, *Tree Physiology*, 25: 915–927.

Van Straten, G. & Keesman, K. J. (1991), Uncertainty propagation and speculation in projective forecasts of environmental change, *J. Forecasting*, 10: 163–190.

Varis, O. (1955), Belief networks for modelling and assessment of environmental change, *Environmetrics*, 6: 439–444.

Varis, O. (1997), Bayesian decision analysis for environmental and resource management, *Environ. Modell. & Software*, 12: 177–185.

Varis, O. & Kuikka, S. (1997), Joint use of multiple environmental assessment models by a Bayesian meta-model: the Baltic salmon case, *Ecol. Modell.*, 102: 341–351.

Ver Hoef, J. M. (1996), Parametric empirical Bayes methods for ecological applications, *Ecol. Applic.*, 6: 1047–1055.

Von Bertalanffy, L. (1968), *General Systems Theory*, Braziller: New York.

Von Neumann, J. & Morgenstern, O. (1944), *Theories of Games and Economics*, Princeton University Press: Princeton, NJ.

Vrugt, J. A., Bouten, W., Gupta, H. V. & Sorooshian, S. (2002), Toward improved identifiability of hydrologic model parameters: the information content of experimental data, *Water Resources Research*, 38(12): doi:10.1029/2001WR001118.

Vrugt, J. A., Gupta, H. V., Bouten, W. & Sorooshian, S. (2003), A shuffled complex evolution Metropolis algorithm for optimization and uncertainty assessment of hydrologic model parameters, *Water Resources Research*, 39: doi:10.1029/2002WR001642.

Vrugt, J. A., Gupta, H. V., O'Nuallain, B., & Bouten, W. (2006), Real-time data assimilation for operational ensemble streamflow forecasting, *J. Hydrometeorol.*, 7: 548–565.

Vrugt, J. A. & Robinson, B. A. (2007a), Improved evolutionary optimisation from genetically adaptive multimethod search, *Proc. Nation. Acad. Sci.*: 104(3): 708–711, doi: 10.1073/pnas0610471104.

Vrugt, J. A. & Robinson, B. A. (2007b), Treatment of uncertainty using ensemble methods: comparison of sequential data assimilation and Bayesian model averaging, *Water Resources Research*, 43: W01411, doi: 10.1029/2005WR004838.

Wade, A. J., Butterfield, D., Griffiths, T. & Whitehead, P. G. (2007), Eutrophication control in river-systems: an application of INCA-P to the River Lugg, *Hydrol. Earth Syst. Sci.*, 11: 584–600.

Wagener, T., McIntyre, N., Lees, M. J., Wheater, H. S., & Gupta, H. V. (2003), Towards reduced uncertainty in conceptual rainfall-runoff modelling: Dynamic identifiability analysis, *Hydrol. Process.*, 17: 455–476.

Walker, J. P., Wilgoose, G. R. & Kalman, J. D., One-dimensional soil moisture profile retrieval by assimilation of near surface observations: a comparison of retrieval algorithms, *Adv. Water Resour.*, 24: 631–650.

Walley, P. (1991), *Statistical Reasoning with Imprecise Probabilities*, Chapman & Hall: London.

Walley, P. (2000), Towards a unified theory of imprecise probabilitiy, *Int. J. Approx. Reason.*, 24: 125–148.

Walters, C. J. (1986), *Adaptive Management of Renewable Resources*, Macmillan: NY.

Wang, Y.-M., Yang, J.-B., Xu, D.-L. & Chin, K.-S. (2005), The evidential reasoning approach for multiple attribute decision analysis using interval belief degrees, *European J. Operational Research*, 182: 1249–1312.

Wang, Y.-M., Yang, J.-B., & Xu, D.-L. (2006), Environmental impact assessment using the evidential reasoning approach, *European J. Operational Research*, 174: 1885–1913.

Wang, Z. & Klir, G. J. (1992), *Fuzzy Measure Theory*, Plenum Press: NY.

Weerts, A. H. & E. l. Serafy, G. Y. H. (2006), Particle filtering and ensemble Kalman filtering for state updating with hydrological conceptual rainfall-runoff models, *Water Resources Research*, 42: W09403, doi: 10.1029/20045WR004903.

Weinberg, A. (1972), Trans-science, *Minerva*, 10: 209–22.

Weiss, C. (2006), Can there be science-based precaution?, *Environ. Res. Lett.*, 1, doi: 10.1088/1748-9326/1/1/014003 (7p).

Whitehead, P. G. & Young, P. C., 1979, Water quality in river systems: Monte-Carlo analysis, *Water Resources Research*, 15: 451–459.

Widrow, B., Rumelhart, D. E. & Lehr, M. A. (1994), Neural Networks – applications in industry, business and science, *Comm. ACM*, 37: 93–105.

Wikle, C. K. (2003), Hierarchical Bayesian models for predicting the spread of ecological processes, *Ecology*, 84: 1382–1394.

Wilby, R. L. and Harris I. (2006), A framework for assessing uncertainties in climate change impacts: low-flow scenarios for the River Thames, UK, *Water Resources Research*, 42: W02419, doi:10.1029/WR2005WR004065.

Williamson, R. C. & Downs, T. (1990), Probabilistic arithmetic I: numerical methods for calculation convolutions and dependency bounds, *Int. J. Approx. Reason.*, 4: 89–158.

Wolfson, L. J., Kadane, J. B. & Mitchell, M. J. (1996), Bayesian environmental policy decisions: two case studies, *Ecological Applications*, 6: 1056–1066.

Wooldridge, S., & Done, T. (2004), Learning to predict large-scale coral bleaching from past events: a Bayesian approach using remotely sensed data, in-situ data, and environmental proxies, *Coral Reefs*, 23: 96–108.

Wynne, B. (1992), Uncertainty and environmental learning: reconceiving science and policy in the preventative paradigm, *Global Environ. Change*, 2: 111–127.

Yadav, M., Wagener, T. & Gupta, H. V. (2007), Regionalization of constraints on expected watershed response behavior for improved predictions in ungauged basins, *Adv. Water Resour.*, 30: 1756–1774.

Yager, R. R. (1986), Arithmetic and other operations on Dempster–Shafer structures, *Int. J. Man-Mach. Stud.*, 25: 357–366.

Yang, J., Reichert, P., Abbaspour, K. C. & Yang, H. (2007), Hydrological modelling of the Chaohe Basin in China: statistical model formulation and Bayesian inference, *J. Hydrol.*, doi: 10.1016/j.jhydrol.2007.04.006.

Yang, Z. & Hamrick, J. M. (2002), Variational inverse parameter estimation in a long-term tidal transport model, *Water Resources Research*, 38(10): W01204, doi:10.1029/20–01WR001121.

Yapo, P. O., Gupta, H. & Sorooshian, S. (1996), Calibration of conceptual rainfall-runoff models: sensitivity to calibration data, *J. Hydrol.*, 181: 23–48.

Yapo, P. O., Gupta, H. & Sorooshian, S. (1998), Multi-objective global optimisation for hydrologic models, *J. Hydrol.*, 204: 83–97.

Ye, M., Neuman, S. P. & Meyer, P. D. (2004), Maximum likelihood Bayesian averaging of spatial variability models in unsaturated fractured tuff, *Water Resources Research*, 40: W05113, doi: 10.1029/2003WR002557.

Yeh, H. D. & Huang, Y. C. (2005), Parameter estimation for leaky aquifers using the extended Kalman filter and considering model and data measurement uncertainties, *J. Hydrol.*, 302: 28–45.

Young, P. C. (1983), The validity and credibility of models for badly-defined systems, in M. B. Beck & G. van Straten (Eds.), *Uncertainty and Forecasting of Water* Quality, Springer-Verlag: Berlin, 69–98.

Young, P. C. (1984), *Recursive Estimation and Time Series Analysis*, Springer: Berlin.

Young, P. C. (1998), Data-based mechanistic modelling of environmental, ecological, economic and engineering systems, *Environmental Modelling & Software*, 13: 105–122.

Young, P. C. (2001), Data-based mechanistic modelling and validation of rainfall-flow processes, in Anderson, M. G. & Bates, P. D. (Eds.), *Model Validation: Perspectives in Hydrological Science*, Wiley: Chichester, 117–161.

Young, P. C. (2002), Advances in Real Time Forecasting, *Phil. Trans. Roy. Soc. Lond.*, A360: 1430–1450.

Young, P. C. (2003), Top-down and data-based mechanistic modelling of rainfall-flow dynamics at the catchment scale, *Hydrol. Process.*, 17: 2195–2217.

Young, P. C., Chotai, A. & Beven, K. J. (2004), Data-based mechanistic modelling and the simplification of environmental systems, in J. Wainwright & M. Mulligan (Eds.), *Environmental Modelling: Finding Simplicity in Complexity*, Wiley: Chichester, 371–388.

Young, P. C. & Parkinson, S. (2002), Simplicity out of complexity, in M. B. Beck (Ed.), *Environmental Foresight and Models: A Manifesto*. Elsevier Science: NY, 251–301.

Young, P. C. and Wallis, S. G. (1993), Solute transport and dispersion in channels, in K. J. Beven and M. J. Kirkby (Eds.), *Channel Network Hydrology*, Wiley: Chichester, 129–175.

Young, R. A. (2001), *Uncertainty and the Environment: Implications for Decision Making and Environmental Policy*, Edward Elgar: Cheltenham.

Zadeh, L. (1965), Fuzzy Sets, *Control*, 8: 338–353.

Zadeh, L. (2004), From imprecise to granular probabilities, *Fuzzy Sets & Systems*, 154: 370–374.

Zadeh, L. (2005), Towards a generalised theory of uncertainty (GTU) – an outline, *Information Sciences*, 172: 1–40.

Zak, S. & Beven, K. J. (1999), Equifinality, sensitivity and uncertainty in the estimation of critical loads, *Science of the Total Environment*, 236: 191–214.

Zhang, D., Beven, K. J. & Mermoud, A. (2006), A comparison of nonlinear least square and GLUE for model calibration and uncertainty estimation for pesticide transport in soils. *Adv. Water Resour.*, 29: 1924–1933.

Index

Indexed terms in bold font will also be found in the glossary

T - #0603 - 071024 - C0 - 246/174/15 - PB - 9780415457590 - Gloss Lamination